# Scientific Computation

Springer-Verlag Berlin Heidelberg GmbH

M. Křížek P. Neittaanmäki
R. Glowinski S. Korotov (Eds.)

# Conjugate Gradient Algorithms and Finite Element Methods

With 93 Figures Including 8 Color Figures, and 24 Tables

Springer

Prof. Michal Křížek
Academy of Sciences
Mathematical Institute
Žitná 25
115 67 Prague, Czech Republic

Prof. Roland Glowinski
University of Houston
Department of Mathematics
University Park
Houston, TX 77004-3008, USA

Prof. Pekka Neittaanmäki
Dr. Sergey Korotov
University of Jyväskylä
Department of Mathematical
Information Technology
P.O. Box 35 (AGORA)
40014 Jyväskylä, Finland

ISBN 978-3-642-62159-8 ISBN 978-3-642-18560-1 (eBook)
DOI 10.1007/978-3-642-18560-1
Library of Congress Control Number: 2004104847



springeronline.com


Originally published by Springer-Verlag Berlin Heidelberg New York in 2004
Softcover reprint of the hardcover 1st edition 2004


Typesetting: by the authors/editors
Final data conversion and figure processing by Frank Herweg, Hirschberg-Leutershausen
Cover design: *design & production* GmbH, Heidelberg
Printed on acid-free paper 55/3141/rm 5 4 3 2 1 0

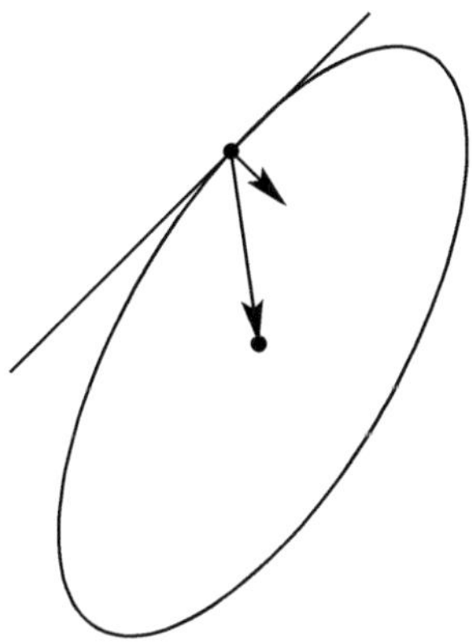

Most of the arts, such as painting, sculpture, and music, have emotional appeal to the general public. This is because these arts can be experienced by one or more of our senses. Such is not true of the art of mathematics; this art can be appreciated only by mathematicians, and to become a mathematician requires a long period of intensive training. The community of mathematicians is similar to an imaginary community of musical composers whose only satisfaction is obtained by the interchange among themselves of the musical scores they compose.

Cornelius Lanczos
(1893–1974)

# Preface

For more than 50 years already, *Conjugate Gradient Algorithms* and *Finite Element Methods* have been around. There is no doubt that these two methods have deeply influenced – not to say revolutionized – the way one solves complicated problems from Science and Engineering. It is interesting to observe that at their birth both methods were considered no more than scientific curiosities: finite element methods were seen as just particular cases of the Galerkin method, while conjugate gradient methods were viewed as something between iterative methods and direct methods for their finite termination properties. Actually, it took just few years to realize that these two methodologies had their own merits and that their capabilities were much beyond what was perceived at their inception. Indeed, applications of both methodologies are everywhere and many of their potentials still have to be discovered, this bold statement being based on the fact that new applications appear almost daily. Having said that, some readers can wonder about the rational of publishing a volume dedicated simultaneously to both methods. The answer lies in the complementarity of both methodologies: Clearly, finite element methods reach maximal efficiency when applied to the solution of problems enjoying a variational formulation in some Hilbert space, directly, or after some manipulations (fortunately, many important problems from Science and Engineering have this property). On the other hand, conjugate gradient algorithms are more than tools for solving linear problems in finite dimension; actually, they provide the most general approach to the solution of linear variational problems in Hilbert spaces of arbitrary dimension (possibly infinite). The destiny of these two methods was to encounter and their meeting took place in the sixties. Via the combination of conjugate gradient algorithms and finite element approximations, practitioners have been able to solve complicated, direct and inverse, multidimensional problems modeled by ordinary or partial differential equations and inequalities, not necessarily linear, optimal control and optimal design being part of these problems. The objectives of this volume are many-fold; among them:

- Provide historical perspective and the foundations of conjugate gradient type methods.
- Show the complementarity of the conjugate gradient and finite element methods.

- Show that both methods apply to the solution of complicated problems modeled by linear and nonlinear partial differential equations.
- Provide an in-depth discussion on the implementation aspects of both methods.
- Provide the readers with numerous references, giving them the possibility to go to further sources of information in order to learn more about both methodologies.
- Show that conjugate gradient and finite element methods apply to the solution of real-life problems.

This volume is structured as follows:

In Part I, one provides the *foundations* and *historical facts* on conjugate gradient methods; their complementarity with finite element methods is also addressed.

Part II is dedicated to the algorithmic aspects and to implementation issues, including parallelization and domain decomposition methods.

Part III addresses mesh qualitity issues for finite element methods. Special geometric conditions on meshes yelding monotone stiffness matrices and guaranteeing the validity of the discrete maximum principle are examined.

In Part IV, one discusses the application of conjugate gradient and related techniques to the solution of initial/boundary value problems for linear and nonlinear partial differential equations (and even inequalities). These are typical situations where the computer implementation of the above iterative methods requires coupling with approximation methods such as finite elements. The conjugate gradient solution of some optimal control problems is also discussed.

Finally, Part V is dedicated to the finite element approximation and the iterative solution of highly non-trivial problems from image processing, physics of nuclear reactors, and computational fluid dynamics, including multiphase flow in chemical reactors and the solution of the Euler equations modeling compressible inviscid fluid flow.

Some of the issues discussed in this volume have been addressed at the International Conference *Finite Element Methods: Fifty Years of Conjugate Gradients*, which took place at the University of Jyväskylä in Finland during Summer 2002, but additional material has been included in order to broaden the historical and scientific perspectives. LOGO of the Conference is marked on page V.

When applying conjugate gradient type methods to the solution of a difficult problem, the difference between success and failure lies, most often, in the choice of the preconditioning operator. From that point of view one of our favorite quotations can be found in J. E. Dennis, R. B. Schnabel: *A view of unconstrained optimization*, In: Handbook in Operation Research and Management Science, Vol. 1, G. L. Newhauser, A. H. G. Rinnooy and M. J. Todd, eds., North-Holland, Amsterdam, 1–66. It reads as follows (in this quotation $p$ is the number of iterations necessary to achieve convergence and $n$ is the dimension of the problem under consideration):

"It is not unusual for strictly convex quadratics arising from discretized partial differential equations to be solved with $p \sim n/10^3$. Such spectacularly successful preconditioning nearly always comes from deep insight into the problem and not from matrix theoretic considerations. They often come from discretizing and solving a simplified problem."

There is little to add to the above quotation, except that one may find in this volume examples of situations where, indeed, the above approach has been quite successful.

The editors would like to acknowledge the assistance of many collaborators at their respective institutions and of Ms. Sabine Lehr at Springer. Our gratitude and warmest thanks go to Professor Wolf Beiglböck for his encouragements, and for accepting the publication of this volume in the prestigious Springer Series in Scientific Computing. We would also like to express our sincere gratitude to Dr. Jan Brandts, Ms. Helena Holovská, Mr. Štěpán Kasal, and Mr. Pavel Křížek for their invaluable help in preparing these proceedings. The financial support of the Academy of Finland and Grant No. A 1019201 of the Grant Agency of the Academy of Sciences of the Czech Republic are gratefully appreciated.

Prague
October 2003

*Michal Křížek*
*Pekka Neittaanmäki*
*Roland Glowinski*
*Sergey Korotov*

# Contents

## Part I. Foundations



## Part II. Aspects of Conjugate Gradient Algorithms



## Part III. Finite Element Meshes

# List of Contributors

**Axelsson Owe**
Department of Mathematics
University of Nijmegen
6525ED Nijmegen, Netherlands
axelsson@sci.kun.nl

**Brandts Jan**
Korteweg-de Vries Institute
Faculty of Science
University of Amsterdam
Plantage Muidergracht 24
1018 TV Amsterdam, Netherlands
brandts@science.uva.nl

**Dacorogna Bernard**
École Polytechnique Fédérale
de Lausanne – DMA
CH–1015 Lausanne, Switzerland
Bernard.Dacorogna@epfl.ch

**Duintjer Tebbens Erik Jurjen**
Department of Numerical
Mathematics
Faculty of Mathematics and Physics
Charles University
Sokolovská 83
CZ–186 75 Praha 8, Czech Republic
jurjend.tebbens@gmx.net

**Faragó István**
Department of Applied Analysis
Eötvös Loránd University
P.O. Box 120
H–1518 Budapest, Hungary
faragois@cs.elte.hu

**Glowinski Roland**
Department of Mathematics
University of Houston
Houston, TX 77204–3008, USA
roland@math.uh.edu

**Horváth Robert**
Institute of Mathematics
and Statistics
University of West-Hungary
Erzsébet u. 9
H–9400 Sopron, Hungary
rhorvath@ktk.nyme.hu

**Karátson János**
Department of Applied Analysis
Eötvös Loránd University
P.O. Box 120
H–1518 Budapest, Hungary
karatson@cs.elte.hu

**Korotov Sergey**
Department of Mathematical
Information Technology
University of Jyväskylä
P.O. Box 35
FIN–40014 Jyväskylä, Finland
korotov@mit.jyu.fi

**Křížek Michal**
Mathematical Institute
Academy of Sciences of the Czech
Republic
Žitná 25
CZ–115 67 Prague 1, Czech Republic
krizek@math.cas.cz

**Křížek Pavel**
Center for Machine Perception
Faculty of Electrical Engineering
Czech Technical University
Karlovo náměstí 13
CZ–121 35 Prague 2, Czech Republic
krizekp1@feld.cvut.cz

**Kuzmin Dmitri**
Institute of Applied Mathematics
University of Dortmund
(LS III)
Vogelpothsweg 87
D–44227 Dortmund, Germany
kuzmin@math.uni-dortmund.de

**Kuznetsov Yuri**
Department of Mathematics
University of Houston
Houston, TX 77204–3008, USA
kuz@math.uh.edu

**Lapin Serguei**
Department of Mathematics
University of Houston
Houston, TX 77204-3008, USA
slapin@math.uh.edu

**Lukšan Ladislav**
Institute of Computer Science
Academy of Sciences of the Czech Republic
Pod vodárenskou věží 2
CZ–182 07 Praha 8, Czech Republic
luksan@cs.cas.cz
and
Technical University of Liberec
Hálkova 6
CZ–461 17 Liberec, Czech Republic

**Martikainen Janne**
Department of Mathematical Information Technology
University of Jyväskylä
P.O. Box 35
FIN–40014 Jyväskylä, Finland
jamartik@mit.jyu.fi

**Matonoha Ctirad**
Institute of Computer Science
Academy of Sciences of the Czech Republic
Pod vodárenskou věží 2
CZ–182 07 Praha 8, Czech Republic
matonoha@cs.cas.cz
and
Technical University of Liberec
Hálkova 6
CZ–461 17 Liberec, Czech Republic

**Möller Matthias**
Institute of Applied Mathematics
(LS III)
University of Dortmund
Vogelpothsweg 87
D–44227 Dortmund, Germany
matthias.moeller@math.uni-dortmund.de

**Neittaanmäki Pekka**
Department of Mathematical Information Technology
University of Jyväskylä
P.O. Box 35
FIN–40014 Jyväskylä, Finland
pn@mit.jyu.fi

**Neumanová Eva**
Department of Technical Mathematics
Faculty of Mechanical Engineering
Czech Technical University
Karlovo náměstí 13
CZ–121 35 Praha 2, Czech Republic
neumanov@marian.fsik.cvut.cz

**Pan Tsorng-Whay**
Department of Mathematics
University of Houston
Houston, TX 77204–3008, USA
pan@math.uh.edu

**Schilders Willy**
Philips Research Laboratories Eindhoven
Prof. Holstlaan 4
5656AA Eindhoven, Netherlands
wil.schilders@philips.com

**Segal Guus**
Department of Applied Mathematical Analysis
Delft University of Technology
Mekelweg 4
2628 CD Delft, Netherlands
g.segal@math.tudelft.nl

**Šolc Jakub**
Department of Mathematics
Faculty of Civil Engineering
Czech Technical University
Thákurova 7
CZ–166 29 Praha 6, Czech Republic
solc@math.cas.cz

**Šolcová Alena**
Department of Mathematics
Faculty of Civil Engineering
Czech Technical University
Thákurova 7
CZ–166 29 Prague 6, Czech Republic
solcova@mbox.cesnet.cz

**Stańdo Jacek**
Institute of Mathematics
Technical University of Łódź
Al. Politechniki 11
90-924 Łódź, Poland
standoj@ck-sg.p.lodz.pl

**Turek Stefan**
Institute of Applied Mathematics
(LS III)
University of Dortmund
Vogelpothsweg 87
D–44227 Dortmund, Germany
ture@featflow.de

**van der Vorst Henk**
Mathematical Institute
Utrecht University
Budapestlaan 6, P.O. Box 80.010
3508 TA Utrecht, Netherlands
vorst@math.uu.nl

**Vassilevski Yuri V.**
Institute of Numerical Mathematics
Russian Academy of Sciences
8, Gubkina str.
119991 Moscow, Russia
vasilevs@dodo.inm.ras.ru

**Vejchodský Tomáš**
Mathematical Institute
Academy of Sciences of the Czech Republic
Žitná 25
CZ–115 67 Prague 1, Czech Republic
vejchod@math.cas.cz

**Vermolen Fred**
Department of Applied Mathematical Analysis
Delft University of Technology
Mekelweg 4
2628 CD Delft, Netherlands
F.J.Vermolen@its.tudelft.nl

**Vlček Jan**
Institute of Computer Science
Academy of Sciences of the Czech Republic
Pod vodárenskou věží 2
CZ–182 07 Praha 8, Czech Republic
vlcek@cs.cas.cz
and
Technical University of Liberec
Hálkova 6
CZ–461 17 Liberec, Czech Republic

**Vuik Kees**
Department of Applied Mathematical Analysis
Delft University of Technology
Mekelweg 4
2628 CD Delft, Netherlands
C.Vuik@its.tudelft.nl

# Part I

# Foundations

# The Founders of the Conjugate Gradient Method

Alena Šolcová

Department of Mathematics, Faculty of Civil Engineering, Czech Technical University, Thákurova 7, CZ–166 29 Prague 6, Czech Republic
solcova@mbox.cesnet.cz

*Better to have an approximate answer to the right question than a precise answer to the wrong question.*

Paraphrased quotation of Tukey J. W. (1915–2000)

## 1 Introduction

The Conjugate Gradient Method was cited among the *"Top 10 Algorithms of the 20th Century"* by Computing in Science & Engineering Magazine, vol. 2, January 2000. It can be considered as a direct method as well as an iterative method for solving systems of linear algebraic equations. In this paper we introduce the fathers of this significant method.

The history started in papers written by Magnus Hestenes and Eduard Stiefel [10], and also by Cornelius Lanczos [13] who cooperated with the Institut of Numerical Analysis at the University of California, Los Angeles. Hestenes and Lanczos were National Bureau of Standards staff members whereas Stiefel was a visitor from the Swiss Eidgenössische Technische Hochschule (ETH) in Zürich.

In December 1952, the paper entitled *Methods of Conjugate Gradients for Solving Linear Systems* by M. Hestenes and E. Stiefel was published (see [10]). This was a starting point of a great revolution in the numerical solution of large systems of simultaneous linear algebraic equations. Lanczos' fundamental paper [13] is quoted in [10] and appeared in the same journal. For a historical survey of the conjugate gradient method we refer to [5].

In the early fourties Hestenes worked in function theory. Let us mention one of his most beautiful results in this field published in [6]. He considers an arbitrary function $f \in C^k(\overline{\Omega})$, where $k \in \{0, 1, 2, \dots\}$ and $\Omega$ is a bounded domain in $R^2$ (without any assumptions on the boundary). For such an $f$ he proves a theorem that states that if it is possible to extend $f$ locally at each point of the boundary keeping it in the class $C^k$, then there exists a global $C^k$-extension of $f$ on the whole space $R^2$.

Journal of Research of the National Bureau of Standards Vol. 49, No. 6, December 1952 Research Paper 2379

# Methods of Conjugate Gradients for Solving Linear Systems[1]

Magnus R. Hestenes[2] and Eduard Stiefel[3]

An iterative algorithm is given for solving a system $Ax=k$ of $n$ linear equations in $n$ unknowns. The solution is given in $n$ steps. It is shown that this method is a special case of a very general method which also includes Gaussian elimination. These general algorithms are essentially algorithms for finding an $n$ dimensional ellipsoid. Connections are made with the theory of orthogonal polynomials and continued fractions.

## 1. Introduction

One of the major problems in machine computations is to find an effective method of solving a system of $n$ simultaneous equations in $n$ unknowns, particularly if $n$ is large. There is, of course, no best method for all problems because the goodness of a method depends to some extent upon the particular system to be solved. In judging the goodness of a method for machine computations, one should bear in mind that criteria for a good machine method may be different from those for a hand method. By a hand method, we shall mean one in which a desk calculator may be used. By a machine method, we shall mean one in which sequence-controlled machines are used.

A machine method should have the following properties:

(1) The method should be simple, composed of a repetition of elementary routines requiring a minimum of storage space.

(2) The method should insure rapid convergence if the number of steps required for the solution is infinite. A method which—if no rounding-off errors occur—will yield the solution in a finite number of steps is to be preferred.

(3) The procedure should be stable with respect to rounding-off errors. If needed, a subroutine should be available to insure this stability. It should be possible to diminish rounding-off errors by a repetition of the same routine, starting with the previous result as the new estimate of the solution.

(4) Each step should give information about the solution and should yield a new and better estimate than the previous one.

(5) As many of the original data as possible should be used during each step of the routine. Special properties of the given linear system—such as having many vanishing coefficients—should be preserved. (For example, in the Gauss elimination special properties of this type may be destroyed.)

In our opinion there are two methods that best fit these criteria, namely, (a) the Gauss elimination method; (b) the conjugate gradient method presented in the present monograph.

There are many variations of the elimination method, just as there are many variations of the conjugate gradient method here presented. In the present paper it will be shown that both methods are special cases of a method that we call the method of conjugate directions. This enables one to compare the two methods from a theoretical point of view.

In our opinion, the conjugate gradient method is superior to the elimination method as a machine method. Our reasons can be stated as follows:

(a) Like the Gauss elimination method, the method of conjugate gradients gives the solution in $n$ steps if no rounding-off error occurs.

(b) The conjugate gradient method is simpler to code and requires less storage space.

(c) The given matrix is unaltered during the process, so that a maximum of the original data is used. The advantage of having many zeros in the matrix is preserved. The method is, therefore, especially suited to handle linear systems arising from difference equations approximating boundary value problems.

(d) At each step an estimate of the solution is given, which is an improvement over the one given in the preceding step.

(e) At any step one can start anew by a very simple device, keeping the estimate last obtained as the initial estimate.

In the present paper, the conjugate gradient routines are developed for the symmetric and nonsymmetric cases. The principal results are described in section 3. For most of the theoretical considerations, we restrict ourselves to the positive definite symmetric case. No generality is lost thereby. We deal only with real matrices. The extension to complex matrices is simple.

The method of conjugate gradients was developed independently by E. Stiefel of the Institute of Applied Mathematics at Zurich and by M. R. Hestenes with the cooperation of J. B. Rosser, G. Forsythe, and L. Paige of the Institute for Numerical Analysis, National Bureau of Standards. The present account was prepared jointly by M. R. Hestenes and E. Stiefel during the latter's stay at the National Bureau of Standards. The first papers on this method were

[1] This work was performed on a National Bureau of Standards contract with the University of California at Los Angeles, and was sponsored (in part) by the Office of Naval Research, United States Navy.

[2] National Bureau of Standards and University of California at Los Angeles.

[3] University of California at Los Angeles, and Eidgenössische Technische Hochschule, Zurich, Switzerland.

409

Figure 1. The first page of the original paper [10] by Hestenes and Stiefel.

**Magnus Hestenes** (1906–1991) was born in Bricelyn, Minnesota, USA. His undergraduate education started at St. Olaf College, and his graduate work took place at the University of Wisconsin. Later he continued at the University of Chicago.

In 1947 he accepted a professorship at the University of California in Los Angeles, where he taught until his retirement. He had 34 Ph.D. students and

Figure 2. Magnus Hestenes.

was a well appreciated well-loved advisor and teacher. The fields of activities of M. Hestenes were variational problems, problems of control theory, and conjugate direction methods. His best known work includes many publications on the problem of Bolza, and a famous paper on quadratic forms in Hilbert spaces [7]. The impact of his scientific output has been enormous. More than 800 papers citing the original paper [10] on the Conjugate Gradient Method appeared in the period 1983–1999. He was a Guggenheim fellow, a Fulbright scholar, and also the vice president of the American Mathematical Society.

Thirty years after the publication of the pioneering paper [10] Magnus Hestenes expressed his respect to the next significant man in the history of the Conjugate Gradient Method – his monograph [8] is devoted to the memory of his colleague E. Stiefel.

**Eduard Stiefel** (1909–1978) was born in Zürich, Switzerland. He spent his youth as a student at the famous ETH, where Albert Einstein worked on the theory of relativity. Stiefel's early work was devoted to topology of Lie groups.

He is connected with the development of one of the first electronic computers Z4 of Konrad Zuse. In 1948 Eduard Stiefel wanted to set up a new Institut for Applied Mathematics in the ETH Zürich. His declared goal was to advance numerical analysis. Accordingly he was looking for a way to gain access to computing power beyond the level that could be performed by simple desktop calculators.

In 1947 Konrad Zuse with his family and the computer named Z4 moved to Hopferau (Bavaria). Two years later Stiefel was informed about the existence of the Z4 and he was told that the machine may possibly be for sale. He inspected this computer in Hopferau in 1949. Konrad Zuse explained him its

Figure 3. Eduard Stiefel.

specifications, which were very convincing. Stiefel gave Konrad Zuse a differential equation to solve it. Zuse wrote a program for the Z4, created a punch tape, and the machine worked. The result was correct. Thus, despite the fact that the Z4 was only barely in operational condition (its construction was not completely finished), Stiefel decided that the idea of transferring it to the ETH Zürich should be by all means tried.

Figure 4. The Z4's input and output devices and the program construction unit in Hopferau.

Installation and operation of the Z4 absorbed all the efforts of Stiefel's small group. Stiefel did not want to build computers, he just wanted to use them in an easy way for complex numerical calculations. For this reason, the speed of the computer was not the most important to Stiefel. Among other aspects, the simplicity of programming was very significant to him. The Z4's attached program construction unit convinced him to get this computer, because it allowed scientists to enter programs in less than a few hours. Also the powerful arithmetic unit with its exceptional handling capability was an ideal prerequisite for the calculations Stiefel would make.

The use of computers stimulated the development of numerical methods. The results of Stiefel indicated that numerical methods are very suitable for high speed machines. The largest linear system with a symmetric matrix that had been solved at that time by the conjugate gradient method contained 106 difference equations. The computations were done on a Zuse relay-computer at ETH in Zürich. The estimate obtained in the 90th step had an acceptable accuracy (see also [w3]).

**Cornelius Lanczos** (1893–1974) was a mathematician and physicist. His papers cover a vast array of disciplines including numerical analysis, applied mathematics, scientific computation, quantum mechanics, and also general relativity.

Figure 5. Cornelius Lanczos.

He was born as Kornél Löwy in a family of Jewish origin in Székesfehérvár, Hungary. His father was a lawyer. Cornelius Lanczos attended a Jewish elementary school, where he learned several foreign languages. Then he entered the local gymnasium which was a catholic school (Order of Cistercians). After graduating in 1915, Lanczos was appointed as an assistant at the Technical University of Budapest. He worked on the theory of relativity and sent a copy of his dissertation: "*Relation of Maxwell's Aether Equations to Functional Theory*" to Albert Einstein, who replied to him:

*I have read your work in as much detail as my present excess of work allows. I can say that it is sound and original thinking. It makes you worthy of the doctorate. I am pleased to give you my permission to honour me by dedicating it to me.*

Lanczos spent the next three years in Freiburg, and then moved to Frankfurt. In the period 1928–1929, Lanczos was Einstein's assistent in Berlin. Then he was as a visiting professor at the University of Purdue in Lafayette, Indiana. His Jewish origin did not allow him to live in Germany and Lanczos quickly returned to Indiana. There, in 1938 he published his first work in numerical analysis. Two years later he introduced a new matrix method for calculating the Fourier coefficients (see [4] and [w2]) which, over 25 years later, was recognized as the *Fast Fourier Transform algorithm* by John W. Tukey (see [2]). The main trick is that a discrete Fourier transform of length $2n$ can be written as the sum of two discrete Fourier transforms, each of length $n$ (see, e.g., Danielson-Lanczos lemma in [15, p. 391]).

Also he continued to work on his first love – the theory of relativity and corresponded with Einstein both on a scientific level and as a friend.

During the Second World War, Lanczos spent the year 1944 working for the Boeing Aircraft Company. Later he got a permanent position at Boeing. There he worked on applications of mathematics to aircraft design and developed new numerical methods to solve associated real-life problems.

In 1949 he moved to the Institute for Numerical Analysis of the National Bureau of Standards, where he worked on developing digital computers. In 1950 he published a fundamental paper [13], which has a close relationship to the conjugate gradient method, see e.g., [1].

During the early fifties the political atmosphere in the USA became unpleasant for Lanczos. He accepted an offer from Erwin Schrödinger to head the Theoretical Physics Department at the Dublin Institut for Advanced Study in Ireland. Lanczos published over 120 papers and books. He died in a hospital in Budapest after a sudden heart attack in 1974. His *Collected Papers with Commentaries* were published by the North Carolina State University [w1].

## Final remarks

The conjugate gradient algorithm of Hestenes and Stiefel and the Lanczos method are examples from a large class of methods, actually a class of methods known today as Krylov subspace methods. At present Krylov subspace methods can be used to solve efficiently very large sparse systems of linear equations and eigenvalue problems. These projection methods are called according to the Russian maritime engineer and applied mathematician A. N. Krylov (1863–1945).

Figure 6. Alexei Nikolaevich Krylov.

It is interesting to hear two distinct voices in the paper [10]. Hestenes came to the algorithm from the point of view of variational methods and optimal control. In 1936 he had developed an algorithm for constructing a set of mutually conjugate basis vectors, but he was advised by a Harvard professor that it was too obvious for publication [6]. Stiefel, on the other hand, had a strong orientation toward relaxation algorithms, continued fractions, the QR-algorithm, and he developed conjugate gradients from this viewpoint. The conjugate gradient method became popular, when G. Meinardus in [14] (see also [3], [11]) derived an upper bound for the rate of convergence of this method. Its proof is based on some old ideas of L. V. Kantorovich (the Nobel Prize Winner for Economy in 1975). It essentially uses properties of the Chebyshev polynomials (for a detailed proof in complex arithmetic see, e.g., [12]). Today the algorithm of the conjugate gradient method belongs to standard topics in numerical analysis textbooks.

**Acknowledgement**: The author thanks Prof. Dr. Lawrence Somer (Washington, D.C.) for valuable suggestions.

## References

[1] J. Brandts, H. A. van der Vorst, *The convergence of Krylov methods and Ritz values*, in: Conjugate Gradient Algorithms and Finite Element Methods, Springer-Verlag, Berlin, 2004, pp. 47–68.

[2] J. W. Cooley, J. W. Tukey, *An algorithm for the machine calculation of complex Fourier series*, Math. Comp. **19** (1965), 297–301.

[3] J. W. Daniel, *The conjugate gradient method for linear and nonlinear operator equations*, SIAM J. Numer. Anal. **4** (1967), 10–26.

[4] G. C. Danielson, C. Lanczos, *Some improvements in practical Fourier analysis and their application to X-ray scattering from liquids*, J. Franklin Inst. **233** (1942), 365–380.

[5] G. H. Golub, D. P. O'Leary, *Some history of the conjugate gradient and Lanczos algorithms: 1948–1976*, SIAM Rev. **31** (1989), 50–102.

[6] M. R. Hestenes, *Extension of the range of a differenciable function*, Duke Math. J. **8** (1941), 183–192.

[7] M. R. Hestenes, *Applications of the theory of quadratic forms in Hilbert space to the Calculus of Variations*, Pacific J. Math. **1** (1951), 525–581.

[8] M. R. Hestenes, *Conjugate direction methods in optimizations*, Springer-Verlag, New York, 1980.

[9] M. R. Hestenes, *Conjugacy and gradients*, Princeton History Conference (Summer 1987).

[10] M. R. Hestenes, E. Stiefel, *Methods of conjugate gradients for solving linear systems*, J. Res. Nat. Bur. Standards **49** (1952), 406–436.

[11] S. Kaniel, *Estimates for some computational techniques in linear algebra*, Mat. Comp. **20** (1966), 369–378.

[12] M. Křížek, P. Neittaanmäki, *Mathematical and numerical modelling in electrical engineering*, Kluwer, Dordrecht, 1996.

[13] C. Lanczos, *An iterations method for the solution of the eigenvalue problem of linear differential and integral operators*, J. Res. Nat. Bur. Standards **45** (1950), 255–282.

[14] G. Meinardus, *Über eine verallgemeinerung einer ungleichung von L. V. Kantorovich*, Numer. Math. **5** (1963), 14–23.

[15] W. H. Press, B. P. Flannery, S. A. Teukolsky, and W. T. Vetterling, *Numerical Recipes: The Art of Scientific Computing*, Cambridge Univ. Press, Cambridge, 1986, 1987, 1988.

[w1] `http://www.physics.ncsu.edu/lanczos/`

[w2] `http://www-groups.dcs.st-and.ac.uk/~history/Mathematicians/Lanczos.html`

[w3] `http://www.epemag.com/zuse/part6c`

[w4] `http://math.nist.gov/mcsd/highlights/cg50.html`

# Conjugate Gradients and Finite Elements – a Golden Jubilee

Jan Brandts

Korteweg-de Vries Institute, Faculty of Science, University of Amsterdam, Plantage Muidergracht 24, 1018 TV Amsterdam, Netherlands

brandts@science.uva.nl

*Dedicated to CG and FE*
*on the occasion of their 50th wedding anniversary*

## 1 Introduction

Both the finite element method for elliptic partial differential equations, and the conjugate gradient method for linear systems with symmetric positive definite system matrix, are instances of *Ritz-Galerkin projection* on a finite dimensional subspace. In the finite element method, the problem under approximation is infinite dimensional, whereas in the conjugate gradient method it is finite, though usually high dimensional. We will briefly recall both methods, and then comment on using conjugate gradients to solve the finite elements system.

### 1.1 The Ritz-Galerkin method for elliptic PDEs

Consider the standard weak formulation of an elliptic second-order partial differential equation on a domain $\Omega$ with homogeneous Dirichlet boundary conditions, which is to find $u \in H_0^1(\Omega)$ such that

$$a(u, y) = \langle f, y \rangle \quad \text{for all } \; y \in H_0^1(\Omega). \tag{1}$$

In this context it is assumed (for more details see also Section 2.1) that

- $a\colon H_0^1(\Omega) \times H_0^1(\Omega) \to \mathbb{R}$ is a continuous, symmetric, coercive bilinear form,
- $f \in H^{-1}(\Omega)$,
- $\langle \cdot, \cdot \rangle\colon H^{-1}(\Omega) \times H^1(\Omega) \to \mathbb{R}$ is the dual pairing.

Given an $n$-dimensional subspace $\mathcal{W} \subset H_0^1(\Omega)$, the Ritz-Galerkin approximation $\omega \in \mathcal{W}$ of $u$ is defined by requiring that the residual $a(\omega - u, \cdot) \in H^{-1}(\Omega)$ is in the annihilator $\mathrm{Ann}(\mathcal{W})$ of $\mathcal{W}$, where

$$\mathrm{Ann}(\mathcal{W}) = \{g \in H^{-1}(\Omega) \mid \langle g, \xi \rangle = 0 \quad \text{for all } \; \xi \in \mathcal{W}\}. \tag{2}$$

This can be expressed as $a(\omega - u, \xi) = 0$ for all $\xi \in \mathcal{W}$, or equivalently in the more familiar form

$$a(\omega, \xi) = \langle f, \xi \rangle \quad \text{for all } \; \xi \in \mathcal{W}. \tag{3}$$

Let $w_1, \ldots, w_n$ be a basis for $\mathcal{W}$ and let $e_1, \ldots, e_n$ be the canonical basis for $\mathbb{R}^n$. By $W\colon \mathbb{R}^n \to \mathcal{W}$ we denote the linear isomorphism defined by $We_j = w_j$ for all $j \in \{1, \ldots, n\}$. Since $\omega \in \mathcal{W}$, there exists a vector $x \in \mathbb{R}^n$ such that $\omega = Wx$. Substituting this into (3) and choosing for $\xi$ the basis functions $w_1, \ldots, w_n$ consecutively, yields $n$ linear equations for the $n$ unknown entries of $x$. These equations can be written as an $n \times n$ matrix-vector equation $Ax = b$, in which the entries $(a_{ij})$ of $A$ are given by $a_{ij} = a(w_i, w_j)$, and $b = (b_i)$ with $b_i = \langle f, w_i \rangle$. In the following diagram, which symbolizes the whole procedure, $G$ stands for Ritz-Galerkin approximation.

$$H_0^1(\Omega) \xrightarrow{\quad G \quad} \mathcal{W} \xleftarrow{\quad W \quad} \mathbb{R}^n$$

### 1.1.1 Finite Elements

In the finite element method, $\mathcal{W}$ is usually chosen as a space of piecewise polynomials relative to a face-to-face partition of the domain, for instance, the space of continuous piecewise linear functions relative to a partition of the domain into simplices. For details on the finite element method, we refer to the three excellent textbooks [3], [6], [7].

## 1.2 The Ritz-Galerkin method for linear systems

Let $A$ be a real symmetric positive definite $n \times n$ matrix and $b \in \mathbb{R}^n$ and consider the linear system $Ax = b$. Even though matrices are commonly identified with linear mappings, it is not often realized that each $n \times n$ matrix $A$ also represents a bilinear form on $\mathbb{R}^n \times \mathbb{R}^n$. In fact, similar as in Section 1.1, $x$ can be seen as the unique solution of a variational formulation of $Ax = b$, which is to find $x \in \mathbb{R}^n$ such that

$$(Ax, z) = (b, z) \quad \text{for all } z \in \mathbb{R}^n, \tag{4}$$

where $(y, z)$ is the Euclidean inner product on $\mathbb{R}^n$ with induced norm $\|z\|^2 = (z, z)$. The similarity with (3) is obscured by the fact that one usually identifies $\mathbb{R}^n$ with its dual $(\mathbb{R}^n)^*$, but notice that we could indeed have worked with the dual pairing on $(\mathbb{R}^n)^* \times \mathbb{R}^n$ in the right-hand side of (4).

**Observation 1.1.** Since $A$ is symmetric positive definite, it has a unique Cholesky factorization $A = U^*U$, where $U$ has positive diagonal entries (see [9]). Hence,

$$(Ax, z) = (U^*Ux, z) = (Ux, Uz) = (b, z). \tag{5}$$

This takes the role of the usual integration by parts in the finite element method for the Poisson equation $-\Delta u = f$ with homogeneous Dirichlet boundary conditions,

$$(-\Delta u, y) = (-\operatorname{div} \nabla u, y) = (\nabla u, \nabla y) = \langle f, y \rangle. \tag{6}$$

Since $U$ is unknown, in practice this step can not be performed. In theory, it does not need to be performed, since contrary to the finite element method, it does not weaken the equation.

Let $\mathcal{V}$ be a $k$-dimensional subspace of $\mathbb{R}^n$. The *Ritz-Galerkin* approximation of $x$ is the unique $x_k \in \mathcal{V}$ such that $(A(x_k - x), \cdot) \in (\mathbb{R}^n)^*$ is in the annihilator $\mathrm{Ann}(\mathcal{V})$ of $\mathcal{V}$. Or equivalently, its Riesz representation $Ax_k - b \in \mathbb{R}^n$ should be orthogonal to $\mathcal{V}$ as depicted in Fig. 1. This orthogonality can be expressed similarly as in the context of elliptic partial differential equations as follows: find $x_k \in \mathcal{V}$ such that

$$(Ax_k, v) = (b, v) \quad \text{for all } v \in \mathcal{V}. \tag{7}$$

Let the columns $v_1, \ldots, v_k$ of the $n \times k$ matrix $V$ form a basis for $\mathcal{V}$, then $x_k = Vq$ for some $q \in \mathbb{R}^k$. Substituting this into (7), it can be seen that $q$ solves the $k \times k$ matrix-vector equation $Tq = V^*b$ with $T = V^*AV$. Notice that similar as in Section 1.1, the entries $t_{ij}$ of $T$ are given by $t_{ij} = (Av_i, v_j)$. Writing $x_k$ as a function of $x$, we have $x_k = Rx$ with $R = VT^{-1}V^*A$. Similar as above, we can depict the method in a diagram as follows.

$$\mathbb{R}^n \xrightarrow{\quad R \quad} \mathcal{V} \xleftarrow{\quad V \quad} \mathbb{R}^k$$

### 1.2.1 Conjugate Gradients

Choosing $\mathcal{V} = \mathcal{K}^k(A, b) = \mathrm{span}\{b, Ab, \ldots, A^{k-1}b\}$, the conjugate gradient method arises. In its implementation, a basis $v_1, \ldots, v_k$ is computed recursively: $v_k$ is defined as the normalized residual $b - Ax_{k-1}$, where $x_{k-1}$ is the conjugate gradient approximation of $x$ in $\mathcal{K}^{k-1}(A, b)$, and without loss of generality we may set $x_0 = 0$. By (7), $v_k$ is orthogonal to $\mathcal{K}^{k-1}(A, b)$. As an additional advantage, all solutions $x_k$ and basis vectors $v_k$, with $k = 1, \ldots, n$, can be computed by means of two coupled two-term recursions in which only one matrix-vector multiplication with $A$ per iteration is involved.

For details on the conjugate gradient method, we refer to the original paper by Hestenes and Stiefel [10], and to [5] in Part II of this volume.

## 1.3 Ritz-Galerkin projection of a Ritz-Galerkin projection

Suppose now that the linear system $Ax = b$ from Section 1.2 is, in fact, the result of the Ritz-Galerkin discretization of the elliptic problem in Section 1.1. The space $\mathcal{V} \subset \mathbb{R}^n$ used in Section 1.2 then corresponds, by means of the isomorphism $W$, to a $k$-dimensional subspace $\mathcal{W}_\mathcal{V} \subset \mathcal{W} \subset H_0^1(\Omega)$. As such, $\mathcal{W}_\mathcal{V}$ can also be used to approximate $u$ using Ritz-Galerkin projection, which results in an approximation $\omega_\mathcal{V} \in \mathcal{W}_\mathcal{V}$ defined by

$$a(\omega_\mathcal{V}, \eta) = \langle f, \eta \rangle \quad \text{for all } \eta \in \mathcal{W}_\mathcal{V}. \tag{8}$$

A basis for $\mathcal{W}_\mathcal{V}$ is already available through the composition $U = WV$, i.e., $u_i = Wv_i = WVe_i$. Using this basis to solve (8) results in a $k \times k$

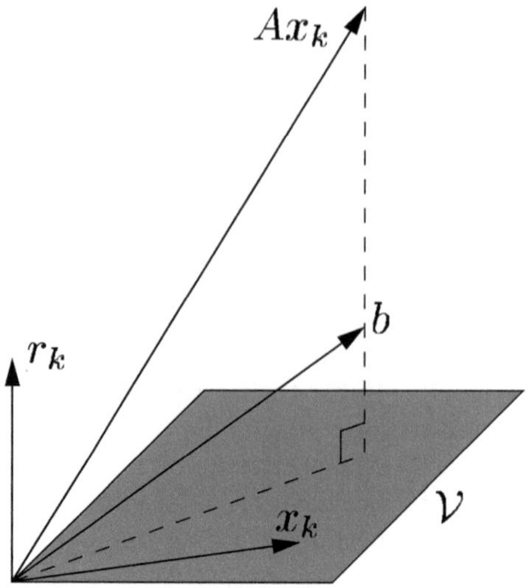


Figure 1. The vector $x_k \in \mathcal{V}$ is such that $r_k = b - Ax_k \perp \mathcal{V}$.

linear system similar as in Section 1.1. The long diagonal in the diagram below symbolizes this procedure. The symbol $H$ stands for the Ritz-Galerkin approximation (8) in $\mathcal{W}_{\mathcal{V}}$. The other edges of the large triangle are the methods from the previous two sections. Additionally, $r\colon \mathcal{W} \to \mathcal{W}_{\mathcal{V}}$ given by $r = WRW^{-1}$ symbolizes $R$ on the functional level.

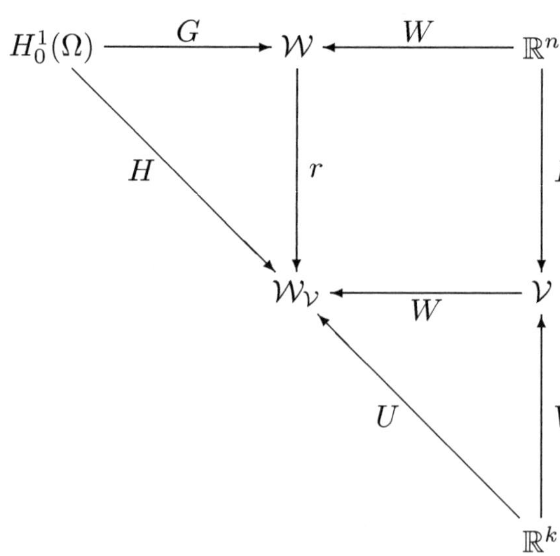


(9)

The following result is well known and should therefore not come as a surprise. Loosely speaking, it states that the Ritz-Galerkin projection of a

Ritz-Galerkin projection equals the Ritz-Galerkin projection on the smallest space.

**Proposition 1.2.** *We have that* $H = rG$. *In other words,* $\omega_{\mathcal{V}} = Wx_k$, *where* $x_k$ *solves* (7) *and* $\omega_{\mathcal{V}}$ *solves* (8).

Proof. Each $\eta \in \mathcal{W}_{\mathcal{V}}$ can be written as $Wv$ with $v \in \mathcal{V}$, and in particular $\omega_{\mathcal{V}} = W\tilde{x}_k$ for some $\tilde{x}_k \in \mathcal{V}$. From Section 1.1 we recall that $\omega = Wx$. Since $\mathcal{W}_{\mathcal{V}} \subset \mathcal{W}$, from (3) and (8) we find that $a(\omega_{\mathcal{V}}, \eta) = a(\omega, \eta)$ for all $\eta \in \mathcal{W}_{\mathcal{V}}$, and thus,

$$a(W\tilde{x}_k, Wv) = a(Wx, Wv) \quad \text{for all } v \in \mathcal{V}. \tag{10}$$

Now, by definition of $A$ we have for all $y, z \in \mathbb{R}^n$ that $(Ay, z) = a(Wy, Wz)$. From this we conclude that (10) is equivalent to $(A\tilde{x}_k, v) = (Ax, v) = (b, v)$ for all $v \in \mathcal{V}$ and since (7) has a unique solution, we find that $\tilde{x}_k = x_k$. □

# 2 Comparing results and theorems in both settings

Both the finite element method and the conjugate gradient method are Ritz-Galerkin projections in the *energy inner product.* Hence we may expect that some of their characteristic properties exhibit similarities. In this section, we will explore those similarities.

**Remark 2.1.** In the sequel, we will denote vector norms by $|||\cdot|||_*$ and function norms by $\|\cdot\|_*$. Contrary to common use, but for conformation of notation, the Sobolev 1-seminorm will be denoted by $\|\cdot\|_1$. See Section 2.1 below.

## 2.1 Céa's Lemma and a priori bounds

Define the *energy norm* in the finite element context for all $w \in H_0^1(\Omega)$ as

$$\|w\|_a^2 := a(w, w). \tag{11}$$

As a special case, denote by $\|\cdot\|_1$ the energy norm corresponding to the Poisson equation, for which the bilinear form takes the simple form $a(w, w) = (\nabla w, \nabla w)$. This latter norm is usually employed to analyze the behaviour of the Ritz-Galerkin method for more general elliptic equations. First, it is proved that the bilinear form $a$ is continuous and coercive in the sense that there exist constants $m > 0$, $\alpha > 0$ such that

$$\forall v, w \in H_0^1(\Omega), \quad |a(v, w)| \leq m\|v\|_1\|w\|_1 \quad \text{and} \quad a(v, v) \geq \alpha^2\|v\|_1^2. \tag{12}$$

Using these properties together with the Galerkin orthogonality $a(u - \omega, \xi) = 0$ for all $\xi \in \mathcal{W}$, we find that

$$\alpha^2\|u - \omega\|_1^2 \leq a(u - \omega, u - \omega) = a(u - \xi, u - \omega) \leq m\|u - \xi\|_1\|u - \omega\|_1, \tag{13}$$

with as a result that

$$\|u - \omega\|_1 \leq \frac{m}{\alpha^2} \inf_{\xi \in \mathcal{W}} \|u - \xi\|_1. \tag{14}$$

So, the Galerkin orthogonality shows that $\omega$ is the energy-orthogonal projection of $u$ onto $\mathcal{W}$, and hence the best approximation of $u$ in the energy-norm, whereas (14) shows that in the $\|\cdot\|_1$-norm we only have *quasi-optimality.* This result is known as Céa's Lemma [3], [6], [7]. In return for the loss of optimality, we have that the $\|\cdot\|_1$-norm is generally much easier to analyze.

**Observation 2.2.** Roughly speaking, Céa's Lemma states that each "decent enough" elliptic equation is "equivalent" to the Poisson equation. This justifies the fact that the Poisson equation is often studied as model problem.

The conjugate gradient analogue is as follows. Define the $A$-norm, sometimes also called *energy norm*, by

$$|||z|||_A^2 := z^* A z = \left(\sqrt{A} z, \sqrt{A} z\right). \tag{15}$$

If $A$ does not result from the discretization of an elliptic PDE, then no specific reference equation (like the Poisson equation above) can be indicated. In that case we merely find, using the Galerkin orthogonality $V^* A(x_k - x) = 0$ and the Cauchy-Schwarz inequality $|y^* A z| \leq |||y|||_A \, |||z|||_A$ that for all $v \in \mathcal{V}$,

$$\begin{aligned} |||x - x_k|||_A^2 = (x - x_k)^* A(x - x_k) &= (x - v)^* A(x - x_k) \\ &\leq |||x - v|||_A \, |||x - x_k|||_A. \end{aligned} \tag{16}$$

The equality $\|y\|_a = |||Wy|||_A$ for all $y \in \mathbb{R}^n$ allows us to define a discrete version of the $\|\cdot\|_1$-norm by

$$|||z|||_1 := \|Wz\|_1. \tag{17}$$

Similar as in (14), we get that

$$|||x - w|||_1 \leq \frac{m}{\alpha^2} \inf_{z \in \mathcal{V}} |||x - z|||_1. \tag{18}$$

This shows that conjugate gradients, which is clearly optimal in the $A$-norm as a result of (16) is quasi-optimal in other "elliptic" norms than the $A$-norm.

### 2.1.1 A priori bounds

In both the finite element method and the conjugate gradient method, the analysis of the error in the energy norm is usually done by explicit construction of an element from the discrete space with suitable approximation properties. In the finite element context, $\xi$ in (14) is chosen as a local interpolant $\pi u$ of the exact solution $u$, turning the problem of finding error bounds for the method into finding error bounds for an interpolation scheme,

$$\|u-\omega\|_1 \leq \frac{m}{\alpha^2}\|(I-\pi)u\|_1. \tag{19}$$

In the conjugate gradient method, it is observed that each $v \in \mathcal{K}^k(A,b)$ can be written as

$$v = P(A)b = P(A)Ax = AP(A)x, \tag{20}$$

where $P$ is a polynomial of degree $k-1$ in $A$. As a consequence of this and (16) we find that

$$|||x-w|||_A \leq |||(I-AP(A))x|||_A. \tag{21}$$

In both (19) and (21), the original problem is decoupled from its context of Ritz-Galerkin projection, and standard approximation theory can be used to analyze the remaining expressions in the respective right-hand sides.

### 2.2 Duality estimate, Nitsche trick

In the continuous setting, a bound in the $L^2$-norm (denoted by $\|\cdot\|_0$) is derived from the energy norm bound by means of the so-called Nitsche trick, which involves a duality argument. Given the error $u-\omega$, an auxiliary elliptic problem

$$a(y,\rho) = (u-\omega, y) \quad \text{for all } y \in H_0^1(\Omega) \tag{22}$$

is introduced. Then, using the Galerkin orthogonality of the error and the Cauchy-Schwarz inequality, we find that

$$\|u-\omega\|_0^2 = (u-\omega, u-\omega) = a(u-\omega,\rho) \leq \|u-\omega\|_a \inf_{\xi\in\mathcal{W}} \|\rho-\xi\|_a, \tag{23}$$

which can be rewritten as

$$\|u-\omega\|_0 \leq \|u-\omega\|_a \left(\inf_{\xi\in\mathcal{W}} \frac{\|\rho-\xi\|_a}{\|u-\omega\|_0}\right). \tag{24}$$

The term between parentheses is the improvement of the $L^2$-bound over the energy-norm bound. It signifies, relative to the given data in $L^2$, how well the auxilliary solution $\rho$ can be approximated in $\mathcal{W}$ in the energy norm. In a typical application with $\Omega$ convex and employing continuous piecewise polynomial finite elements, this term is asymptotically proportional to $h(\mathcal{W})$, which, together with $d(\mathcal{W})$, is defined as

$$h = h(\mathcal{W}) = \inf_{0\neq y\in\mathcal{W}} \frac{\|y\|_0}{\|y\|_a} \quad \text{and} \quad d = d(\mathcal{W}) = \sup_{0\neq y\in\mathcal{W}} \frac{\|y\|_0}{\|y\|_a} \tag{25}$$

and which are called the *meshsize* and *diameter* of $\mathcal{W}$, respectively. The Poincaré inequality shows that $d_0 := d(H_0^1(\Omega)) < d(H^1(\Omega)) = \infty$, whereas $h(H_0^1(\Omega)) = 0$.

**Remark 2.3.** In the finite element context, $h$ is usually defined as the largest diameter of an element from the partition of the domain, after which the *discrete inverse inequality* $\|\xi\|_a \leq Ch^{-1}\|\xi\|_0$ for all $\xi \in \mathcal{W}$, is derived from this definition for some constant $C > 0$. Here, conversely, we axiomatized the discrete inverse inequality, in order to be able to associate a parameter $h$ to arbitrary subspaces $\mathcal{W} \subset H_0^1(\Omega)$.

**Proposition 2.4.** *Let $\mathcal{W}_0 \subset \mathcal{W}_1 \subset \cdots \subset H_0^1(\Omega)$ and set $h_\ell = h(\mathcal{W}_\ell)$. Then,*

$$\sup_{u \in H_0^1(\Omega)} \left( \lim_{\ell \to \infty} \inf_{y \in \mathcal{W}_\ell} \frac{\|u - y\|_a}{\|u\|_a} \right) = 0 \Rightarrow \lim_{\ell \to \infty} h_\ell = 0, \tag{26}$$

*showing that $h_\ell \to 0$ is a necessary condition for convergence in the energy norm of the Ritz-Galerkin method to any possible solution $u \in H_0^1(\Omega)$.*

Proof. The sequence $(h_\ell)_{\ell \geq 0}$ is decreasing and must have a limit $h_\infty$. Thus, for all $\ell \geq 0$ and all $y \in \mathcal{W}_\ell$ we have that $\|y\|_0 \geq h_\infty \|y\|_a$. Since $h(H_0^1(\Omega)) = 0$ we may choose $u \in H_0^1(\Omega)$ such that $\|u\|_0 \leq \frac{1}{2} h_\infty \|u\|_a$. Using the triangle inequality twice, we find that for all $\ell \geq 0$ and all $y \in \mathcal{W}_\ell$,

$$\begin{aligned} d_0 \|u - y\|_a \geq \|u - y\|_0 \geq \|y\|_0 - \|u\|_0 &\geq h_\infty \|y\|_a - \frac{h_\infty}{2} \|u\|_a \\ &\geq \frac{h_\infty}{2} \|u\|_a - h_\infty \|u - y\|_a, \end{aligned}$$

and hence, after some rearrangement of terms, that for all $\ell \geq 0$,

$$\inf_{y \in \mathcal{W}_\ell} \frac{\|u - y\|_a}{\|u\|_a} \geq \frac{h_\infty}{2(h_\infty + d_0)}. \tag{27}$$

This means that if $h_\infty$ does not equal zero, there exists a function $u \in H_0^1(\Omega)$ such, that no sequence $(\omega_\ell)_{\ell \geq 0}$ with $\omega_\ell \in \mathcal{W}_\ell$ converges to $u$ in the energy norm. □

**Observation 2.5.** We have just shown that $h_\ell \to 0$ is a necessary condition for convergence in the energy norm of the finite element method for both the problem in $H_0^1(\Omega)$ and the auxilliary problem using the sequence $\mathcal{W}_\ell$. Since the factor between parentheses in (24) is often proportional to $h_\ell$, the $L^2$ errors for both problems are even of higher order than the energy norm errors.

We wish to emphasize that with the definition $h_\ell = h(\mathcal{W}_\ell)$ as in (25), the condition number of the matrix $A_\ell$ resulting from discretization of a second order elliptic equation in $\mathcal{W}_\ell$ is of magnitude $\mathcal{O}(h_\ell^{-2})$. This shows that the notorious ill-conditioning of the finite element method for such problems results, in fact, from the wish that the method should converge for all possible solutions $u \in H_0^1(\Omega)$.

In the finite dimensional context, and in particular the context of conjugate gradients, this is different. Given the equation $Ax = b$, the Ritz-Galerkin subspace $K^k(A,b)$ explicitly depends on $b$, and this makes conjugate gradients into an *adaptive method.* There is no specific reason why $K^k(A,b)$ should contain good approximations of the solution $c$ of the auxilliary problem $Ac = x - x_k$. Of course, we can still write

$$|||x - x_k|||_0^2 = (Ac, x - x_k) \leq \inf_{v \in \mathcal{V}} |||c - v|||_A \, |||x - x_k|||_A, \tag{28}$$

and similar as before, we find that

$$|||x - x_k|||_0 \leq \left( \inf_{v \in \mathcal{V}} \frac{|||c - v|||_A}{|||Ac|||_0} \right) |||x - x_k|||_A. \tag{29}$$

However, since conjugate gradients *specializes* in the approximation of $Ax = b$, the Krylov subspace generally does not have approximation quality for the equation $Ac = x - x_k$ as well.

## 3 Multilevel methods and conjugate gradients

Even though the conjugate gradient method is our main interest, it is clear from Proposition 1.2 that no specific property of the conjugate gradient method is used. Hence, each Ritz-Galerkin projection of the linear system $Ax = b$ on a subspace $\mathcal{V}$ gives rise to the relation $H = rG$ in diagram (9). We recommend Chapter V in [3] for more details on what follows in this section, including a lucid exposition of the space decomposition analysis by Bramble et al. [4].

### 3.1 Coarse grid corrections in the multigrid method

A special case of projection on a subspace is projection of a finite element system $Ax = b$ on the space $\mathcal{V}$ for which $\mathcal{W}_\mathcal{V}$ is a subspace of finite element functions belonging to a coarser grid. From Proposition 1.2 it can be concluded that this results in the finite element system *belonging* to that coarser grid. This is a central property in the multigrid method. In the multigrid setting, one works explicitly with a *restriction* operator $r\colon \mathcal{W} \to \mathcal{W}_\mathcal{V}$ and a *prolongation* operator $p\colon \mathcal{W}_\mathcal{V} \to \mathcal{W}$. Denoting their algebraic counterparts by

$$R\colon \mathbb{R}^n \to \mathcal{V}, \;\; z \mapsto W^{-1}rWz, \quad \text{and} \quad P\colon \mathcal{V} \to \mathbb{R}^n, \;\; v \mapsto W^{-1}pWv, \tag{30}$$

a coarse grid correction aims to find an element $x_k \in \mathbb{R}^k$ for which

$$RAPx_k = Rb. \tag{31}$$

For simplicity of implementation as well as for ease of formulation of convergence proofs, $R$ is usually chosen to be equal to $P^*$, where $P$ is chosen such that $p$ is the trivial inclusion of $\mathcal{W}_\mathcal{V}$ in $\mathcal{W}$. In that case, (31) is easily seen to be a valid Ritz-Galerkin projection on $\mathcal{V}$. But even if $P$ and $R$ are chosen differently, as long as the column span of both $P$ and $R^*$ equal $\mathcal{V}$, the requirement $R(APx_k - b) = 0$ still orthogonalizes the residual to $\mathcal{V}$, hence $x_k$ is the same as when $R = P^*$. Notice, however, that (31) needs to be solved somehow, and that in general only the choice $R = P^*$ results in a symmetric positive definite matrix $RAP$. Upon completion of a coarse-grid correction, either *smoothing steps* with a classical solver can be applied, in order to prepare the resulting error for another coarse grid correction, or a correction in a *complementary subspace* can be performed, as we will briefly discuss now.

### 3.2 Schwarz alternating method

Given a subspace $\mathcal{V}$ and its functional counterpart $\mathcal{W}_\mathcal{V}$, suppose that a basis is available for a complementary space $\mathcal{Z}$, i.e., a space for which

$$\mathcal{W} = \mathcal{W}_\mathcal{V} \oplus \mathcal{Z}. \tag{32}$$

If $\mathcal{Z}$ happens to be orthogonal to $\mathcal{W}_\mathcal{V}$ with respect to the energy inner product $a$, i.e., if

$$\forall w \in \mathcal{W}_\mathcal{V}, \quad \forall z \in \mathcal{Z}, \quad a(w,z) = 0, \tag{33}$$

then the difference $\zeta := \omega - \omega_\mathcal{V} \in \mathcal{Z}$, and since we assumed that a basis is available for $\mathcal{Z}$, it can be solved from

$$a(\zeta, z) = a(\omega - \omega_\mathcal{V}, z) = \langle f, z\rangle - a(\omega_\mathcal{V}, z) \ \text{ for all } \ z \in \mathcal{Z}, \tag{34}$$

where in fact, in the case of an $a$-orthogonal $\mathcal{Z}$, the term $a(\omega_\mathcal{V}, z)$ vanishes. Although a basis for the orthogonal complement of $\mathcal{W}_\mathcal{V}$ is generally not available, it is often not difficult to construct a basis for a non-orthogonal complementary space $\mathcal{Z}$. Notice that since $\omega - \omega_\mathcal{V}$ remains $a$-orthogonal to $\mathcal{W}_\mathcal{V}$, it is not an element of $\mathcal{Z}$ anymore. If, however, for such a $\mathcal{Z}$ one is able to prove the following *strengthened Cauchy-Schwarz inequality* with $\gamma < 1$,

$$\forall w \in \mathcal{W}_\mathcal{V}, \ \forall z \in \mathcal{Z}, \quad a(w,z) \le \gamma \|w\|_a \|z\|_a, \tag{35}$$

the $\zeta \in \mathcal{Z}$ satisfying (34) yields an improved approximation $\omega_\mathcal{V} + \zeta$ of $\omega$ in the sense that

$$\|\omega - (\omega_\mathcal{V} + \zeta)\|_a \le \gamma \|\omega - \omega_\mathcal{V}\|_a. \tag{36}$$

The case $\gamma = 0$ corresponds to $\mathcal{Z}$ being the orthogonal complement of $\mathcal{W}_\mathcal{V}$ in $\mathcal{W}$. Repeating this procedure by solving a correction to $\omega_\mathcal{V} + \zeta$ in $\mathcal{W}_\mathcal{V}$ again, results in the Schwarz alternating method, which therefore consists of iterated projections on complemetary subspaces. This idea can be applied recursively.

**Remark 3.1.** In the finite element method, many results are available on strengthened Cauchy-Schwarz inequalities in different settings. As an example, the splitting of the continuous piecewise linear functions into a space of coarse grid functions and their hierarchical surplus resulting from uniform refinement leads to a strengthened Cauchy-Schwarz inequality, and so does the splitting of the continuous piecewise quadratic functions into linears and quadratic bubbles. See [1] for details and references.

### 3.3 Cascadic algorithms: CG and FEM in perfect harmony

A relatively recent combination of conjugate gradients and the use of discretizations in nested subspaces, is the cascade algorithm of Bornemann and Deuflhard [2]. The main idea can be illustrated as follows. Suppose that

$$\cdots \subset \mathcal{W}_{8h} \subset \mathcal{W}_{4h} \subset \mathcal{W}_{2h} \subset \mathcal{W}_{h} \subset H_0^1(\Omega) \tag{37}$$

and that $\mathcal{W}_{jh}$ consists of continuous piecewise linear functions relative to a partition with mesh size $jh$. For example, the spaces could be constructed by uniform refinement of an initial mesh. Denote the matrix corresponding to $\mathcal{W}_{jh}$ by $A_{jh}$ and let $\kappa_{jh}$ be its condition number, i.e., the ratio of its largest and smallest singular value. Write $\omega_{jh}$ for the Ritz-Galerkin approximation of the exact solution $u \in H_0^1(\Omega)$ in $\mathcal{W}_{jh}$. Hence, $\omega_h$ is the approximation on the finest mesh, $\omega_{2h}$ on the next finest mesh, and so on. Typically, we have that

- $\lim_{h\to 0} \|u - \omega_{2jh}\|_a / \|u - \omega_{jh}\|_a = 2$,
- $\lim_{h\to 0} \kappa_{jh}/\kappa_{2jh} = 4$.

The main idea in cascade is, to compute approximations $\chi_{jh}$ of $\omega_{jh}$ recursively as given by steps (C1)–(C4), where $(\varepsilon_{jh})_j$ is a sequence of tolerances with $\varepsilon_h = \mathcal{O}(h)$.

(C1) Assume that $\chi_{2jh} \in \mathcal{W}_{2jh}$ is such that $\|\omega_{2jh} - \chi_{2jh}\|_a \le \varepsilon_{2jh}$.
(C2) Use conjugate gradients to approximate $\omega_{jh}$, with $\chi^0 = \chi_{2jh}$ as start function.
(C3) Stop as soon as the $k$-th iterate $\chi^k$ satisfies $\|\omega_{jh} - \chi^k\|_a \le \varepsilon_{jh}$.
(C4) Set $\chi_{jh} = \chi^k$.

The question is, if this sequence can be designed such that the overall computational complexity needed to arrive at the approximation $\chi_h$ of $\omega_h$ is optimal. The answer is affirmative, even though at first glance this seems impossible. Indeed, suppose that we have only two grids, that $\chi_{2h} = \omega_{2h}$, and that this has been achieved in optimal complexity. If we use $\omega_{2h}$ as start of a conjugate gradient iteration to compute $\omega_h$, then the standard a priori bound for conjugate gradients tells us that the iterates $\chi^k$, with $\chi^0 = \omega_{2h}$, satisfy

$$\|\omega_h - \chi^k\|_a \le 2\left(\frac{\sqrt{\kappa_h}-1}{\sqrt{\kappa_h}+1}\right)^k \|\omega_h - \chi^0\|_a. \tag{38}$$

The initial error $\|\omega_h - \chi^0\|_a$ then needs to be decreased roughly by a factor of three, since the mutual distance between $\omega_h$ and $\omega_{2h}$ may be three times as big as the error in $\omega_h$. Moreover, this should be achieved in a number $k$ of iterations that is independent of the level (and hence of $h$). However, the requirement

$$2\left(\frac{\sqrt{\kappa_h}-1}{\sqrt{\kappa_h}+1}\right)^k \leq \frac{1}{3} \tag{39}$$

already leads to $k = \mathcal{O}(h)$. This does not give us what we would like, but that is due to a shortcoming of (38) and not of the cascade idea. The point is that (38) holds for arbitrary start function $\chi^0$, whereas the $\chi^0$ in cascade is very special: it is equal, or close to, the exact solution on the next coarser grid. As such, the difference $\omega_h - \chi^0 \in \mathcal{W}_h$ is almost orthogonal to $\mathcal{W}_{2h}$. An analysis of the convergence of conjugate gradients, starting with the exact solution on the next coarser level can be performed similarly as for the general start, using Chebyshev polynomials with carefully designed properties. As a result, a sequence $(\varepsilon_{jh})_j$ can be found with the desired properties. Relatively high accuracy is required on the coarser levels. However, smaller matrices and condition numbers on the coarser levels guarantee the overall optimal complexity. We refer to [2], [8] for details.

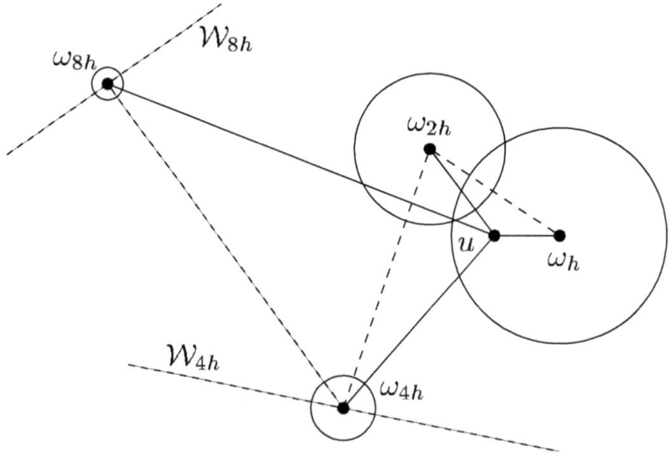


Figure 2. Cascadic conjugate gradients: Use conjugate gradients to get close enough to the solution on a certain level, and then go to the next finer level.

**Remark 3.2.** Since $\mathcal{W}_{2h} \perp (\omega_h - \omega_{2h}) \in \mathcal{W}_h$, this initial error is highly oscillatory in $\mathcal{W}_h$. The fact that conjugate gradients quickly reduces this error shows that it is a good *smoother*. Also other smoothers can be used in combination with the multilevel approach, giving rise to other cascadic algorithms.

### 3.4 Conjugate gradients as adaptive method

In the multilevel methods as described above, the subspace $\mathcal{V} \subset \mathbb{R}^n$ does not depend on the operator $A$ nor on the right-hand side $b$. In the conjugate gradient method, this is completely the opposite. After all, it uses the Krylov subspace $\mathcal{K}^k(A,b)$, which is impossible to construct before discretization in a larger finite element space. As such, conjugate gradients may be seen as an *adaptive* method: the residual corresponding to the current approximation is used to construct an additional dimension of the subspace. Since

$$r_k = b - Ax_k = A(x - x_k), \tag{40}$$

the residual contains information about the error, and hence, expansion of the subspace with $r_k$ may be interpreted as an unsuccessful though not unnatural attempt to expand the space with the error: hopefully, $r_k$ will approximate $x - x_k$ in some sense. Notice that if the subspace is enriched with the *exact* error, by linearity it will then also contain the exact solution $x$, hence $x_{k+1} = x$.

Summarizing, we might say that the finite element method and conjugate gradients are two aspects of the same problem, that of solving a continuous problem. The finite element method *a priori* requires a space with sufficient approximation quality, hence a good approximation exists in this space and will be the solution of the finite element linear system. The conjugate gradient method tries to find this solution as quickly as possible, by adaptively building a subspace of the finite element space that contains good approximations of the finite element solution.

**Acknowledgments.** The author was supported by a Research Grant (1999–2004) of the *Royal Netherlands Academy of Arts and Sciences.* He thanks Michal Křížek and Tomáš Vejchodský from the Academy of Sciences of the Czech Republic for their comments on an earlier version of this paper.

## References

[1] O. Axelsson, R. Blaheta, *Two simple derivations of universal bounds for the CBS inequality constant,* Appl. Math. (2003), accepted.

[2] F. Bornemann, P. Deuflhard, *The cascadic multigrid method for elliptic problems,* Numer. Math. **75** (1996), 135–152.

[3] D. Braess, *Finite Element: Theory, Fast Solvers, and Applications in Solid Mechanics,* University Press, Cambridge; Second Edition, 2001.

[4] J. H. Bramble, J. Pasciak, J. Wang, and J. Xu, *Convergence estimates for multigrid algorithms without regularity assumptions,* Math. Comp. **57** (1991), 23–45.

[5] J. H. Brandts, H. A. van der Vorst, *The convergence of Krylov methods and Ritz values*, in: Conjugate Gradient Algorithms and Finite Element Methods, Springer-Verlag, Berlin, 2004, pp. 47–68.

[6] S. C. Brenner, L. R. Scott, *The Mathematical Theory of Finite Element Methods*, Springer Verlag, New York; Second Edition, 2002.

[7] P. Ciarlet, *The Finite Element Method for Elliptic Problems*, SIAM Classics in Applied Mathematics, 2002; Second Edition.

[8] S.K. Godunov, G.P. Prokopov, *On the solution of the Laplace difference equation*, Zh. Vychisl. Mat. Mat. Fiz. **9** (1969), 462–468.

[9] G.H. Golub, C.F. van Loan, *Matrix computations*, The John Hopkins University Press; Third Edition, 1996.

[10] M.R. Hestenes, E. Stiefel, *Methods of conjugate gradients for solving linear systems*, J. Res. Nat. Bur. Standards **49** (1952), 409–436.

# Geometric Interpretations of Conjugate Gradient and Related Methods

Michal Křížek,[1] Sergey Korotov[2]

[1]Mathematical Institute, Academy of Sciences of the Czech Republic, Žitná 25, CZ–115 67 Prague 1, Czech Republic krizek@math.cas.cz

[2]Department of Mathematical Information Technology, University of Jyväskylä, P. O. Box 35, FIN–40014 Jyväskylä, Finland korotov@mit.jyu.fi

*A picture is worth a thousand words.*

An old Chinese proverb

## 1 Introduction

In 1952, M. R. Hestenes and E. Stiefel published the pioneering paper [14], in which they introduced the conjugate gradient method for solving the system of linear algebraic equations

$$Ax = b, \tag{1}$$

where $A = (a_{ij})$ is a real symmetric and positive definite $n \times n$ matrix, $x \in \mathbb{R}^n$ is the vector of unknowns, and $b = (b_1, \dots, b_n)^{\mathrm{T}} \in \mathbb{R}^n$ is a given right-hand side. This method can be considered as direct as well as iterative. It is similar to the Lanczos method for finding eigenvalues presented in [24] (which is mentioned in [14, p. 410]).

The conjugate gradient has been unnoticed for a long time, since for dense matrices it requires $2n^3 + \mathcal{O}(n^2)$ arithmetic operations (for the dense direct case), whereas the standard Gaussian elimination requires only $\frac{1}{3}n^3 + \mathcal{O}(n^2)$ operations as $n \to \infty$.

Later it was found out that for matrices $A$ coming especially from technical applications (e.g., from the use of finite difference or finite element methods) the conjugate gradient method converges very quickly if it is used as an iterative method – cf. (18). This observation is based on some old ideas of L. V. Kantorovich[1] – see [23] and also [10], [17]. The iteration process is usually stopped when the iteration error of the conjugate gradient method is approximately equal to the discretization error of the finite element method (see [3]). At present, the fast convergence of the conjugate gradient method

[1]L. V. Kantorovich is the Nobel prize winner for economy in 1975.

enables us to solve sparse systems (1) with a sufficiently high accuracy for $n$ up to $10^8$ in a reasonable computational time. Therefore, the associated algorithm is often considered to be the "algorithm of the century" (see [8]). It is also commonly used in various optimization techniques (see, e.g., the monograph by Hestenes [13], which is devoted to Stiefel).

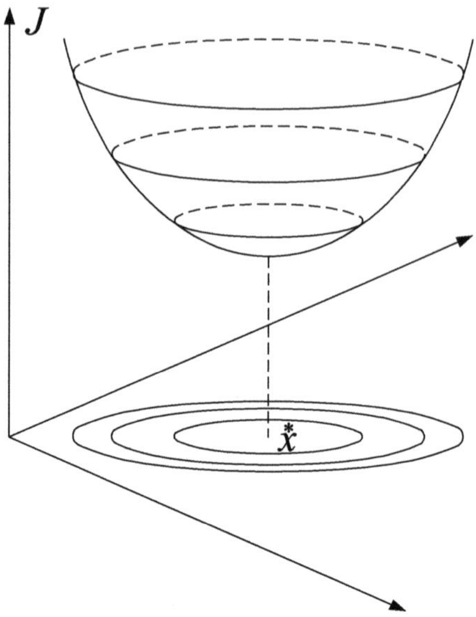


Figure 1

It is well known that problem (1) is equivalent to the minimization of the quadratic functional (see Fig. 1)

$$J(x) = \frac{1}{2}x^{\mathrm{T}}Ax - b^{\mathrm{T}}x \tag{2}$$

over the space $\mathbb{R}^n$. Geometric interpretations of our considerations and proposed methods will be illustrated mostly for $n = 2$.

Let $\overset{*}{x}$ be the (unique) solution of (1) and let us set $x = \overset{*}{x} + s$. Then we get

$$J(x) + \frac{1}{2}b^{\mathrm{T}}\overset{*}{x} = \frac{1}{2}(\overset{*}{x} + s)^{\mathrm{T}}A(\overset{*}{x} + s) - b^{\mathrm{T}}(\overset{*}{x} + s) + \frac{1}{2}b^{\mathrm{T}}\overset{*}{x} = \frac{1}{2}s^{\mathrm{T}}As.$$

The hypersurface (compare with Fig. 2)

$$S = \{s = (s_1, \ldots, s_n) \mid \tfrac{1}{2}s^{\mathrm{T}}As = \text{const.}\}$$

is a hyperellipsoid in the variables $s_1, \ldots, s_n$ with the centre at $s = 0$, i.e., at $\overset{*}{x}$. Therefore, the equation $J(x) = \text{const.}$ also represents a hyperellipsiod.

Since $A$ is symmetric, there exists an orthogonal matrix $P$ (i.e., $P^{\mathrm{T}} = P^{-1}$) such that

$$\Lambda = P^{\mathrm{T}} A P = \begin{pmatrix} \lambda_1 & & 0 \\ & \ddots & \\ 0 & & \lambda_n \end{pmatrix} \qquad (\lambda_1 \leq \lambda_2 \leq \ldots \leq \lambda_n)$$

is a diagonal matrix with the eigenvalues $\lambda_i$ of $A$ on its diagonal. Setting $z = (z_1, \ldots, z_n)^{\mathrm{T}} = P^{\mathrm{T}} s$, we immediately observe that the columns of $P$ are eigenvectors of $A$ and

$$s^{\mathrm{T}} A s = z^{\mathrm{T}} \Lambda z = \sum_{i=1}^{n} \lambda_i z_i^2.$$

From this it follows that the hyperellipsoid $S$ has its axes in the directions of eigenvectors of $A$. The lengths of these axes are proportional to $\dfrac{1}{\sqrt{\lambda_i}}$, $i = 1, \ldots, n$, where $\lambda_i > 0$, since $A$ is positive definite.

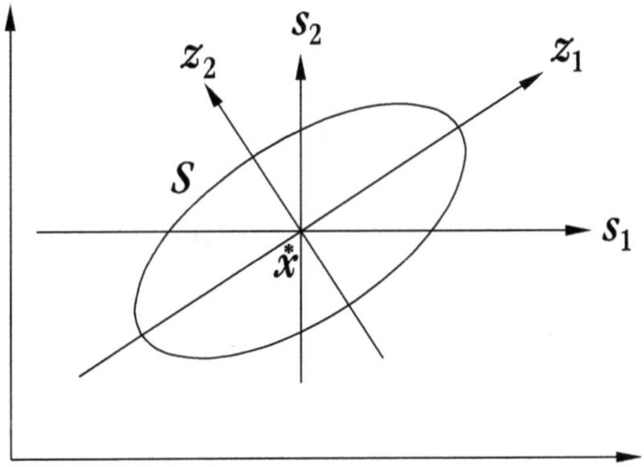


Figure 2

Introduce the *condition number* of $A$,

$$\kappa(A) = \frac{\max_i \lambda_i}{\min_i \lambda_i} = \frac{\lambda_n}{\lambda_1}.$$

We see that if $\kappa(A) \to 1$ then the hyperellipsoid $S$ transforms to a hypersphere and we say that $A$ is *well-conditioned.* On the other hand, if $\kappa(A) \gg 1$ then the hyperellipsoid $S$ is more "elongated" and we say that $A$ is *ill-conditioned.*

## 2 Geometric interpretation of classical iterative methods

Geometric interpretation of iterative methods for solving (1) helps us in better understanding the behaviour of the "iteration trajectory" (cf. Figs. 3–10) and the efficiency of numerical processes, namely, how to compare the computational complexity of particular methods, how to improve or accelerate their convergence, etc. – see, e.g., [7], [26].

In this section we survey some geometric ideas from [21].

**The Jacobi method.** Suppose that $x_k$, $k \in \{0, 1, 2, \dots\}$, is already known. We define the $i$th component $x_{k+1}^{(i)}$ of the vector $x_{k+1}$ so that it is the point at which the convex quadratic function

$$\begin{aligned} J_k^{(i)}(y) &= J(x_k^{(1)}, \dots, x_k^{(i-1)}, y, x_k^{(i+1)}, \dots, x_k^{(n)}) \\ &= \frac{1}{2} a_{ii} y^2 + y \sum_{\substack{j=1 \\ j \neq i}}^{n} a_{ij} x_k^{(j)} - b_i y + \text{ const.} \end{aligned}$$

attains its minimum over the space $\mathbb{R}^1$ (see Fig. 3).

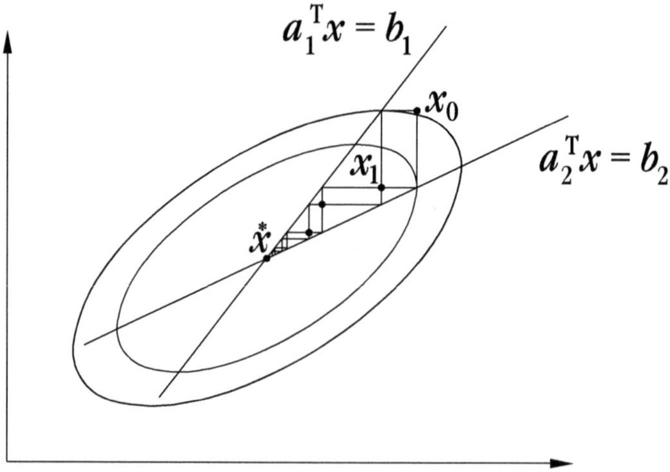


Figure 3

Clearly, $a_{ii} > 0$, since $A$ is positive definite and $a_{ii} = \mathrm{e}_i^{\mathrm{T}} A \mathrm{e}_i$, where $\mathrm{e}_i = (0, \dots, 0, 1, 0, \dots, 0)^{\mathrm{T}}$ with 1 on the $i$th position. Setting $\partial J_k^{(i)} / \partial y = 0$, we get

$$x_{k+1}^{(i)} = \frac{1}{a_{ii}} \Bigl( b_i - \sum_{\substack{j=1 \\ j \neq i}}^{n} a_{ij} x_k^{(j)} \Bigr), \quad i = 1, \dots, n.$$

From this we see that the point $(x_k^{(1)}, \dots, x_k^{(i-1)}, x_{k+1}^{(i)}, x_k^{(i+1)}, \dots, x_k^{(n)})^{\mathrm{T}}$ lies on the affine hyperplane

$$a_i^{\mathrm{T}} x = b_i$$

for all $i \in \{1, \dots n\}$, where $a_i$ is the $i$th row of $A$ (compare with Fig. 3). This method can be interpreted as a simultaneous minimization of $J$ in the directions of the coordinate axes, which is especially suitable for computers with parallel architecture.

Note that the Jacobi method can diverge if $A$ is not irreducibly diagonally dominant even though it is positive definite (see [31, p. 73]). In this case the convergence can be achieved by introducing a suitable underrelaxation factor $\omega \in (0, 1)$ like in the SOR method.

**The Gauss-Seidel method.** The main difference between the Jacobi method and the Gauss-Seidel method is that in the latter we already use the new values $x_{k+1}^{(1)}, \dots, x_{k+1}^{(i-1)}$ to compute $x_{k+1}^{(i)}$, i.e.,

$$x_{k+1}^{(i)} = \frac{1}{a_{ii}} \left( b_i - \sum_{j=1}^{i-1} a_{ij} x_{k+1}^{(j)} - \sum_{j=i+1}^{n} a_{ij} x_k^{(j)} \right)$$

for $i = 1, \dots, n$. This is, in fact, a successive minimization of $J$ in the direction of the coordinate axes (which can be observed in Fig. 4).

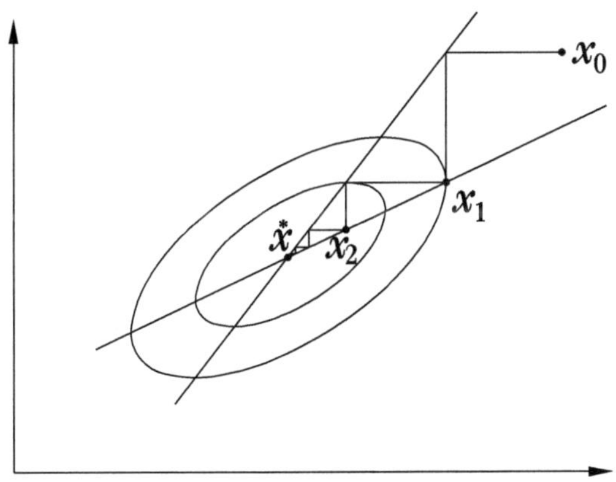


Figure 4

It is well known (see [31, p. 78]) that the Gauss-Seidel method converges independently of the initial vector $x_0$ if $A$ is symmetric and positive definite. Moreover, it converges faster than the Jacobi method if it is applicable (except for the trivial case, when $A$ is a diagonal matrix).

Next, we will introduce a famous geometric interpretation of a method for accelerating the convergence of the Gauss-Seidel method.

**The method of successive overrelaxation (SOR).** This method is defined by the relation (see [31])

$$x_{k+1}^{(i)} = x_k^{(i)} + \frac{\omega}{a_{ii}}\Big(b_i - \sum_{j=1}^{i-1} a_{ij} x_{k+1}^{(j)} - \sum_{j=i}^{n} a_{ij} x_k^{(j)}\Big),$$

where $\omega \in (0,2)$. We call $\omega$ the *overrelaxation* (*underrelaxation*) *factor* if $\omega > 1$ ($\omega < 1$). Comparing the above equation with the Gauss-Seidel method, we find that

$$x_{k+1}^{(i)} = x_k^{(i)} + \omega \left( {}^{\mathrm{GS}}x_{k+1}^{(i)} - x_k^{(i)} \right),$$

where ${}^{\mathrm{GS}}x_{k+1}^{(i)}$ denotes iterates obtained by the Gauss-Seidel method. If $\omega = 1$ we get the Gauss-Seidel method. If $\omega \neq 1$ we do not stop in the point of minimum of $J$ in the direction $(0,\dots,0,1,0,\dots,0)^{\mathrm{T}}$ (the digit 1 stands in the $i$th position). In particular, for $\omega \in (1,2)$ we shall "go" a little bit further and for $\omega \in (0,1)$ we stop before the point of minimum. The associated iteration trajectory is illustrated in Fig. 5.

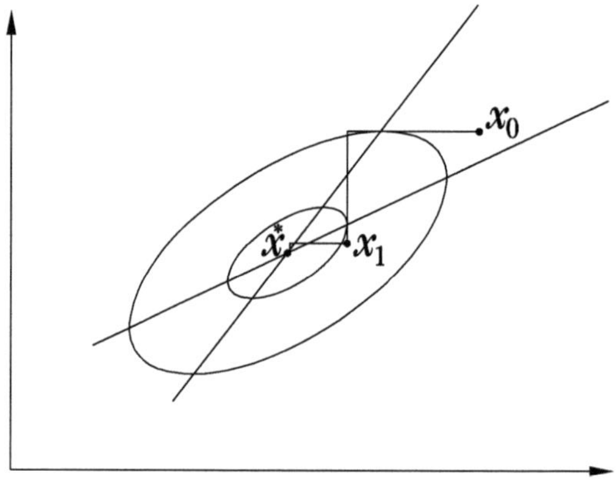


Figure 5

**The Kaczmarz method.** This method is also called the method of orthogonal projections [19]. It has a very simple geometric interpretation (see Fig. 6) and is defined as follows:

Let $k \in \{0,1,2,\dots\}$ be fixed and assume that $x_0,\dots,x_k$ are already given. We shall construct $x_{k+1}$. First we set $y_0 = x_k$ and define

$$y_i = \Big(I - \frac{a_i a_i^{\mathrm{T}}}{a_i^{\mathrm{T}} a_i}\Big) y_{i-1} + \frac{b_i}{a_i^{\mathrm{T}} a_i} a_i$$

successively for $i = 1,\dots,n$, where $I$ is the identity matrix. Then we put

$$x_{k+1} = y_n.$$

We can easily check that $y_i$ is the orthogonal projection of the point $y_{i-1}$ to the affine hyperplane $a_i^{\mathrm{T}} x = b_i$. Although the convergence of this method is, in general, very slow, it converges for any nonsingular matrix $A$ and any initial vector $x_0$.

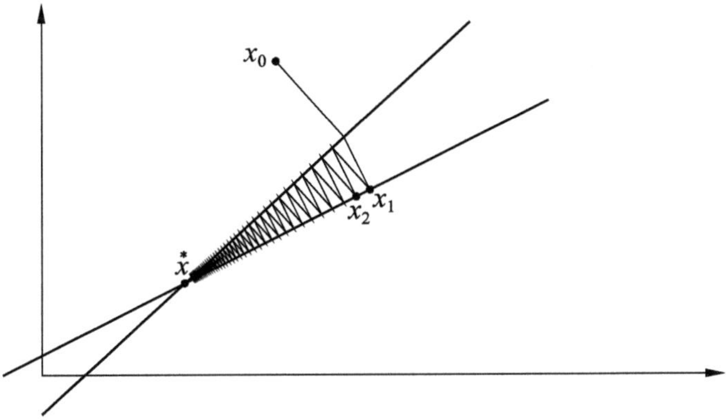


Figure 6

The convergence of the methods mentioned above can essentially be increased by Aitken's post-processing, since the corresponding iterates form "almost geometric sequences" (see [19] for details).

## 3 Geometric interpretation of gradient iterative methods

In this section we present further iterative methods in which the functional (2) is minimized in a given direction $v_k \neq 0$. First we establish a formula for the value $\alpha_k \in \mathbb{R}^1$ such that $J(x_k + \alpha_k v_k)$ is minimal and then we put

$$x_{k+1} = x_k + \alpha_k v_k.$$

From (2) we find that

$$J\left(x_k + \alpha_k v_k\right) = \frac{1}{2} x_k^{\mathrm{T}} A x_k + \alpha_k x_k^{\mathrm{T}} A v_k + \frac{\alpha_k^2}{2} v_k^{\mathrm{T}} A v_k - b^{\mathrm{T}} x_k - \alpha_k b^{\mathrm{T}} v_k.$$

Since this function is convex and quadratic in $\alpha_k$, from the relation $\dfrac{\partial J}{\partial \alpha_k} = 0$, we obtain

$$\alpha_k = \frac{b^{\mathrm{T}} v_k - x_k^{\mathrm{T}} A v_k}{v_k^{\mathrm{T}} A v_k} = \frac{r_k^{\mathrm{T}} v_k}{v_k^{\mathrm{T}} A v_k}, \tag{3}$$

where

$$r_k = b - A x_k \tag{4}$$

is the so-called *residual vector.*

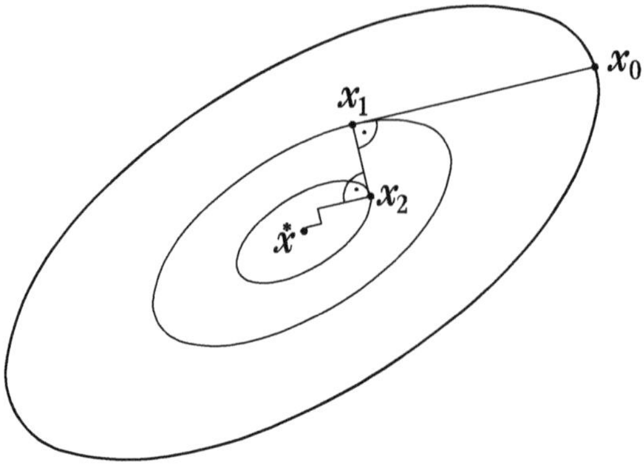


Figure 7

**The method of steepest descent.** This method originates from the 19th century by A. L. Cauchy and also by B. Riemann (cf. [27]). It is based on the natural idea of choosing $v_k$ in the direction of the greatest rate of change of $J$, namely, in the direction opposite to the gradient (see Fig. 7). From (2) and (4) we deduce that the gradient of $J$ at $x = x_k$ is

$$\text{grad } J(x_k) = Ax_k - b = -r_k.$$

Putting $v_k = r_k$, we obtain from (3) that

$$\alpha_k = \frac{r_k^{\mathrm{T}} r_k}{r_k^{\mathrm{T}} A r_k} \tag{5}$$

when $r_k \neq 0$, i.e., the steepest descent algorithm is then defined as follows:

$$x_{k+1} = x_k + \frac{r_k^{\mathrm{T}} r_k}{r_k^{\mathrm{T}} A r_k} r_k, \qquad k = 0, 1, \dots . \tag{6}$$

Consequently,

$$Ax_{k+1} - b = Ax_k - b + \alpha_k A r_k$$

and by (4) we find the recurrence formula for the residual vectors

$$r_{k+1} = r_k - \alpha_k A r_k. \tag{7}$$

From this and (5) we easily derive that

$$r_{k+1}^{\mathrm{T}} r_k = 0 \ , \qquad k = 0, 1, \dots ,$$

i.e., subsequent residual vectors are orthogonal.

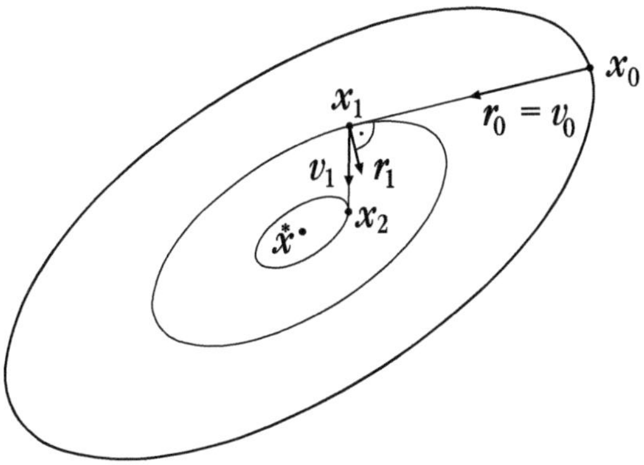


Figure 8

**The heavy ball method.** The movement of a heavy ball on a surface (cf. Fig. 1) in the gravity field is not, in general, in the direction of steepest descent due to inertial forces. The instantaneous direction depends on "old" directions. We use this fact to define the direction $v_k \neq 0$ in the following manner (see Fig. 8):

$$v_k = r_k + \beta_{k-1} v_{k-1} \qquad (v_{-1} = 0). \tag{8}$$

According to [30], the parameter $\beta_{k-1}$ should be non-negative (appropriately chosen – cf. e.g. (13)). If $\beta_{k-1} = 0$ we get (6), otherwise

$$x_{k+1} = x_k + \frac{r_k^{\mathrm{T}} v_k}{v_k^{\mathrm{T}} A v_k} v_k, \qquad k = 0, 1, \ldots.$$

Analogously to the method of steepest descent (cf. (7)) we can derive the recurrence formula for the residual vectors, i.e.,

$$r_{k+1} = r_k - \alpha_k A v_k,$$

where $\alpha_k$ is defined by (3) and $v_k$ by (8). Therefore, we see that

$$r_{k+1}^{\mathrm{T}} v_k = r_k^{\mathrm{T}} v_k - \frac{r_k^{\mathrm{T}} v_k}{v_k^{\mathrm{T}} A v_k} v_k^{\mathrm{T}} A v_k = 0, \qquad k = 0, 1, \ldots.$$

**The conjugate gradient method.** In this method the direction $v_k$ is usually denoted by $p_k$. It is chosen so that the minimum of $J$ is attained in the plane passing through $x_k$ and parallel with $r_k$ and $p_{k-1}$ (see Fig. 9 for $k = 1$). It is well known that the conjugate gradient method requires only a finite number of iterations (in exact arithmetic) to reach the true solution, which immediately follows from (16) below. The method is defined as follows:

Let

$$p_0 = r_0 = b - Ax_0, \tag{9}$$

where $x_0$ is any vector such that $Ax_0 \neq b$, and let

$$\left.\begin{aligned} \alpha_k &= \frac{r_k^{\mathrm{T}} r_k}{p_k^{\mathrm{T}} A p_k} && (10)\\ x_{k+1} &= x_k + \alpha_k p_k && (11)\\ r_{k+1} &= r_k - \alpha_k A p_k && (12)\\ \beta_k &= \frac{r_{k+1}^{\mathrm{T}} r_{k+1}}{r_k^{\mathrm{T}} r_k} && (13)\\ p_{k+1} &= r_{k+1} + \beta_k p_k && (14) \end{aligned}\right\} \quad k = 0, 1, \ldots$$

until $r_{k+1} = 0$. When the algorithm stops, we set

$$m = k + 1.$$

Clearly, $\overset{*}{x} = x_m$ is the true solution of (1) and from relation (12) we get

$$x_m = x_0 + \sum_{k=0}^{m-1} \alpha_k p_k. \tag{15}$$

The vectors $p_k$ are called the *conjugate directions* (see Fig. 9).

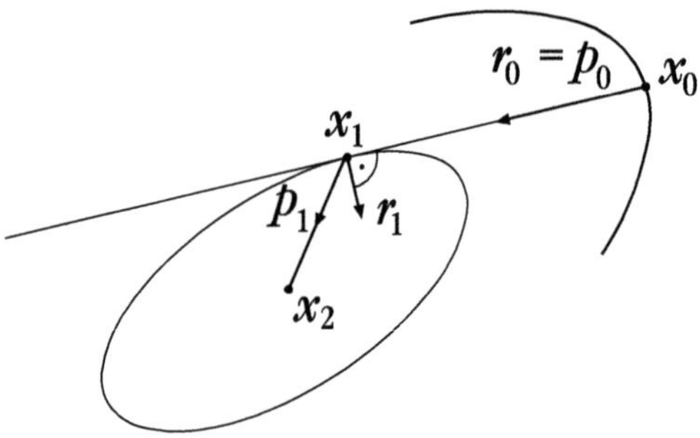


Figure 9

From (4) and (9) we see that $r_0$ is the residual vector. Using (12), we can easily derive by induction that the whole sequence $r_1, r_2, \ldots, r_{m-1}$ is formed by residual vectors. Moreover, it is well-known that all the vectors $r_k$ are mutually orthogonal, i.e.,

$$r_k^{\mathrm{T}} r_\ell = 0 \quad \text{for } k \neq \ell, \tag{16}$$

and the vectors $p_k$ are $A$-orthogonal (conjugate), i.e.,

$$p_k^{\mathrm{T}} A p_\ell = 0 \quad \text{for } k \neq \ell.$$

To reduce round-off errors and guarantee the orthogonality (16) in practical calculations, a multiple-precision arithmetic is sometimes used. It is based on the Fast Fourier Transform, which requires $\mathcal{O}(N \log N \log\log N)$ bit operations for multiplication of two $N$-digit binary numbers, whereas the number of bit opeartions for standard straightforward multiplication is $\mathcal{O}(N^2)$.

Notice that the heavy ball method contains its two important special cases, namely the method of steepest descent (with $\beta_k = 0$) and the conjugate gradient method (with $\beta_k$ defined as the ratio of squared sizes of subsequent residua – see (13)).

The conjugate gradient method can be interpreted as an iterative process of two-dimensional minimizations. It gives us the true solution after a finite number of steps (at most $n$), whereas all methods based on *one-dimensional minimizations* yield the true solution after infinitely many steps, in general.

The conjugate gradient method is also a special case of the *method of conjugate directions*. This method was introduced in [14]. It chooses $x_{k+1} = x_k + \alpha_k v_k$, where $\alpha_k$ is given by (3) and $v_1, v_2, \ldots$ is an arbitrary set of nonzero conjugate directions (i.e., $v_k^{\mathrm{T}} A v_\ell = 0$ for $k \neq \ell$). By [14, p. 412] the conjugate direction algorithm terminates in at most $n$ iterations for any initial vector $x_0$.

Further, let us recall the rate of convergence of the conjugate gradient method in the *energy norm*

$$\|x\|_A = \sqrt{x^{\mathrm{T}} A x}. \tag{17}$$

First notice that the points $x$, for which $\|\overset{*}{x} - x\|_A$ is constant lie on a hyperellipsoid (compare with Fig. 9), since by (2) and (17) we have $J(x) + \frac{1}{2}\overset{*}{x}{}^{\mathrm{T}} A \overset{*}{x} = \|x - \overset{*}{x}\|_A^2$, where $\overset{*}{x}$ is the true solution of (1).

If $x_0$ is arbitrary and $A$ is symmetric and positive definite then for the algorithm (9)–(14) we have

$$\|\overset{*}{x} - x_k\|_A \le 2\left(\frac{\sqrt{\kappa(A)} - 1}{\sqrt{\kappa(A)} + 1}\right)^k \|\overset{*}{x} - x_0\|_A \quad \forall k \in \{0, 1, \ldots, m\}. \tag{18}$$

A detailed proof is given, e.g., in [22]. Notice that relation (18) represents only an upper bound of the error. Let us point out (see [15], [16], [33]) that the convergence of the conjugate gradient method is generally much faster, which is called *superlinear*. The superlinear convergence for general Krylov subspace methods is investigated in the recent paper [5].

For $k \in \{1, \ldots, m\}$ introduce the so-called *Krylov space*

$$\mathcal{K}_k = \operatorname{span}(r_0, \ldots, A^{k-1} r_0),$$

where span denotes the linear hull. It can be shown that

$$\begin{aligned} \mathcal{K}_k &= \operatorname{span}(p_0, \ldots, p_{k-1}) = \operatorname{span}(r_0, r_1, \ldots, r_{k-1}) \\ &= \operatorname{span}(A(\overset{*}{x} - x_0), \ldots, A^k(\overset{*}{x} - x_0)), \end{aligned}$$

where $\overset{*}{x}$ is the true solution of (1). The following minimization property of the conjugate gradient method is also well known

$$\|\overset{*}{x} - x_k\|_A = \min_{x \in x_0 + \mathcal{K}_k} \|\overset{*}{x} - x\|_A \quad \forall k \in \{1, 2, \ldots, m\}. \tag{19}$$

For geometric aspects of Krylov subspace methods, namely a detailed analysis of angles between particular vectors, see [11]

**Example.** Let us illustrate a geometric interpretation of a possible breakdown of the conjugate gradient algorithm when the matrix $A$ is not positive definite. In Fig. 10 we see parabolic contour lines of functional (2) for

$$A = \begin{pmatrix} 1 & 2 & 2 \\ 2 & 4 & -2 \\ 2 & -2 & 0 \end{pmatrix}$$

and $b = 0$. Taking $x_0 = (4,4,3)^{\mathrm{T}}$, we easily find by (9)–(14) that $p_0 = r_0 = (-18,-18,0)^{\mathrm{T}}$, $x_1 = (0,0,3)^{\mathrm{T}}$, $r_1 = (-6,6,0)^{\mathrm{T}}$, $p_1 = (-8,4,0)^{\mathrm{T}}$ and $p_1^{\mathrm{T}} A p_1 = 0$, i.e., the algorithm breaks down. In Fig. 10 we see that there is no minimum of $J$ in the direction $p_1$.

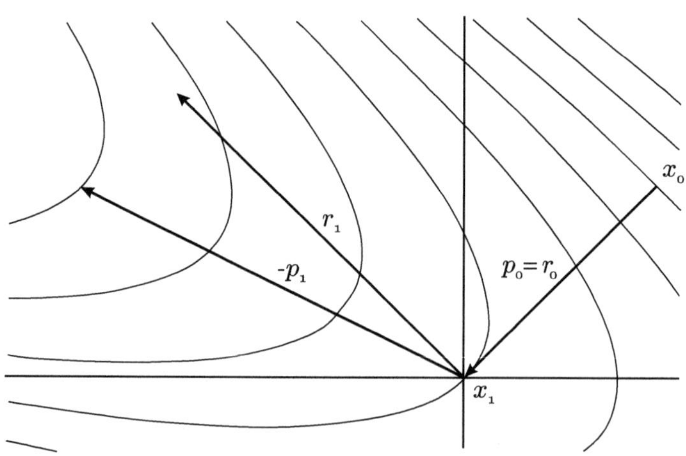


Figure 10

## 4 Convergence of the conjugate gradient method in the Euclidean norm

First note that the value of functional (2) at $x_k$ decreases as $k$ increases to $m$, since in each iteration we perform one-dimensional minimization of $J$ in the line $x_k + \alpha_k p_k$. From (18) we know an upper bound for the rate of convergence of the *error*

$$e_k = \overset{*}{x} - x_k \tag{20}$$

in the energy norm. In the next theorem we show (cf. [22, p. 118]) that the error $e_k$ decreases in the Euclidean norm, i.e., in the norm $\|e_k\| = \sqrt{e_k^{\mathrm{T}} e_k}$.

**Theorem.** *Let $\{x_k\}$ be the sequence defined by (9)–(14). Then there exists a constant $C \in (0,1)$ such that*

$$\|\overset{*}{x} - x_{k+1}\| < C\|\overset{*}{x} - x_k\| \quad \forall k \in \{0, 1, \ldots, m-1\}.$$

Proof. Using (14) repeatedly, we come to

$$p_k = r_k + \beta_{k-1} r_{k-1} + \beta_{k-1}\beta_{k-2} r_{k-2} + \cdots + \left(\prod_{j=0}^{k-1} \beta_j\right) r_0, \tag{21}$$

which implies that

$$r_\ell^{\mathrm{T}} p_k = 0 \quad \text{for} \quad 0 \le k < \ell \le m. \tag{22}$$

From this and (14),

$$r_k^{\mathrm{T}} p_k = r_k^{\mathrm{T}} r_k + \beta_{k-1} r_k^{\mathrm{T}} p_{k-1} = r_k^{\mathrm{T}} r_k. \tag{23}$$

According to (14) again and (22) we find that

$$p_k^{\mathrm{T}} p_j = r_k^{\mathrm{T}} p_j + \beta_{k-1} p_{k-1}^{\mathrm{T}} p_j = \beta_{k-1} p_{k-1}^{\mathrm{T}} p_j \quad \text{for } 0 \le j < k, \tag{24}$$

Thus by induction,

$$p_k^{\mathrm{T}} p_j = \beta_{k-1} \ldots \beta_j \|p_j\|^2 > 0. \tag{25}$$

Moreover, using (20), (11), and (25), we get

$$\begin{aligned} p_k^{\mathrm{T}} e_{k+1} &= p_k^{\mathrm{T}} (\overset{*}{x} - x_{k+1}) \\ &= p_k^{\mathrm{T}} \left( x_0 + \sum_{j=0}^{m-1} \alpha_j p_j - x_0 - \sum_{j=0}^{k} \alpha_j p_j \right) \\ &= p_k^{\mathrm{T}} \sum_{j=k+1}^{m-1} \alpha_j p_j = \sum_{j=k+1}^{m-1} \alpha_j p_k^{\mathrm{T}} p_j > 0. \end{aligned} \tag{26}$$

Employing (20) again and (12), we obtain

$$e_k = \overset{*}{x} - x_k = \overset{*}{x} - x_{k+1} + \alpha_k p_k = e_{k+1} + \alpha_k p_k. \tag{27}$$

From this and (25) we see that

$$\|e_k\|^2 = \|e_{k+1}\|^2 + 2\alpha_k p_k^{\mathrm{T}} e_{k+1} + \alpha_k^2 \|p_k\|^2 \tag{28}$$

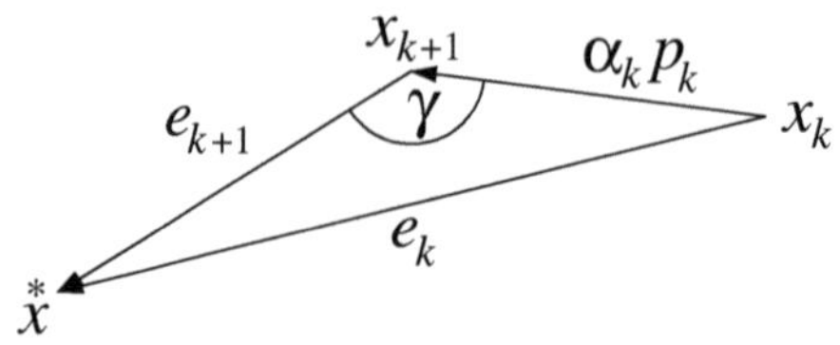


Figure 11

and the last two terms are positive. Hence,

$$\|\overset{*}{x} - x_{k+1}\| < C_k \|\overset{*}{x} - x_k\|$$

for some $C_k \in (0,1)$. Since the sequence $\{x_k\}$ terminates in a finite number of iterations, we can set $C = \max_k C_k$. □

**Remark 1.** The geometric meaning of relation (27) is illustrated in Fig 11. By to (26), the angle between the vectors $p_k$ and $e_{k+1}$ is acute. Therefore, the angle $\gamma$ in Fig. 11 is obtuse, and thus $\|e_{k+1}\| < \|e_k\|$ (cf. (28)).

**Remark 2.** According to (4) and (22), we find that

$$e_k = \overset{*}{x} - x_k = A^{-1}b - x_k = A^{-1}r_k.$$

Thus, by Theorem 1, we get the monotonic decrease $r_{k+1}^{\mathrm{T}} A^{-2} r_{k+1} < r_k^{\mathrm{T}} A^{-2} r_k$ for every $k \in \{0, 1, \ldots, m-1\}$. Similarly, (18) gives an upper bound for the sequence $r_k^{\mathrm{T}} A^{-1} r_k = e_k^{\mathrm{T}} A e_k$, which also monotonically decreases due to (19). On the other hand, $r_k^{\mathrm{T}} r_k$ is not monotone, in general.

**Remark 3.** An upper bound for the rate of convergence of the method of steepest descent in the Euclidean norm is derived in [18] and in the energy norm in [3, p. 14].

## 5 Geometric interpretation of preconditioning

From (18) it is evident that the convergence of the conjugate gradient method will be fast when the condition number $\kappa(A)$ is close to one. Then the iteration process may be stopped much earlier than after $n$ steps. The condition number can be reduced by so-called *preconditioning*, which means that instead of system (1) we solve another system

$$\tilde{A}\tilde{x} = \tilde{b},$$

where

$$\tilde{A} = P^{-1} A P^{-\mathrm{T}}$$

is symmetric and positive definite, $P$ is a nonsigular matrix, $P^{-\mathrm{T}} = (P^{-1})^{\mathrm{T}}$,

$$\tilde{x} = P^{\mathrm{T}}x, \quad \tilde{b} = P^{-1}b,$$

and where $P$ is appropriately chosen such that

$$\kappa(\tilde{A}) < \kappa(A).$$

Here $\tilde{A}$ is called the *preconditioned matrix* and $P$ *preconditioning matrix* or just *preconditioner.* In the left part of Fig. 12, we observe a "typical" spectrum of the matrix $A$, whereas the right part illustrates the spectrum of the associated matrix $\tilde{A}$.

Figure 12

The matrix $P$ is often chosen so that $\kappa(\tilde{A}) < \sqrt{\kappa(A)}$, which is considered as a good preconditioning.

Another geometric interpretation of preconditioning is as follows: We minimize (instead of functional (2)) over the space $\mathbb{R}^n$ the following functional

$$\tilde{J}(\tilde{x}) = \frac{1}{2}\tilde{x}^{\mathrm{T}}\tilde{A}\tilde{x} - \tilde{b}^{\mathrm{T}}\tilde{x},$$

whose contour lines (see the right part of Fig. 13) are less elongated hyperellipsoids than the contour lines of the functional $J$ (see the left part of Fig. 13). Straight lines in Fig. 13 correspond to particular equations of systems $Ax = b$ and $\tilde{A}\tilde{x} = \tilde{b}$.

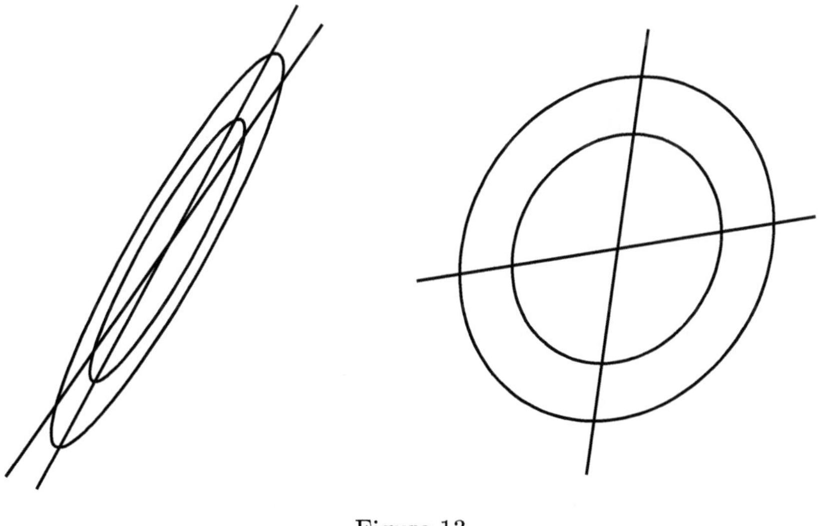

Figure 13

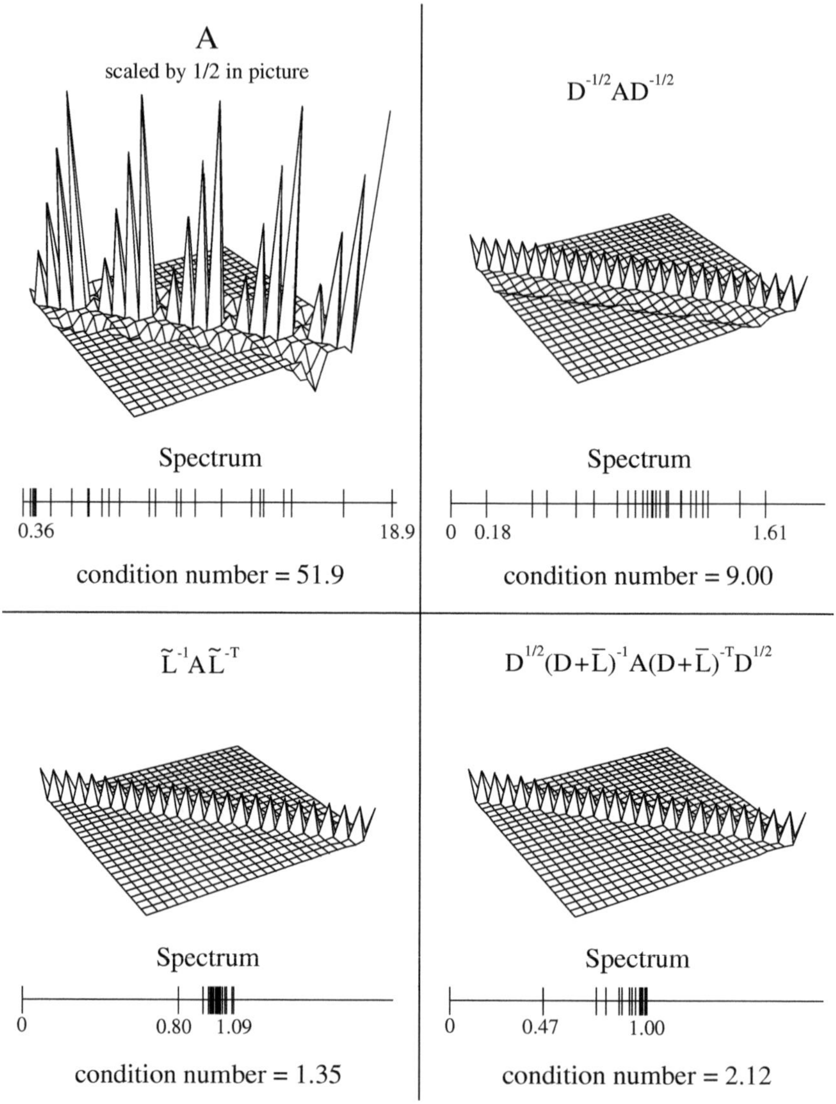


Figure 14

The effect of three standard preconditioning techniques is sketched in Fig. 14 for

$$A = \left( \int_{\Omega} \Big( \frac{\partial v^i}{\partial r} \frac{\partial v^j}{\partial r} + \frac{\partial v^i}{\partial z} \frac{\partial v^j}{\partial z} \Big) r \, \mathrm{d}r \, \mathrm{d}z \right)_{i,j=1}^{25},$$

where $\Omega = (0,1) \times (0,1)$ and $v^i$ are the Courant basis functions corresponding to $5 \times 5$ degrees of freedom. The matrix $A$ arises by discretization of the Poisson equation in an axially symmetric domain (with homogeneous Dirichlet boundary conditions) using cylindrical coordinates. For a graphical illustration of the entries of $A$ (see Fig. 14) we used bilinear elements over a square mesh with $25 \times 25$ nodal points.

For the first preconditioner we take $P = D^{\frac{1}{2}}$ (see [321, p. 20]), where $D$ is a diagonal matrix containing the main diagonal of $A$. Each diagonal entry of the associated preconditioned matrix $\tilde{A} = D^{-\frac{1}{2}} A D^{-\frac{1}{2}}$ is thus equal to 1 (compare with Fig. 14).

The second preconditioner is constructed by means of an incomplete Cholesky factorization (see [3, p. 41]). In this case we decompose $A$ so that $A = \tilde{L}\tilde{L}^{\mathrm{T}} + R$, where $\tilde{L}_{ij} = 0$ whenever $A_{ij} = 0$, and $R$ is a small residual matrix. Then we set $P = \tilde{L}$. We observe from Fig. 14 that the preconditioned matrix $\tilde{A} = \tilde{L}^{-1} A \tilde{L}^{-\mathrm{T}}$ is almost the identity matrix.

Finally, in the third case we set

$$P = \left(\frac{1}{\omega}D + \overline{L}\right)\left(\frac{1}{\omega}D\right)^{-\frac{1}{2}},$$

where $\omega$ is an overrelaxation parameter ($\omega = 1$ in Fig. 14) and $\overline{L}$ is a lower triangular matrix such that $A = \overline{L} + D + \overline{L}^{\mathrm{T}}$. Such a choice of $P$ leads to the so-called SSOR preconditioning technique (see [3, p. 51]).

For further details concerning geometric interpretations of preconditioning (even for nonsymmetric matrices) we refer to [2], [3], [12], [20], [25], [26]. Preconditioning by multilevel or (algebraic) multigrid methods can be found in [1], [4], [6]. The biconjugate gradient method for systems of linear algebraic equations in complex variable is investigated in [20]. Generalizations of the conjugate gradient method for nonlinear problems can be found, e.g., in [9], [21], [28], [29].

**Acknowledgement.** This paper was supported by Grant no. 201/02/0595 of the Grant Agency of the Czech Republic and by Grant no. 49051 of the Academy of Finland. The authors are grateful to J. Brandts, A. V. Knyazev, P. Neittaanmäki, and J. Toivanen for valuable suggestions and help with the preparation of this paper.

## References

[1] O. Axelsson, *Preconditioned conjugate gradient methods; Preface*, BIT **29** (1989), 577–582.

[2] O. Axelsson, *Iterative solution methods*, Cambridge Univ. Press, Cambridge, 1994.

[3] O. Axelsson, V. A. Barker, *Finite element solution of boundary value problems. Theory and computation*, Academic Press, New York, 1984, 2001.

[4] O. Axelsson, P. S. Vassilevski, *Algebraic multilevel preconditioning methods, Part I*, Numer. Math. **56** (1989), 157–177; *Part II*, SIAM J. Numer. Anal. **27** (1990), 1569–1590.

[5] B. Beckermann, A. B. J. Kuijlaars, *Superlinear convergence of conjugate gradients*, SIAM J. Numer. Anal. **39** (2001), 300–329.

[6] J. H. Bramble, J. E. Pasciak, J. Xu, *Parallel multilevel preconditioner*, Third Internat. Sympos. on Domain Decomposition Methods for Partial Differential Equations, SIAM, Philadelfia, 1990, 341–357.

[7] J. Brandts, *Explanation of a phenomenon witnessed in pre-processed GMRES*, ENUMATH 99, Proc. of the 3rd European Conf. on Numer. Math. and Advanced Applications, Jyväskylä, Finland, 1999, ed. by P. Neittaanmäki et al., World Scientific, Singapore, 2000, 440-447.

[8] J. H. Brandts, M. Křížek, *Fifty years of conjugate gradients (in Czech)*, Pokroky Mat. Fyz. Astronom. **47** (2002), 103–112.

[9] Y. H. Dai, *Convergence of nonlinear conjugate gradient methods*, J. Comput. Math. **19** (2001), 539–548.

[10] J. W. Daniel, *The conjugate gradient method for linear and nonlinear operator equations*, SIAM J. Numer. Anal. **4** (1967), 10–26.

[11] M. Eiermann, O. G. Ernst, *Geometric aspects of the theory of Krylov subspace methods*, Acta Numerica **10** (2001), 251–312.

[12] I. Faragó, J. Karátson, *Numerical solution of nonlinear elliptic problems via preconditioning operators. Theory and applications*, Nova Science Publisher, New York, 2002.

[13] M. R. Hestenes, *Conjugate direction methods in optimization*, Springer, New York, 1980.

[14] M. R. Hestenes, E. Stiefel, *Methods of conjugate gradients for solving linear systems*, J. Res. Nat. Bur. Standards **49** (1952), 409–436.

[15] A. Jennings, *Influence of the eigenvalue spectrum on the convergence rate of the conjugate gradient method*, J. Inst. Maths Applics **20** (1977), 61–72.

[16] A. Jennings, G. M. Malik, *The solution of sparse linear equations by the conjugate gradient method*, Internat. J. Numer. Methods Engrg. **12** (1978), 141–158.

[17] S. Kaniel, *Estimates for some computational techniques in linear algebra*, Math. Comp. **20** (1966), 369–378.

[18] A. V. Knyazev, A. L. Skorokhodov, *The rate of convergence of the method of steepest descent in the Euclidean norm (in Russian)*, Zh. Vychisl. Mat. i Mat. Fiz. **28** (1988), 1577–1578, 1599; U.S.S.R. Comput. Math. and Math. Phys. **28** (1988), 195–196 (1990).

[19] M. Křížek, L. Liu, P. Neittaanmäki, *Postprocessing of Gauss-Seidel iterations*, Numer. Linear Algebra Appl. **6** (1999), 147–156.

[20] M. Křížek, J. Mlýnek, *On the preconditioned biconjugate gradients for solving linear complex equations arising from finite elements*, Banach Center Publ. **29** (1994), 195–205.

[21] M. Křížek, P. Neittaanmäki, *Finite element approximation of variational problems and applications*, Pitman Monographs and Surveys in Pure and Applied Mathematics vol. 50, Longman Scientific & Technical, Harlow; copublished in the United States with John Wiley & Sons, New York, 1990.

[22] M. Křížek, P. Neittaanmäki, *Mathematical and numerical modelling in electrical engineering: theory and applications*, Kluwer, Dordrecht, 1996.

[23] G. Meinardus, *Über eine Verallgemeinerung einer Ungleichung von L. V. Kantorovitch*, Numer. Math. **5** (1963), 14–23.

[24] C. Lanczos, *An iteration method for the solution of the eigenvalue problem of linear differential and integral operators*, J. Res. Nat. Bur. Standards **45** (1950), 255–282.

[25] G. I. Marchuk, Yu. A. Kuznetsov, *On optimal iteration processes*, Dokl. Akad. Nauk SSSR **181** (1968), 1331–1334; Soviet Math. Dokl. **9** (1968), 1041–1045.

[26] K. Neymeyr, *A geometric theory for preconditioned inverse iteration, I: Extrema of the Rayleigh quotient, II: Convergence estimates*, Linear Algebra Appl. **322** (2001), 61–85, 87–104.

[27] S. S. Petrova, A. D. Solov'ev, *The origin of the method of steepest descent*, Historia Math. **24** (1997), 361–375.

[28] O. Pironneau, *Finite element methods in fluids*, Springer, New York, 1984.

[29] E. Polak, *Computational methods in optimization*, Academic Press, New York, 1971.

[30] B. T. Poljak, *Introduction to optimization (in Russian)*, Nauka, Moscow, 1983.

[31] R. S. Varga, *Matrix iterative analysis*, Prentice-Hall, New Jersey, 1962.

[32] A. van der Sluis, *Condition numbers and equilibrium of matrices*, Numer. Math. **14** (1969), 14–23.

[33] A. van der Sluis, H. van der Vorst, *The rate of convergence of conjugate gradients*, Numer. Math. **48** (1986), 543–560.

Part II

# Aspects of Conjugate Gradient Algorithms

# The Convergence of Krylov Methods and Ritz Values

Jan Brandts,[1] Henk van der Vorst[2]

[1]Korteweg-de Vries Institute, Faculty of Science, University of Amsterdam, Plantage Muidergracht 24, 1018 TV Amsterdam, Netherlands
brandts@science.uva.nl
[2]Mathematical Institute, Utrecht University, Budapestlaan 6, P.O. Box 80.010, 3508 TA Utrecht, Netherlands vorst@math.uu.nl

## 1 Introduction

We start with recalling one of the many definitions of the conjugate gradient method for the approximation of the solution $x$ of a linear system $Ax = b$ with symmetric and positive definite system matrix $A$. An advantage of this definition is that it is easy to state, and that it immediately shows the connection with the Lanczos method for approximation of eigenvalues of $A$. A disadvantage is that the actual *algorithms* for both the conjugate gradients and the Lanczos method do not follow too easily and require clever combination of several ingredients.

### 1.1 Basic relations and definitions

Given a non-singular $n \times n$ matrix $A$ and a vector $v \in \mathbb{R}^n$, we denote by $\mathcal{K}^i(A, v)$ the $i$th Krylov subspace for $A$ with initial vector $v$, which is defined for $i \geq 1$ as

$$\mathcal{K}^i(A, v) = \text{span}\{v, Av, A^2v, \ldots, A^{i-1}v\},$$

and which consists of the nullvector for $i = 0$. Let $x_0$ be a given initial approximation for the solution $x$ of $Ax = b$, and $r_0 = b - Ax_0$ the *residual* corresponding to this approximation. The conjugate gradient method is an iterative procedure that results in a sequence $(x_i)_{i\geq 1}$ such that

$$x_0 - x_i \in \mathcal{K}^i(A, r_0) \quad \text{and} \quad b - Ax_i = r_i \perp \mathcal{K}^i(A, r_0).$$

Even though these two relations already determines $(x_i)_{i\geq 1}$ uniquely, we will give some more details on how it may be obtained. For ease of presentation, we will assume that all residuals $r_i$ are non-zero, which corresponds to not considering the lucky situation that we find the exact solution.

Since $r_i \in \mathcal{K}^{i+1}(A, r_0)$ but also $r_i \perp \mathcal{K}^i(A, r_0)$, it is clear that $r_0, \ldots, r_{i-1}$ form an orthogonal basis for $\mathcal{K}^i(A, r_0)$. Defining the vectors $v_i = r_i/\|r_i\|$, we get an orthonormal basis. Let for all $i \geq 0$, the $n \times i$ matrix $V_i$ be the defined by $V_i e_{j+1} = v_j$, for $j \in \{0, \ldots, i-1\}$, where $e_j$ is the $j$th standard basis vector of $\mathbb{R}^i$. So, the columns of $V_i$ are $v_0, \ldots, v_{i-1}$, and consequently, any element from $\mathcal{K}^i(A, r_0)$ can be written as a linear combination of $v_0, \ldots, v_{i-1}$. In particular,

$$x_0 - x_i = V_i y_i \tag{1}$$

for some $y_i \in \mathbb{R}^i$. As a result,

$$r_i = b - Ax_i = b - Ax_0 + A(x_0 - x_i) = r_0 + AV_i y_i,$$

and since $r_i$ is orthogonal to $\mathcal{K}^i(A, r_0)$, we have that $V_i^* r_i = 0$, and therefore $y_i$ satisfies

$$V_i^* A V_i y_i = -V_i^* r_0 = -e_1 \|r_0\|. \tag{2}$$

Since $r_0$ is given, it is now clear (at least in theory) how to compute the approximations $x_i$ for $i \geq 1$. First, we set $i = 1$, then we can compute $\|r_0\|$ and $V_1^* A V_1$, solve $y_1$ from (2), after which $x_1 = x_0 + V_1 y_1$ according to (1). The residual for this approximation can be computed and the process can be continued. As long as the residuals remain non-zero, the matrix $V_i$ will be full rank and the fact that $A$ is positive definite guarantees that $y_i$ can always be solved from (2).

Because $A$ maps $\mathcal{K}^i(A, r_0)$ into $\mathcal{K}^{i+1}(A, r_0)$ for all $i \geq 1$, it follows that $Av_i$ is a linear combination of $v_0, \ldots, v_{i+1}$. This results in the observation that for all $i \geq 1$,

$$AV_i = V_{i+1} T_{i+1,i}, \tag{3}$$

and that the entries $t_{ij}$ of the $(i+1) \times i$ matrix $T_{i+1,i}$ are zero for all $i > j+1$. As a consequence, the system matrix from (2) can be found as

$$V_i^* A V_i = V_i^* V_{i+1} T_{i+1,i} = I_{i,i+1} T_{i+1,1} =: T_{ii}, \tag{4}$$

where $I_{i,i+1}$ represents the trivial embedding $\mathbb{R}^{i+1} \to \mathbb{R}^i$. Since the left-hand side in (4) is symmetric, also $T_i$ must be symmetric. So, $t_{ij} = 0$ if $|i - j| > 1$, or, in other words, $T_{ii}$ (and also $T_{i+1,i}$) is tridiagonal. This fact lies at the basis of the efficiency of the conjugate gradient method. The procedure described below (2) can be reduced to a three-term recursion, and using that $T_i$ is positive definite, even be decoupled into two two-term recursions using Cholesky factorization. We will not go into detail how this is done, but refer to other papers in this volume for the actual algorithm.

**The Lanczos method.** As a suitable conclusion of this section with basic definitions and relations, we now define approximations of the eigenpairs of $A$ according to the Lanczos method. For this, we associate to each pair $(z, \theta) \in \mathbb{C}^n \times \mathbb{C}$ with $\|z\| = 1$ an *eigenvalue residual* $r = Az - \theta z$. Given an arbitrary, so-called *initial vector* $v \in \mathbb{R}^n$, the Lanczos method aims to find $z \in \mathcal{K}^i(A, v)$ and $\theta \in \mathbb{C}$ such that

$$Az - \theta z \ = r \perp \mathcal{K}^i(A, v). \tag{5}$$

Let us now suppose that we apply the Lanczos method with the choice $v = r_0$, the initial residual of the conjugate gradient method described above. Since $z$ is supposed to be in $\mathcal{K}^i(A, r_0)$, it can be written as $V_i y$ for some $y \in \mathbb{R}^i$. Then, demanding orthogonality of the residual to $\mathcal{K}^i(A, v)$ results in the quest for pairs $y_j^{(i)}, \theta_j^{(i)}$ such that

$$V_i^* \left(AV_i y_j^{(i)} - \theta_j^{(i)} V_i y_j^{(i)}\right) = 0, \quad \text{or equivalently,} \quad T_{ii} y_j^{(i)} = \theta_j^{(i)} y_j^{(i)},$$

which shows that the Lanczos approximations for eigenpairs of $A$, using $r_0$ as startvector are the pairs $V_i y_j^{(i)}, \theta_j^{(i)}$ derived from the matrix $T_{ii}$ from the conjugate gradient method. The approximate eigenvalues $\theta_j^{(i)}$ are called *Ritz-values*, and the approximate eigenvectors $V_i y_j^{(i)}$ *Ritz vectors*. We will elabo rate on the convergence of Ritz values in Section 3. Before that, we put the Lanczos method and conjugate gradients in the broader context of *subspace methods*.

## 1.2 CG and Lanczos as instances of more general methods

Suppose $\mathcal{V} \subset \mathbb{R}^n$ and $\dim \mathcal{V} = i < n$, and let $\mathcal{W} = A\mathcal{V}$. Let $V$ be a full rank matrix with column span $\mathcal{V}$ and set $W = AV$. Then both the linear system $Ax = b$ and the eigenvalue problem $Az = \lambda z$ can be approximated in $\mathcal{V}$. We discuss here two general selection principles, by which we mean strategies to determine the $i$ free variables that we have to our disposal. The first one is demanding orthogonality of the residual with respect to $\mathcal{V}$, whereas the second demands orthogonality with respect to $\mathcal{W}$. In both cases, this gives $i$ equations for the $i$ variables.

**$\mathcal{V}$-orthogonal methods.** The first principle is to find $v \in \mathcal{V}$ such that the residual $r$, where $r = b - Av$ in case of the linear system, and $r = Av - \theta v$ in case of the eigenvalue problem, is orthogonal to $\mathcal{V}$. As a result, one is interested in solving

$$V^* W y = V^* b \quad \text{and} \quad V^* W y = \theta V^* V y, \tag{6}$$

respectively. Indeed, in the first case, $Vy$ can be interpreted as to approximate $x$, whereas in the second case, $(Vy, \theta)$ can be seen as an approximate eigenpair. Note that CG and Lanczos belong to this class of methods, with $\mathcal{V} = \mathcal{K}^i(A, b)$, whereas coarse grid corrections in MultiGrid [2] and the Jacobi-Davidson method [12] are instances of such methods in which $\mathcal{V}$ is *not* a Krylov subspace. The class is generally known as *Rayleigh-Ritz* methods, or $\mathcal{V}$-orthogonal methods.

**$\mathcal{W}$-orthogonal methods.** A second principle is given by the following symmetry-argument. If $A$ is invertible, then the eigenvalue problem $Az = \lambda z$ can be formulated equivalently as $A^{-1}z = \lambda^{-1}z$. Since we have subspaces $\mathcal{V}$ and $\mathcal{W}$ at our disposal together with the action of both $A$ and $A^{-1}$ between these spaces, there is no mathematical reason to prefer either formulation of the eigenvalue problem. Therefore, interchanging the role of $\mathcal{V}$ and $\mathcal{W}$, we may as well look for $w \in \mathcal{W}$ and scalars $\mu \in \mathbb{C}$ such that $A^{-1}w - \mu w \perp \mathcal{W}$. Of course, $w$ can be written as $Wy$ for some $y$, and using that $W = AV$ the orthogonality constraint results in

$$W^*Vy = \mu W^*Wy. \tag{7}$$

In case of the linear system $Ax = b$, the corresponding formulation becomes to find $w \in \mathcal{W}$ such that $w - b \perp \mathcal{W}$, which reduces to

$$W^*Wy = W^*b. \tag{8}$$

In this case, $Wy$ can be seen as an approximation of $b$, and hence $A^{-1}Wy = Vy$ is an approximation of $x$. In correspondence to the $\mathcal{V}$-orthogonal methods above, this class of methods is sometimes referred to as $\mathcal{W}$-orthogonal methods, but also as *minimal residual* (in case of linear systems) methods or *harmonic Rayleigh-Ritz* (in case of eigenvalue approximation) methods.

**Similarities in appearance.** Notice that we have the freedom to choose $b \in \mathcal{V}$. For ease of presentation, we will assume that this is the case. Thus, there exists an $e \in \mathbb{R}^i$ such that $b = Ve$, and using this, a more clear relation between the two different linear system- and eigenvalue approximation methods becomes visible:

$$\begin{aligned} V^*Wy &= V^*Ve, & V^*Wy &= V^*V\theta y, \\ W^*W\mu y &= W^*Vy, & W^*Wy &= W^*Ve. \end{aligned} \tag{9}$$

If $A$ is non-singular and $V$ has full rank, then also $W$ has full rank and $W^*W$, being a Gramm matrix, is non-singular. Therefore, only the first of the four problems in (9) needs extra assumptions to result in a well-defined approximation, of which symmetry and positive definiteness of $A$ is sufficient.

**Krylov subspace methods.** Choose $\mathcal{V} = \mathcal{K}^k(A, b)$. Refraining from any extra assumptions, the first method in (9) leads to the full orthogonalization method (FOM) for general matrices $A$, whereas for symmetric $A$, we arrive at the Lanczos method for linear systems. Assuming that $A$ is symmetric and positive definite, which is sufficient for a unique solution in $\mathcal{V}$, results in the conjugate gradient (CG) method. If $A$ is merely symmetric, then the second method is the Lanczos method, whereas for general $A$ it leads to the Arnoldi method. The third method in (9), corresponding to the second approach, leads to harmonic Lanczos and harmonic Arnoldi. Finally, the fourth method in (9) leads for general $A$ to the generalized minimal residuals method (GMRES), and for symmetric $A$ to minimal residuals (MinRes). See [5] for references in on Krylov subspace methods. We stress that we have only classified the methods according to their mathematical definition. We did not explain how from these definitions efficient algorithms can be derived. Generally, in case $A$ is symmetric, algorithms can be derived that infer short recursions and low, constant memory costs, whereas for general matrices, the length of the recursions grows linearly with $i$ and the memory costs quadratically in $i$. Finally, we emphasize that different implementations of the same method may lead to differently behaving algorithms in finite precision arithmetic.

## 1.3 Minimization properties for linear systems

Notice that for the linear system $Ax = b$, the two methods just described are linear mappings of the exact solution into the approximating space $\mathcal{V}$. Due to the left equation in (6), the corresponding operator, suggestively called $\mathcal{CG}$, can be written as

$$\mathcal{CG}\colon \mathbb{R}^n \to \mathcal{V}\colon x \mapsto V(V^*W)^{-1}V^*Ax \tag{10}$$

which can be checked to be idempotent and self-adjoint with respect to the bilinear form defined by $(\cdot,\cdot)_A\colon \mathbb{R}^n \times \mathbb{R}^n \to \mathbb{R}\colon (x,y)_A \mapsto (Ax,y)$, hence, if $A$ is symmetric positive definite, $\mathcal{CG}(x)$ is the best approximation of $x$ within $\mathcal{V}$ with respect to the associated $A$-norm. If $A$ is non-symmetric but coercive, the Lax-Milgram lemma gives quasi-optimality, but we will not go into detail here.

Secondly, for arbitrary non-singular $A$, orthogonalization of the residual to $\mathcal{W}$ as in (8), corresponds to an operator, suggestively called $\mathcal{MR}$, that can be written as

$$\mathcal{MR}\colon \mathbb{R}^n \to \mathcal{V}\colon x \mapsto V(W^*W)^{-1}W^*Ax, \tag{11}$$

which can be checked to be idempotent and self-adjoint with respect to the $A^*A$-inner product defined by $(\cdot,\cdot)_{A^*A}\colon \mathbb{R}^n \times \mathbb{R}^n \to \mathbb{R}\colon (x,y)_{A^*A} \mapsto (A^*A\ x, y)$, hence, $\mathcal{MR}(x)$ is the best approximation of $x$ within $\mathcal{V}$ with respect to the associated $A^*A$-norm. Since the $A^*A$ norm of the error $x - \mathcal{MR}(x)$ equals the $L^2$-norm of $A(x - \mathcal{MR}(x)) = b - A\mathcal{MR}(x)$, the name minimal residuals is explained.

**A priori bounds.** Similar as, for example, in finite element methods, the error of the method can now be bounded by the error of another element in the same subspace $\mathcal{V}$ that is easier to analyze. In finite element methods, one usually studies the error of some locally defined nodal interpolant. Also in Krylov methods, an interpolation problem can be formulated and solved in order to obtain upper bounds on the error. We will show how to do this in Section 2.

### 1.4 Circumventing the lack of symmetry

Before concluding the introduction, we would like to comment on several ways to avoid the long recursions and higher memory requirements that come with non-symmetric matrices $A$. The most obvious solution for linear systems is to consider the normal equations $A^*A\,x = A^*b$ and to use a solver for symmetric positive definite system matrices. Unfortunately, the convergence of conjugate gradients, as we will see further on, depends heavily on the condition number of the system matrix, and the normal equations have a squared condition number, compared to the original ones. Another way to symmetrize the matrix would be to consider the system

$$\begin{pmatrix} M & A^* \\ A & 0 \end{pmatrix} \begin{pmatrix} x \\ y \end{pmatrix} = \begin{pmatrix} c \\ b \end{pmatrix}, \tag{12}$$

for some symmetric $M$. This corresponds to solving both the systems $Mx + A^*y = c$ and $Ax = b$ simultaneously. An immediate consequence is that something is computed that may have no use at all. The obvious choice $M = 0$ together with the use of conjugate gradients leads to the bi-conjugate gradient (BiCG) algorithm, which allows short recursions and a low use of memory. Since for all $M$, and hence in particular for $M = 0$ the resulting matrix is not positive definite, BiCG may break down.

An reasonable alternative seems to apply the approach from (8) to the larger symmetric system (12). This would result in short recurrences, and would minimize the residual over $\mathcal{V}$. Notice that for given $(v^*, w^*)^* \in \mathbb{R}^{2n}$, the squared residual norm equals

$$\left\| \begin{pmatrix} c \\ b \end{pmatrix} - \begin{pmatrix} M & A^* \\ A & 0 \end{pmatrix} \begin{pmatrix} v \\ w \end{pmatrix} \right\|^2 = \|b - Av\|^2 + \|c - Mv - A^*w\|^2. \tag{13}$$

Not even for the choice $M = 0$, much can be concluded about the convergence behaviour of the single component $\|b - Av\|$. But although monotonicity cannot be guaranteed, clearly, any bound for the minimal residual method can be applied to the residual in (13). We will discuss this method briefly in Section 4.2.

**A strategy motivated by mixed finite element methods.** Another interesting choice in (12) is $M = I$ together with $c = 0$. Again, the resulting system is symmetric indefinite. Choosing subspaces $\mathcal{V} \subset \mathbb{R}^n$ and $\mathcal{W} \subset \mathbb{R}^n$, we can aim to solve $(\widehat{x}, \widehat{y}) \in \mathcal{V} \times \mathcal{W}$ from the following system

$$\begin{aligned} (\widehat{x}, v) + (\widehat{y}, Av) &= 0 \qquad \text{for all } v \in \mathcal{V} \\ (A\widehat{x}, w) &= (b, w) \quad \text{for all } w \in \mathcal{W}. \end{aligned} \tag{14}$$

It is well-known from the theory of mixed finite element methods [3] for which choices there exists a unique solution of this discretized problem. For example, given $\mathcal{V}$, choosing $\mathcal{W} = A\mathcal{V}$ is a possibility. Note that we are solving the "second order" equation $AA^*y = b$ by splitting it up into two "first order" systems $x = A^*y$ and $Ax = b$, and "testing" the equations in $\mathcal{V}$ and $\mathcal{W}$ respectively, using "integration by parts" in the first. Instead of (12), and thinking back to the normal equations, we may also formulate $A^*Ax = A^*b$ in this way, by discretizing

$$\begin{pmatrix} I & A \\ A^* & 0 \end{pmatrix} \begin{pmatrix} y \\ x \end{pmatrix} = -\begin{pmatrix} 0 \\ A^*b \end{pmatrix}. \tag{15}$$

Here too, the link with mixed finite element methods is obvious, although at present, we know of no results in this direction.

**Outline of the rest of this paper.** In Section 2, we will concentrate on deriving a priori bounds for the conjugate gradient method, as a prototype of Krylov subspace methods in general. In Section 3 we examine the convergence of Ritz values in the Lanczos method. Both sections are strongly related. Finally, in Section 4 we reflect on the application of the techniques developed in the previous sections to more general Krylov subspace methods.

## 2 The convergence of conjugate gradients

The conjugate gradient method constructs in the $i$th iteration step an approximation $x_i$ of $x$ such that $x_0 - x_i \in \mathcal{K}^i(A, r_0)$. Since each element from $\mathcal{K}^i(A, r_0)$ can be written as $Q_{i-1}(A)r_0$, where $Q_{i-1}$ is a polynomial of degree $i-1$, we have

$$x_0 - x_i = Q_{i-1}(A)r_0 = Q_{i-1}(A)(Ax - Ax_0) = AQ_{i-1}(A)(x - x_0). \tag{16}$$

Adding $x - x_0$ to both sides, we see that

$$x_i - x = P_i(A)(x_0 - x), \tag{17}$$

for some polynomial $P_i$ of degree $i$ with $P_i(0) = 1$. By our observations in Section 1.3, $P_i$ is such that $\|x_i - x\|_A$ is minimal over all polynomials of degree $i$ for which $P_i(0) = 1$. If we denote the eigenvalues and the orthonormalized eigenvectors of $A$ by $\lambda_j$, $z_j$, where $0 < \lambda_1 \le \cdots \le \lambda_n$, and write $r_0 = \sum_j \gamma_j z_j$, then we find that

$$r_i = b - Ax_i = A(x - x_i) = P_i(A)r_0 = \sum_j \gamma_j P_i(\lambda_j) z_j, \tag{18}$$

and, hence,

$$\|x_i - x\|_A^2 = \sum_j \frac{\gamma_j^2}{\lambda_j} P_i^2(\lambda_j). \tag{19}$$

Notice that only those $j$ for which $\gamma_j \neq 0$ give a positive contribution to this sum. In particular, if $A$ would have been semidefinite with $\lambda_j = 0$ for all $j \leq J \leq n$ for some $J \leq n$, then the minimization problem is still well-defined as long as the corresponding coefficients $\gamma_j, j \leq J$ in the expansion of the initial residual are also zero. The situation in which these $\gamma_j$ are small, for example due to rounding errors, is discussed in [6].

### 2.1 A priori bounds for the error

A priori bounds for the error in the $A$-norm are obtained by observing that the right-hand side of (19) can be bounded as follows. Using (19) itself with $i = 0$, we find that for each polynomial $R_i$ of degree $i$ with $R_i(0) = 1$,

$$\begin{aligned} \sum_j \frac{\gamma_j^2}{\lambda_j} P_i^2(\lambda_j) \leq \sum_j \frac{\gamma_j^2}{\lambda_j} R_i^2(\lambda_j) &\leq \max_j R_i^2(\lambda_j) \sum_j \frac{\gamma_j^2}{\lambda_j} \\ &= \max_j R_i^2(\lambda_j) \|x_0 - x\|_A^2, \end{aligned} \tag{20}$$

where the maximum, in principle, only needs to be taken over those $j$ for which $\gamma_j \neq 0$. Since generally we do not know the location of the eigenvalues of $A$, but may have some information on its extremal eigenvalues, and hence on its condition number $\kappa \equiv \lambda_n/\lambda_1$, we will continue our analysis with the simple bound

$$\max_j R_i^2(\lambda_j) \leq \max_{x \in [\lambda_1, \lambda_n]} R_i^2(x). \tag{21}$$

A well-known and descriptive upper bound now results when we choose for $R_i$ the $i$th degree *Chebyshev polynomial* $C_i$ transformed to the interval $[\lambda_1, \lambda_n]$ and scaled such that its value at 0 is equal to 1, or in other words,

$$R_i(x) = C_i\left(\frac{2x - (\lambda_1 + \lambda_n)}{\lambda_n - \lambda_1}\right) \Big/ C_i\left(\frac{-(\lambda_1 + \lambda_n)}{\lambda_n - \lambda_1}\right).$$

Uniformly, $R_i(x)$ is the best approximation of the zero function on $[\lambda_1, \lambda_n]$ under the restriction that it has value one at zero. The resulting upper bound for the $A$-norm of the error is (see [1], [4], [5]):

$$\|x_i - x\|_A \leq 2\left(\frac{\sqrt{\kappa} - 1}{\sqrt{\kappa} + 1}\right)^i \|x_0 - x\|_A, \quad \text{for all } i \geq 1, \tag{22}$$

which shows that we have fast convergence for small condition numbers.

**Upper bounds for special situations.** Other, though similar bounds can be obtained by choosing tailor-made polynomials when extra information is available on the eigenvalue distribution. For instance Axelsson [1] considered the specific case in which $\lambda_n$ is much larger than $\lambda_{n-1}$. Then the polynomial

$$R_i(x) = \left(1 - \frac{x}{\lambda_n}\right) C_{i-1}\left(\frac{2x - (\lambda_1 + \lambda_{n-1})}{\lambda_{n-1} - \lambda_1}\right) / C_{i-1}\left(\frac{-(\lambda_1 + \lambda_{n-1})}{\lambda_{n-1} - \lambda_1}\right)$$

satisfies $R_i(0) = 1$, $R(\lambda_n) = 0$, and if $\lambda_n$ is relatively much larger than $\lambda_{n-1}$, then on $[\lambda_1, \lambda_{n-1}]$, it is almost equal to the polynomial

$$C_{i-1}\Big(\frac{2x - (\lambda_1 + \lambda_{n-1})}{\lambda_{n-1} - \lambda_1}\Big)/C_{i-1}\Big(\frac{-(\lambda_1 + \lambda_{n-1})}{\lambda_{n-1} - \lambda_1}\Big).$$

This results in an improved upper bound, that approximately lags one step behind the upper bound (22) for a process with a condition number $\lambda_{n-1}/\lambda_1$.

These types of upper bounds show that it is important to have small condition numbers, or, in the case of larger condition numbers, to have eigenvalue distributions for which the eigenvalues that cause these large condition numbers are well separated. In that case we say that the (remaining) part of the spectrum is (relatively) clustered. The purpose of preconditioning is to reduce the condition number $\kappa$ and/or clustering of the eigenvalues. In [1] the situation is analysed where the eigenvalues of $K^{-1}A$ are in disjoint a priori known intervals.

## 2.2 Local effects in the convergence behavior

Upper bounds as in (22) show that we have global convergence, but they do not help us to explain all sorts of local effects in the convergence behavior of CG. A very well-known effect is the so-called superlinear convergence: in many situations one observes that the average speed of convergence seems to increase as the iteration proceeds. As we have seen in Section 1.1, the conjugate gradient algorithm and the Lanczos algorithm are closely related. The eigenvalues of the implicitly generated tridiagonal matrix $T_{ii}$ are the Ritz values of $A$ with respect to the current Krylov subspace. It is known from Lanczos theory that these Ritz values converge to the eigenvalues of $A$, and that in general the extreme eigenvalues of $A$ are well approximated already during early iterations [7], [9], [10]. Furthermore, the speed of convergence depends on how well these eigenvalues are separated from the others (gap ratio) [10]. This will help us to understand the superlinear convergence behavior of the conjugate gradient method (as well as other Krylov subspace methods). It can be shown that as soon as one of the extreme eigenvalues is modestly well approximated by a Ritz value, the procedure converges from then on as a process in which this eigenvalue is absent, i.e., as a process with a reduced condition number. Note that superlinear convergence behavior in this connection is used to indicate linear convergence with a factor that is gradually decreasing during the process, as more and more of the extreme eigenvalues are sufficiently well approximated.

**Example.** We consider a linear system of order $n = 100$ with one isolated eigenvalue 0.005 and the other 99 eigenvalues equally distributed over the interval $[0.08, 1.21]$. The right-hand side was chosen such that the initial residual has equal components in all eigenvector directions. The $A$-norms of the error $x - x_i$ are plotted on a logarithmic scale in Fig. 1. We observe in this figure, that the convergence for the first 8 or so iteration steps is relatively slow, in agreement with the condition number of $A$, which is 242.0. After iteration 10 the convergence is noticeably faster: about 4 times faster than in the earlier phase. This corresponds to a condition number of about 15.1 and that is the condition number for a matrix with eigenvalue spectrum $[0.08, 1.21]$, that is, the smallest eigenvalue 0.005 seems to have lost its influence on the convergence behaviour after 10 iteration steps. The smallest Ritz value at iteration step 10 is $\theta_1^{(10)} \approx 0.0061$, at iteration step 11: $\theta_1^{(11)} \approx 0.0054$, and at step 12: $\theta_1^{(12)} \approx 0.0052$. Clearly, it is converging to $\lambda_1 = 0.005$, but we see that the faster convergence of CG already sets in when the smallest Ritz value is still a relatively crude approximation (at step 10 we have an error of about 20 %). We will now analyze this behaviour in more detail.

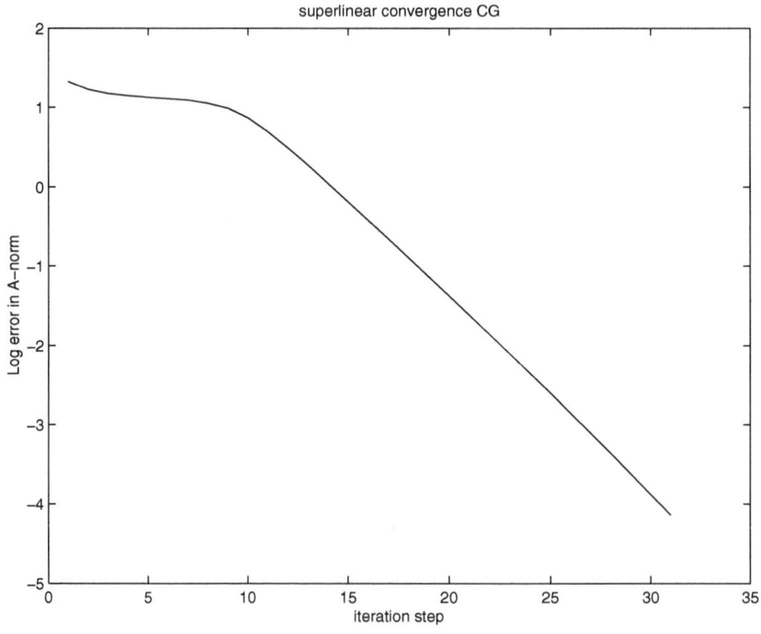


Figure 1. The convergence of CG for one isolated small eigenvalue.

### 2.3 Derivation of superlinear bounds

The local convergence behavior of CG, and especially the occurrence of superlinear convergence, was first explained in a qualitative sense in [4], and later in a quantitative sense in [13]. In both papers it was linked to the convergence of eigenvalues (Ritz values) of $T_{ii}$ in (4) towards eigenvalues of $A$, for increasing $i$. We will follow here the exposition given in [13]. The main idea in that paper is to link the convergence of CG in a certain phase with a process in which one or more of the Ritz values are replaced by eigenvalues of the matrix. To that end we first have to make a link with the Ritz values.

As we have seen in (17), the residual $r_i$ can be expressed as

$$r_i = P_i(A)r_0, \tag{23}$$

with $r_i \perp \mathcal{K}^i(A, r_0)$. It can be shown [13], that $P_i$ is the characteristic polynomial of $T_{ii}$ from (4), and hence,

$$\det(T_{ii})P_i(t) = (\theta_1^{(i)} - t)(\theta_2^{(i)} - t) \cdots (\theta_i^{(i)} - t). \tag{24}$$

We will exploit this characterization to study the convergence of CG from a certain iteration $i$ onwards. Writing the initial error $x - x_0$ as

$$x - x_0 = \sum_1^n \mu_j z_j, \tag{25}$$

on account of (17) we have that

$$x - x_i = \sum_1^n \mu_j P_i(\lambda_j) z_j.$$

Now, define $\bar{x}_0 = x_i + \mu_1 P_i(\lambda_1) z_1$, and use $\bar{x}_0$ as the starting vector of an auxilliary CG process (with the same $A$ and $b$), of which the initial error obviously satisfies

$$x - \bar{x}_0 = \sum_2^n \mu_j P_i(\lambda_j) z_j.$$

This new process generates a sequence $(\bar{x}_k)_{k \geq 0}$, for which the errors are characterized by polynomials $\bar{P}_k$ of degree $k$ as follows,

$$x - \bar{x}_k = \bar{P}_k(A)(x - \bar{x}_0) = \sum_2^n \mu_j \bar{P}_k(\lambda_j) P_i(\lambda_j) z_j. \tag{26}$$

For given $k$, the product of $\bar{P}_k$ and the polynomial $q_i$ defined by

$$q_i(t) = \frac{(\lambda_1 - t)(\theta_2^{(i)} - t) \cdots (\theta_i^{(i)} - t)}{\lambda_1 \theta_2^{(i)} \cdots \theta_i^{(i)}} = \frac{\theta_1^{(i)}(\lambda_1 - t)}{\lambda_1(\theta_1^{(i)} - t)} P_i(t) \tag{27}$$

is of degree $k + i$ and has value one at $t = 0$. Therefore, it may be used to bound the error in the $A$-norm of the $(k + i)$-th iterate of the original CG process, hence

$$\|x - x_{i+k}\|_A^2 \leq \|\bar{P}_k(A) q_i(A)(x - x_0)\|_A^2 = \sum_2^n \lambda_j \bar{P}_k(\lambda_j)^2 q_i(\lambda_j)^2 \mu_j^2. \tag{28}$$

Consider the rational factor in (27) representing the quotient $q_i/P_i$. Assuming that $\theta_1^{(i)}$ does not equal any of the eigenvalues $\lambda_2, \ldots, \lambda_n$, the following maximum is well-defined,

$$F_i \equiv \frac{\theta_1^{(i)}}{\lambda_1} \max_{j \geq 2} \left| \frac{\lambda_j - \lambda_1}{\lambda_j - \theta_1^{(i)}} \right|,$$

with as a consequence that $|q_i(\lambda_j)| \leq F_i |P_i(\lambda_j)|$ for all $j \geq 2$. We can now further simplify the upper bound in (28) as

$$\begin{aligned} \|x - x_{i+k}\|_A^2 &\leq F_i^2 \sum_2^n \lambda_j \bar{P}_k(\lambda_j)^2 P_i(\lambda_j)^2 \mu_j^2 \qquad (29) \\ &= F_i^2 \|x - \bar{x}_k\|_A^2 \\ &\leq F_i^2 \frac{\|x - \bar{x}_k\|_A^2}{\|x - \bar{x}_0\|_A^2} \|x - x_i\|_A^2. \end{aligned}$$

We have now proved a result stated in Theorem 3.1 from [13]. Its interpretation is as follows. First note that when $\theta_1^i$ is close to $\lambda_1$ for some $i$ then $F_i \approx 1$. This is illustrated in Fig. 2, where we have sketched the rational factor from (27). In fact, if $\lambda_1 \leq \theta_1^i < \lambda_2$, then $F_i$ equals the product of two factors with a simple interpretation: $\theta_1^i/\lambda_1$, which shows how good the approximation $\theta_1^i$ is, and $\frac{\lambda_2 - \lambda_1}{\lambda_2 - \theta_1^i}$, which measures the relative position of $\theta_1^i$ in the interval $[\lambda_1, \lambda_2]$. We conclude that as soon as $\lambda_1$ is sufficiently well approximated, we may expect a reduction in the next steps that, apart from a small multiplicative factor, is bounded by the reduction that we obtain for

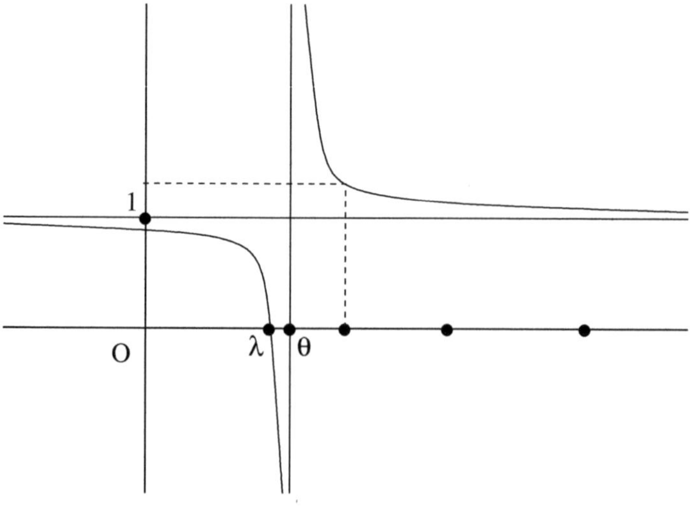


Figure 2. Qualitative sketch of $t \mapsto \frac{\theta(\lambda - t)}{\lambda(\theta - t)}$.

a CG process for $Ax = b$ in which $\lambda_1$ is missing. If we define the effective condition number for that process as $\kappa_2 = \lambda_n/\lambda_2$, we have that

$$\|x - x_{i+k}\|_A \le 2F_i \left( \frac{\sqrt{\kappa_2} - 1}{\sqrt{\kappa_2} + 1} \right)^k \|x - x_i\|_A.$$

Hence, if $\theta_1^{(i)} \approx \lambda_1$, then after the $i$-th step we have a reduction that is bounded by an expression in which we see the condition number for the remaining part of the spectrum. In fact, (29) reveils that the approximation of $\lambda_1$ by the smallest Ritz value does not need to be extremely accurate for the superlinear convergence to become apparent. For instance, if

$$\frac{\theta_1^{(i)} - \lambda_1}{\lambda_1} < 0.1 \quad \text{and} \quad \frac{\theta_1^{(i)} - \lambda_1}{\lambda_2 - \lambda_1} < 0.1,$$

then we already have that $F_i < 1.25$, which shows that we have from then on a reduction as for the reduced process except for a very modest factor. Things cannot be worse from then on, because it can be proved (see also Section 3) that $\theta_1^{(i)}$ is strictly decreasing towards $\lambda_1$ for increasing $i$. This means that also $F_i$ becomes smaller in further iterations.

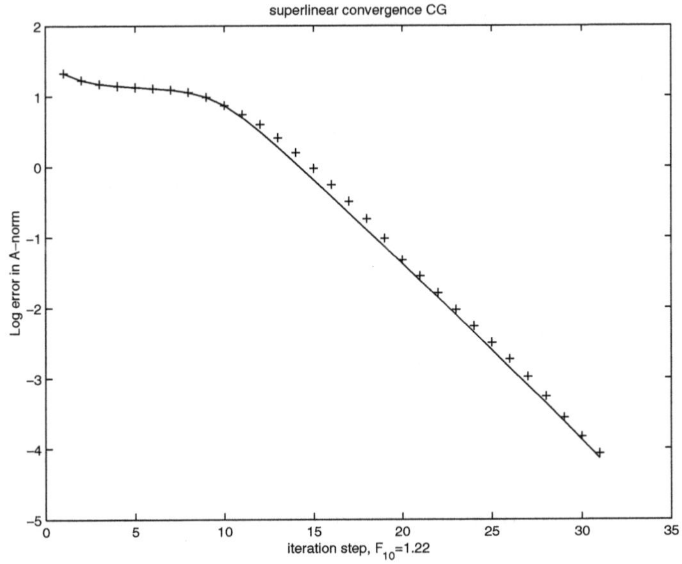


Figure 3. CG (−) and the comparison process (+++)

**Example (continued).** We now return to our earlier example of Fig. 1. At iteration step 10 we have a value $F_{10} \approx 1.22$. In Fig. 3 we have plotted the errors in $A$-norm for the CG process (the drawn line), and we have also marked with + the results for the comparison process. In this comparison process, we have removed at iteration step 10 the error component in the eigenvector direction corresponding to $\lambda_1 = 0.005$, and from then on we have multiplied the error-norms for the modified process by $F_{10}$. We see that this comparison process, which has only eigenvalues in the interval $[0.08, 1.21]$, describes the observed convergence behaviour for the original CG process (with all eigenvalues present) rather well.

In [13] other results are presented that shed more light on all sorts of special situations, such as almost double eigenvalues in the spectrum of $A$, or situations in which more Ritz values have converged to a number of successive eigenvalues. Not surprisingly, these results show that after (modest) approximations of Ritz values for the smallest $m$ eigenvalues, we will see convergence of the CG process from then on, governed by a condition number $\kappa_m \equiv \lambda_n/\lambda_{m+1}$. Of course, similar results can be derived for convergence of Ritz values towards the largest eigenvalues, but usually removal of small eigenvalues leads to much smaller condition numbers.

# 3 The convergence of Ritz values

We have introduced the Lanczos method as an algorithm based on finding pairs $(z, \theta) \in \mathcal{K}^i(A, v) \times \mathbb{C}$ such that the corresponding residual $r = Az - z\theta$ is orthogonal to the Krylov subspace $\mathcal{K}^i(A, v)$, which was spanned by the columns of the matrix $V_i$. The defining *Rayleigh-Ritz* orthogonality relation

$$V_i^*(Az - \theta z) = 0, \tag{30}$$

led, since each $z \in \mathcal{K}^i(A, v)$ can be written as $z = V_i y$ for some $y \in \mathbb{R}^i$, to the smaller eigenvalue problem

$$T_{ii} y = \theta y, \tag{31}$$

for the tridiagonal matrix $T_{ii}$ from (4). The eigenpairs of $T_{ii}$ are usually referred to as *Ritz values* and *Ritz vectors*, and are considered as approximations of eigenpairs of $A$. Strictly speaking, the first basis vector $v$ of $\mathcal{K}^i(A, v)$ may be chosen arbitrarily and need not be an initial residual of an explicit conjugate gradient process, but note that it can always be interpreted as one, simply for solving $Ay = v$ with initial approximation $y_0 = 0$. This is what we will do in the sequel in order to analyze the convergence of the Ritz values.

## 3.1 Monotonicity through the minimax characterization

Notice that for given non-zero $z \in \mathcal{K}^i(A, r_0)$ we may write the *Rayleigh quotient* $\mathcal{R}(A, z)$ for $z$ as

$$\mathcal{R}(A, z) = \frac{z^* A z}{z^* z} = \frac{y^* V_i^* A\, V_i y}{y^* V_i^* V_i y} = \frac{y^* T_{ii} y}{y^* y} = \mathcal{R}(T_{ii}, y), \tag{32}$$

where $y \in \mathbb{R}^i$ is the vector for which $z = V_i y$. This gives some insight into how the Ritz values are related to the eigenvalues of $A$ by means of their *minimax characterization.* Indeed, according to the Minimax Theorem, see, e.g., [10], the eigenvalues $\lambda_j$ of $A$ are equal to

$$\lambda_j = \min_{S \subset \mathbb{R}^n, \dim(S)=j} (\max_{0 \neq z \in S} \mathcal{R}(A, z)), \tag{33}$$

and the maximum is taken at the corresponding eigenvector(s) $z_j$ of $A$. Similarly, the eigenvalues of $T_{ii}$ (the Ritz values) are equal to

$$\theta_j^{(i)} = \min_{U \subset \mathbb{R}^i, \dim(U)=j} (\max_{0 \neq y \in U} \mathcal{R}(T_{ii}, y)), \tag{34}$$

and the maximum is taken at the corresponding eigenvector(s) $y_j^{(i)}$ of $T_{ii}$. Combining (32) with (33) and (34), leads to the interesting observation that the ordered Ritz values $\theta_1^{(i)} \leq \theta_2^{(i)} \leq \cdots \leq \theta_i^{(i)}$ are monotonically decreasing to the ordered eigenvalues $\lambda_1 \leq \lambda_2 \leq \cdots \leq \lambda_n$ of $A$. To be more explicit, with $j \leq i$ we have that

$$\begin{aligned} \lambda_j &= \min_{S \subset \mathbb{R}^n, \dim(S)=j} (\max_{0 \neq z \in S} \mathcal{R}(A, z)) \\ &\leq \min_{S \subset \mathcal{K}^i, \dim(S)=j} (\max_{0 \neq z \in S} \mathcal{R}(A, z)) = \theta_j^{(i)}, \end{aligned} \tag{35}$$

where the latter equality results from the application of the change in coordinates (32) to (34). This leads, for increasing $i$, to

$$\lambda_1 \leq \theta_1^{(1)},\ \lambda_1 \leq \theta_1^{(2)} \text{ and } \lambda_2 \leq \theta_2^{(2)},\ \lambda_1 \leq \theta_1^{(3)},\ \lambda_2 \leq \theta_2^{(3)}, \text{ and } \lambda_3 \leq \theta_3^{(3)},\ \ldots.$$

This shows that the $\theta_j^{(i)}$ with $j$ fixed and $i \geq j$ running, are bounded from below by $\lambda_j$, whereas similar observations can be made for the other extremal end of the spectrum by considering the maximin characterization (we will not go into detail). The right-hand side of (35) also shows that for fixed $j$ we have

$$\lambda_j = \theta_j^{(n)} \leq \theta_j^{(n-1)} \leq \cdots \leq \theta_j^{(j+1)} \leq \theta_j^{(j)}. \tag{36}$$

Since, in exact arithmetic, the Lanczos process is a finite process, we can strictly speaking, not regard the sequence of smallest Ritz values as a converging process. However, in finite precision, the process will usually deliver accurate approximations that cannot be essentially improved, long before the finiteness of the process would play a role. In this sense, it behaves as a converging process that is terminated when finite precision effects inhibit further convergence. We will therefore, in the context of Ritz values and Ritz vectors speak of convergence, and we will consider an eigenpair as (practically) converged when its residual is in norm smaller than some given tolerance.

### 3.2 The polynomial characterization

We have seen that the basis vectors $v_i$ for the Krylov subpace are scalar multiples of the conjugate gradient residuals $r_i$. This means that they can be expressed in terms of the initial vector $v_0 = r_0/\|r_0\|$ as

$$v_i = c_i P_i(A) v_0,$$

where $P_i$ is the polynomial from (17), (23) and (24), and $c_i = \|r_0\|/\|r_i\|$ to assure that $\|v_i\|_2 = 1$. This polynomial characterization permits nice error bounds. In order to show this, we define $R_{i,1}$ as the unique polynomial for which $P_i(t) = (t - \theta_1^{(i)})R_{i,1}$ and write $P_i(A)v_0$ accordingly as

$$P_i(A)v_0 = \frac{1}{\|r_0\|} P_i(A) r_0 = \frac{1}{\|r_0\|} \sum_j (\lambda_j - \theta_1^{(i)}) R_{i,1}(\lambda_j) \gamma_j z_j,$$

where we have written $r_0 = \sum_j \gamma_j z_j$ as before. Because $P_i(A)v_0$ is orthogonal to all lower degree polynomials in $A$ applied to $v_0$, we have that

$$0 = (P_i(A)v_0)^T R_{i,1}(A) v_0 = \sum_j (\lambda_j - \theta_1^{(i)}) R_{i,1}^2(\lambda_j)\gamma_j^2.$$

This gives the following relation for $\theta_1^{(i)}$:

$$\theta_1^{(i)} = \frac{\sum_j \lambda_j R_{i,1}^2(\lambda_j)\gamma_j^2}{\sum_j R_{i,1}^2(\lambda_j)\gamma_j^2} = \frac{z^* A z}{z^* z}, \quad \text{where } z = R_{i,1}(A) v_0. \tag{37}$$

By combining this with our knowledge that the Ritz values are the (local) extrema of the Rayleigh quotient, we conclude that this gives us two nice results:

1. The Ritz vector corresponding to $\theta_1^{(i)}$ is (apart from a scalar factor) given by $z = R_{i,1}(A)v_0$.
2. Because of the minimax property, the above characterization gives us an excellent means to find *a priori* upper bounds for $\theta_1^{(i)}$, namely, any polynomial of degree $i-1$ may replace $R_{i,k}$ and this will give an upper bound.

### 3.3 Construction of a related CG-process

From now on we will assume that $\gamma_1 \neq 0$, which is no loss of generality. Note that expression (37) is scaling invariant with respect to the polynomial $R_{i,1}$, and we scale it such that $\widetilde{R}_{i,1}(\lambda_1) = 1$, that is we replace $R_{i,1}$ by

$$\widetilde{R}_{i,1}(\lambda) = R_{i,1}(\lambda)/R_{i,1}(\lambda_1).$$

Then we obtain the following expression for the distance of the smallest Ritz value $\theta_1^{(i)}$ to the smallest eigenvalue $\lambda_1$ (with $\widetilde{\gamma}_j = \gamma_j/\gamma_1$):

$$0 \leq \theta_1^{(i)} - \lambda_1 = \frac{\sum_j (\lambda_j - \lambda_1) R_{i,1}^2(\lambda_j)\gamma_j^2}{\sum_j R_{i,1}^2(\lambda_j)\gamma_j^2} = \frac{\sum_{j>1} (\lambda_j - \lambda_1) \widetilde{R}_{i,1}^2(\lambda_j)\widetilde{\gamma}_j^2}{1 + \sum_{j>1} \widetilde{R}_{i,1}^2(\lambda_j)\widetilde{\gamma}_j^2}. \quad (38)$$

Since the right-hand side of (38) is, minus a fixed number $\lambda_1$, a Rayleigh quotient of which we know from (35) that it takes its minimum (over the Krylov subspace) at precisely the Ritz vector belonging to the smallest Ritz value, we know that $\widetilde{R}_{i,1}$ is the polynomial that minimizes the right-hand side of (38) over all polynomials of degree $i-1$ having value one at $\lambda_1$.

As we will show now, the error in the Ritz value can be interpreted as the square of the norm of the error of some associated conjugate gradient process for a system $By = d$, where $d$ remains to be specified, and with starting vector $y_0 = 0$. For this process, in analogy to (17), we have a relation of the form

$$y - y_i = (I - BQ_{i-1}(B))(y - y_0) \equiv G_i(B)(y - y_0) = G_i(B)y, \quad (39)$$

where $G_i$ is such that

$$\|y - y_i\|_B = \|G_i(B)y\|_B$$

is minimal over all polynomials of degree $i$ that have value one at zero. Hence, comparing this with (38), we see that the numerator in the right-hand of (38) can be interpreted as the error, in the $B$ norm, of a CG-process for an $(n-1)\times(n-1)$ matrix $B$ with eigenvalues $\lambda_1 - \lambda_j$ $(j > 1)$ and with eigenvectors $w_j$ that are unknown, but this is irrelevant since they do not appear explicitly in the expressions. The solution $y$ can be written as

$$y = \sum_{j\geq 2} \widetilde{\gamma}_j w_j.$$

With $Q_{i-1}$ the polynomial from (39) that minimizes the $B$-norm of this CG-process in step $i-1$, this yields an upper bound for the approximation error in the smallest Ritz value simply by neglecting the denominator, which is larger than one. In fact, we then have that

$$\theta_1^{(i)} - \lambda_1 \leq \sum_{j>1} (\lambda_j - \lambda_1) Q_{i-1}^2(\lambda_j)\widetilde{\gamma}_j^2 = \|y_{i-1} - y\|_B^2. \quad (40)$$

This shows that $\theta_1^{(i)} - \lambda_1$ is bounded from above by the square of the $B$-norm of the error in a related conjugate gradient process. As soon as $\theta_1^{(i)}$ has converged to some modest degree, then (38) implies that

$$1 + \sum_{j>1} \widetilde{R}_{i,1}^2(\lambda_j)\widetilde{\gamma}_j^2 \approx 1,$$

and then the approximation error almost coincides with the CG-error. Or said in a slightly different way, the same kind (only marginally better) of convergence behavior that we observe for conjugate gradients (with a reduced and shifted spectrum) can be observed for the convergence behavior of the smallest Ritz value.

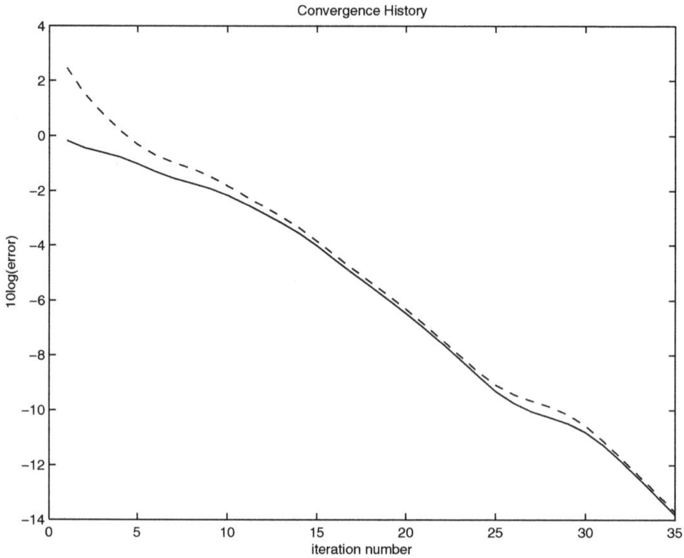


Figure 4. Ritz error (*line*) and CG-error (*dashed*).

**Example.** We consider the diagonal matrix of order 400 with eigenvalues $\lambda_1 = -0.034$, $\lambda_2 = -0.081$, $\lambda_3 = -0.0811$, $\lambda_4 = -0.127$, $\lambda_5 = -0.155$, and $\lambda_6 = -0.190$. The other 394 eigenvalues of this matrix are uniformly distributed in $[-1.2, -0.2]$. We select a $v_0$ that has equal components in all eigenvector directions: $\gamma_j = 1$. In Fig. 4 we see the values of the error $\theta_1^{(i)} - \lambda_1$, and the (squares of the) errors in $B$-norm of the associated CG-process. We see that the Ritz errors are nicely below the CG-errors, and ultimately they almost coincide.

We also see that temporary slow-downs in the convergence behaviour of CG is reflected by a similar slowdown of the Ritz value, although not at the same iteration numbers and at the same level. We would like to stress that the associated CG process is not the CG process that defines the same Ritz values as the Lanczos process that we have used.

In Section 2.3 we have argued that the conjugate gradient process converges superlinearly, or more precisely, that conjugate gradients converge as a process from which an extremal eigenvalue has been removed, as soon as this eigenvalue is modestly well approximated by the Ritz value of the corresponding Lanczos process (that is indeed the Lanczos process defined by the iteration coefficients of CG).

In a similar way, we have superlinear convergence for the smallest Ritz value, as soon as the largest eigenvalue of the shifted reduced problem is modestly well approximated by a Ritz value. Since the Ritz values are shift invariant, this implies that we may expect faster convergence in average, for the smallest Ritz value, as soon as the next Ritz value gets modestly close to the next eigenvalue.

Proceeding in this way, we see that we may expect an increase of the average speed of convergence, each time when the next smallest eigenvalue gets a Ritz value in its neighbourhood. This explains nicely what we see in practice.

This interpretation also shows that the error can be bounded as an error in a CG-process, and for this upper bounds have been derived that decay with a linear factor that depends on the condition number of $B$, or, in our case $(\lambda_n - \lambda_1)/(\lambda_2 - \lambda_1)$.

The similarity between CG errors and Ritz errors had already been noticed and analyzed in [11, Section 6]. There it has been shown that the CG-error for a process with matrix $A$, characterized by the eigenvalues $\lambda_j$ and weights $\gamma_j$, can be interpreted as a Ritz value for a related process that is characterized by eigenvalues $\bar{\lambda}_j$, augmented with $\bar{\lambda}_0 = 0$, and a set of weights $\bar{\gamma}_j$. They give explicit formulas for these eigenvalues and weights, and also lower bounds for the smallest Ritz value of this scheme. The approach in [11] is essentially the same as ours, but gives slightly more detailed results (at the cost of a more complicated analysis). We see that this correspondence between Ritz value convergence and CG-convergence allows us to transfer all the insights that result from analysis that has been carried out for convergence aspects of Ritz values (that is, for instance, misconvergence or stagnation, see [14]) to CG-behavior.

## 4 The convergence of minimal residuals

We introduced the minimal residual method in Section 1.2 as a specific $\mathcal{W}$-orthogonal method, i.e., the one in which $\mathcal{V} = \mathcal{K}^i(A, r_0)$. In Section 1.3 we explained its name by showing that $\mathcal{W}$-orthogonality coincides with minimizing the $A^*A$-norm of the error, or equivalently, the $L^2$-norm of the residual. Here we will briefly discuss its convergence behavior, following an exposition by Wathen et al. [15].

### 4.1 Bounds for symmetric indefinite problems

Considering the problem $Ax = b$ with $A$ symmetric indefinite. The minimal residual method constructs in the $i$-th iteration step an approximation $x_i$ of $x$ such that $x_0 - x_i \in \mathcal{K}^i(A, r_0)$, and $r_i = b - Ax_i$ has minimal Euclidean norm. As in (16), we have $x_i = x_0 + Q_{i-1}(A)r_0$ for some polynomial $Q_{i-1}$ of degree $i-1$ and hence,

$$r_i = b - Ax_i = b - Ax_0 - AQ_{i-1}(A)r_0 = (I - AQ_{i-1}(A))r_0 = P_i(A)r_0, \quad (41)$$

where $P_i(A)$ is, in this case, the polynomial of degree $i$ with $P_i(0) = 1$ such that $\|P_i(A)r_0\|$ is minimal over all such polynomials. Following the same approach as in Section 2, and writing $r_0 = \sum_j \gamma_j z_j$, we get

$$\|r_i\|^2 = \sum_{j=1}^{n} \gamma_j^2 P_i(\lambda_j)^2 \leq \max_j |P_i(\lambda_j)|^2 \|r_0\|^2. \tag{42}$$

The location of the eigenvalues is unknown, but it may be able to show, in specific situations, that they lie in a set of the form

$$E = [-a, -b] \cup [c, d]$$

with $a, b, c, d$ positive real numbers. In that case, it will be interested in finding a polynomial $R_i$ of degree $i$ with $R_i(0) = 1$ which is uniformly small on $E$. Similar as in conjugate gradients, this would lead to upper bounds of the error, though this time in the $A^2$-norm instead of the $A$-norm (which, for indefinite $A$, is not a norm anymore).

**Comparison with CG applied to $A^2x = Ab$.** It is reasonable to aim for convergence bounds of the minimal residual method that are better than the bounds one would obtain by solving the squared system $A^2x = Ab$ by the conjugate gradient method. For this reason, Wathen et al. [15] remark that a bound based on *symmetric* Chebyshev polynomials (see [8]) for the special case that the two intervals containing the eigenvalues have equal length, i.e., $a - b = d - c$, gives no advantage compared to conjugate gradients applied to the squared system. At the same time, a construction is presented in [15] showing that whenever the intervals have *different* lengths, minimal residuals is to be preferred. This construction is however not within the scope of this paper.

### 4.2 Non-symmetric systems

With the above observations in mind, we will briefly return to the methods proposed in Section 1.4 to circumvent the lack of symmetry. In case $A$ is not symmetric, we noticed that it may be worthwhile to apply the minimal residual method to the double-sized symmetric indefinite linear system

$$\begin{pmatrix} M & A^* \\ A & 0 \end{pmatrix} \begin{pmatrix} x \\ y \end{pmatrix} = \begin{pmatrix} c \\ b \end{pmatrix}. \tag{43}$$

Denote the double-sized matrix by $\mathcal{C}(M)$, then it is easily seen that $\mathcal{C}(M)$ and $\mathcal{C}(0)$ are congruent,

$$\mathcal{C}(M) = \begin{pmatrix} I & \frac{1}{2}MA^{-1} \\ 0 & I \end{pmatrix} \mathcal{C}(0) \begin{pmatrix} I & 0 \\ \frac{1}{2}A^{-*}M & I \end{pmatrix}, \tag{44}$$

and hence, by Sylvester's law, the inertia of the double-sized matrix is independent of the choice of $M$. Now, consider $M = \alpha I$ for some real number $\alpha$, possibly zero, then

$$\mathcal{C}(\alpha I)\begin{pmatrix} x \\ y \end{pmatrix} = \lambda \begin{pmatrix} x \\ y \end{pmatrix} \Leftrightarrow \mathcal{C}(\alpha I)\begin{pmatrix} -y \\ x \end{pmatrix} = (\alpha - \lambda)\begin{pmatrix} -y \\ x \end{pmatrix}, \tag{45}$$

which can be easily checked by writing out the equations. Therefore, $\lambda$ is an eigenvalue of $\mathcal{C}(\alpha I)$ if and only if $\alpha - \lambda$ is. Hence, the eigenvalues of $\mathcal{C}(\alpha I)$ are symmetrically distributed around $\frac{1}{2}\alpha$. Since this is the case also for $\alpha = 0$, and taking into account the constant inertia in $\alpha$, this means that there cannot be eigenvalues between zero and $\alpha$. Or, in other words, $\lambda$ and $\alpha - \lambda$ must be on different sides of the origin.

With respect to the convergence of minimal residuals, the eigenvalues of $\mathcal{C}(0)$ are distributed as badly as possible, i.e., symmetrically around the origin. According to [15], application of minimal residuals to solve the system would, in fact, be equally expensive as applying conjugate gradients to the squared system with system matrix $\mathcal{C}(0)^2$. This squared system involves the solution of both the systems $A^*Ax = A^*b$ and $AA^*y = Ac$. This, in turn, is more expensive than solving only $A^*Ax = A^*b$ by conjugate gradients.

For nonzero choices of $\alpha$, the two intervals containing the eigenvalues are also of equal length, although it is not clear a priori of which length. It may be that there exists an optimal value for it, but this issue needs further research.

**Acknowledgments.** Jan Brandts acknowledges the support of a Research Fellowship (1999–2004) of the *Royal Netherlands Academy of Arts and Sciences*, whereas both authors thank Michal Křížek of the Academy of Science of the Czech Republic and the anonymous referee for commenting on earlier versions of this paper.

## References

[1] O. Axelsson, *Solution of linear systems of equations: Iterative methods*, in: Sparse Matrix Techniques, Copenhagen 1976 (V. A. Barker, ed.), Springer Verlag, Berlin, 1977, pp. 1–51.

[2] D. Braess, *Finite Elements. Theory, Fast Solvers, and Applications in Solid Mechanics*, Cambridge University Press, Second Edition, 2001.

[3] F. Brezzi, *On the existence, uniqueness and approximation of saddle-point problems arising from Lagrangian multipliers*, RAIRO Anal. Numér. **8** (1974), 129–151.

[4] P. Concus, G. H. Golub, and D. P. O'Leary, *A generalized conjugate gradient method for the numerical solution of elliptic partial differential equations*, in: Sparse Matrix Computations (J. R. Bunch, D. J. Rose, eds.), Academic Press, New York, 1976.

[5] G. H. Golub, C. F. Van Loan, *Matrix Computations*, The Johns Hopkins University Press, Baltimore, 1996.

[6] E. F. Kaasschieter, *A practical termination criterion for the Conjugate Gradient method*, BIT **28** (1988), 308–322.

[7] C. Lanczos, *An iteration method for the solution of the eigenvalue problem of linear differential and integral operators*, J. Res. Natl. Bur. Standard **45** (1950), 225–280.

[8] V. I. Lebedev, *An iteration method for the solution of operator equations with their spectrum lying on several intervals*, USSR Comput. Math. and Math. Phys. **9** (1969), 17–24.

[9] C. C. Paige, *Computational variants of the Lanczos method for the eigenproblem*, J. Inst. Math. Appl. **10** (1972), 373–381.

[10] B. N. Parlett, *The Symmetric Eigenvalue Problem*, Prentice-Hall, Englewood Cliffs, N.J., 1980.

[11] G. L. G. Sleijpen, A. van der Sluis, *Further results on the convergence of conjugate gradients and Ritz values*, Linear Algebra Appl. **246** (1996), 233–278.

[12] G. L. G. Sleijpen, H. A. van der Vorst, *A Jacobi Davidson method for the eigenvalue problem*, SIAM J. Matrix Anal. Appl. **17** (1996), 401–425.

[13] A. van der Sluis, H. A. van der Vorst, *The rate of convergence of conjugate gradients*, Numer. Math. **48** (1986), 543–560.

[14] A. van der Sluis, H. A. van der Vorst, *The convergence behavior of Ritz values in the presence of close eigenvalues*, Lin. Algebra and Appl. **88/89** (1987), 651–694.

[15] A. J. Wathen, B. Fischer, and D. J. Silvester, *The convergence of iterative solution methods for symmetric and indefinite linear systems*, in: 'Numerical Analysis 1997' (D. F. Griffiths, G. A. Watson, eds.), Pitman Research Notes in Mathematics Series, Addison Wesley Longman, Harlow, England, 1997, pp. 230–243.

# An Application of the Shermann-Morrison Formula to the GMRES Method

Erik Jurjen Duintjer Tebbens

Department of Numerical Mathematics, Faculty of Mathematics and Physics, Charles University, Sokolovská 83, CZ–186 75 Praha 8, Czech Republic
jurjend.tebbens@gmx.net

## 1 Introduction

Solving with the restarted GMRES method the system of linear equations

$$Ax = b, \tag{1.1}$$

where $A \in \mathbb{R}^{n \times n}$ is a nonsingular matrix and the right-hand side $b$ lies in $\mathbb{R}^n$, is in some cases a laborious task. In the worst case, convergence of the GMRES method restarting after $m$ iterations, GMRES($m$), is not guaranteed at all for $m < n$ (for some global convergence criteria we refer to [10] and [11], for criteria in dependency of the restart parameter $m$ of GMRES($m$) see for example [18]). If the process does converge, decreasing of the residual norms can still be very slow. This is in particular so when during several steps residual norms remain stationary. If the process becomes unsuitable for practical use, we shall call this phenomenon stagnation. Popular techniques to overcome stagnation include preconditioning, especially through spectral deflation ([2], [3]), augmentation of Krylov Subspaces ([7]) or polynomial filtration ([13]).

These techniques are based on the idea that the eigenvalue distribution of the system matrix has an influence on the convergence of GMRES. In particular, it is assumed that eigenvalues lying comparatively close to the origin hamper convergence and the mentioned techniques all try, in some way or another, to eliminate unwanted eigenvalues from the spectrum. Though many well-known bounds for GMRES generated residual norms do in fact refer to the spectrum (see, for example, [11]) and in most cases the eigenvalue distribution does have an influence on the residual norms, it is also known this is not always true. In [14] it has even been proven one can construct for whatever spectrum and whatever convergence curve a linear system (1.1) such that $A$ has that specific spectrum and GMRES applied to (1.1) yields the prescribed convergence curve.

In this paper, we will find for the given right-hand side of (1.1) modifications of $A$ that yield any desired convergence curve. Moreover, we try to overcome the problem of a stagnating GMRES($m$) by switching to a parameter-dependent linear system that yields for appropriate parameters if not any prescribed convergence curve, then at least optimal residual norm decreasing at every step. The proposed process does not seek to modify the spectrum of the involved matrix but it consists of an optimization procedure obtained with the help of the Sherman-Morrison formula.

In the following section we describe the transition from the original system to the auxiliary system, Section 3 investigates the convergence properties of the auxiliary system when the GMRES method is applied to it. In the next section we describe options to make use of the convergence behavior of the auxiliary system in order to improve the convergence of the original system. Section 5 proposes a similar but different approach of applying the Sherman-Morrison formula and we construct an algorithm from it. In the last section we present some numerical experiments. Throughout the paper, $I_n$ will denote the $n$-dimensional identity matrix, with $I_0$ being the zero-dimensional matrix.

## 2 The Sherman-Morrison theorem

One of the most powerful techniques to improve poor properties of a linear system when a certain method is applied to it consists of passing to a preconditioned system. Preconditioning is based on the simple idea of multiplying both matrix and right-hand side with a specially constructed matrix and the solution of the preconditioned system is the same as the solution of the original system. The situation gets slightly more complicated when we subtract a specially constructed matrix from the system matrix instead of multiplying with it. The solution of the auxiliary system is not anymore equal to the one of the first system and, as we will see later, finding the wanted solution can become somehow problematic. But these problems can be overcome by appropriate choice of the matrix to subtract. The formula that enables us to change the linear system by matrix subtraction is the Sherman-Morrison formula for inversion of rank-$m$ updated matrices:

**Theorem 2.1** (Sherman-Morrison-Woodbury). *Let the matrix $A \in \mathbb{R}^{n \times n}$ be non-singular and $U, V \in \mathbb{R}^{n \times m}$ be matrices with $m \leq n$. Then the rank-$m$ updated matrix $A + UV^T$ is non-singular if and only if the $m$-dimensional matrix $I_m + V^T A^{-1} U$ is non-singular and its inverse equals*

$$(A + UV^T)^{-1} = A^{-1} - A^{-1}U(I_m + V^T A^{-1} U)^{-1} V^T A^{-1}. \tag{2.1}$$

Proof. For the proof see [8]. □

In order to switch to an auxiliary system we change the original matrix $A$ of (1.1) to

$$\hat{A} := A - UV^T,$$

for some $U, V \in \mathbb{R}^{n\times m}$. Then the solution of (1.1) can be written as

$$A^{-1}b = (\hat{A} + UV^T)^{-1}b = (\hat{A}^{-1} - \hat{A}^{-1}U(I_m + V^T\hat{A}^{-1}U)^{-1}V^T\hat{A}^{-1})b.$$

For $m = 1$, $U := b$ and $V := y \in \mathbb{R}^n$, this equation changes to

$$A^{-1}b = (\hat{A} + by^T)^{-1}b = \hat{A}^{-1}b - \hat{A}^{-1}b(1 + y^T\hat{A}^{-1}b)^{-1}y^T\hat{A}^{-1}b. \tag{2.2}$$

With the auxiliary system being defined by

$$\hat{A}x = b, \tag{2.3}$$

where

$$\hat{A} := A - by^T,$$

it is clear that we can find the solution of the original system (1.1) by solving the auxiliary system (2.3) and substituting $\hat{A}^{-1}b$ in (2.2).

One could also choose $m > 1$ and thus $V$ would consist of several columns $v_i$, $i \leq m$. On the one side such choice augments the number of free parameter vectors $v_i$, but on the other hand solving $A^{-1}b$ implies the computation of $\hat{A}^{-1}U$, that is of more than one linear system. In this paper, we restrict ourselves to the case $m = 1$. The one parameter vector $y \equiv v_1$ gives enough possibilities to construct an auxiliary system (2.3) having better convergence properties then has the original system when GMRES is applied to it.

## 3 The convergence curve of GMRES applied to systems with matrices $A - by^T$

We will try to find matrices $\hat{A} = A - by^T$ such that the GMRES method applied to (2.3) has highest possible convergence speed. An interesting result in this context can be found in [14]. From this paper it follows that given a non-increasing positive sequence $f_0 \geq f_1 \geq \ldots \geq f_{n-1} > 0$ and a right-hand side $b$ with $\|b\| = f_0$, the residual vectors $r_k$ at each step of the GMRES method applied to $\hat{A}x = b$ with vanishing initial guess satisfy $\|r_k\| = f_k$, $0 \leq k \leq n-1$, if and only if $\hat{A}$ is of the form

$$\hat{A} = WR\tilde{H}W^T, \tag{3.1}$$

where $R$ is a nonsingular upper triangular matrix, $W = (w_1, \ldots, w_n)$ is an orthogonal matrix such that

$$W^T b = \begin{pmatrix} \pm\sqrt{f_0^2 - f_1^2} \\ \vdots \\ \pm\sqrt{f_{n-2}^2 - f_{n-1}^2} \\ \pm\sqrt{f_{n-1}^2}, \end{pmatrix},$$

and $\tilde{H}$ is given by

$$\tilde{H} = \begin{pmatrix} 0 & \dots & 0 & 1/(b^T w_n) \\ 1 & & 0 & -(b^T w_1)/(b^T w_n) \\ \vdots & \ddots & \vdots & \vdots \\ 0 & \dots & 1 & -(b^T w_{n-1})/(b^T w_n) \end{pmatrix}.$$

We are particularly interested in the question whether or not it is possible to find a vector $y \in \mathbb{R}^n$ such that $\hat{A} = A - by^T$ belongs to the class of matrices given by (3.1). In other words choose $\hat{A}$ such that it can assume the form $\hat{A} = WR\tilde{H}W^T$ for some upper triangular matrix $R$ and with $W$ and $\tilde{H}$ satisfying the above equations. This is in fact possible, as we will now prove. The proof is not based on the above mentioned theory of Strakoš et al., but it works with the Givens rotations used to calculate the approximations of the GMRES method. The parameters of these rotations have an immediate influence on the convergence speed of the method.

**Definition 3.1.** A Givens rotation for the $i$th row, $i > 1$, is a rotation represented by left multiplication with an orthogonal matrix of the form

$$\begin{pmatrix} I_{i-2} & 0 & \dots & 0 \\ 0 & c_{i-1} & s_{i-1} & \vdots \\ \vdots & -s_{i-1} & c_{i-1} & 0 \\ 0 & \dots & 0 & I_{n-i} \end{pmatrix}.$$

Consequently, $c_{i-1}^2 + s_{i-1}^2 = 1$, and $c_{i-1}$ is called *Givens cosine* and $s_{i-1}$ *Givens sine.*

Givens rotations for the $i$th row can be used to zero out the $i$th element of a vector $g = (g_1, \dots, g_n)^T$. To do so, we will throughout this paper use the choices

$$c_{i-1} := \frac{g_{i-1}}{\sqrt{g_{i-1}^2 + g_i^2}}, \qquad s_{i-1} := \frac{g_i}{\sqrt{g_{i-1}^2 + g_i^2}}.$$

**Lemma 3.2.** *The action of $k$ consecutive Givens rotations, from a rotation for the 2nd row till a rotation for the $(k+1)$st row, on a given vector $g = (g_1, \dots, g_{k+1})^T$ yields a vector $(g_1', \dots, g_{k+1}')^T$ where*

$$g_i' = s_i g_{i+1} + c_i \sum_{j=1}^{i} c_{j-1} \left( \prod_{l=j}^{i-1} (-s_l) \right) g_j, \quad i \le k, \tag{3.2}$$

$$g_{k+1}' = \sum_{j=1}^{k+1} c_{j-1} \left( \prod_{l=j}^{k} (-s_l) \right) g_j,$$

*with $c_0 := 1$.*

Proof. At first, note that a Givens rotation for the $i$th row changes only rows $(i-1)$ and $i$. After application of the Givens rotation for the second row to $g$, the first element has the form

$$g_1' := c_1 g_1 + s_1 g_2,$$

and the second element will change to

$$g_2^* := -s_1 g_1 + c_1 g_2.$$

Therefore, the third element of this column changes to

$$g_3^* := (-s_2) g_2^* + c_2 g_3 = (-s_2)(-s_1 g_1 + c_1 g_2) + c_2 g_3$$

after the rotation for the third row. Simultaneously $g_2'$ assumes its final form

$$g_2' := c_2 g_2^* + s_2 g_3 = (-s_1 g_1 + c_1 g_2) c_2 + s_2 g_3.$$

It is readily seen that $k$ rotations give a $(k+1)$th element of the form

$$g_{k+1}^* = g_{k+1}' := \sum_{j=1}^{k+1} c_{j-1} \left( \prod_{l=j}^{k} (-s_l) \right) g_j,$$

and $g_k$ can be expressed as

$$g_k' := \left( \sum_{j=1}^{k} c_{j-1} \left( \prod_{l=j}^{k-1} (-s_l) \right) g_j \right) c_k + s_k g_{k+1},$$

which proves the lemma. □

The Givens cosines and sines involved in the computation of a GMRES approximation can be expressed in values that were calculated during the Arnoldi orthogonalization process and previously executed Givens rotations:

**Corollary 3.3.** *After the kth step of the* GMRES *method applied to a linear system*

$$Ax = b$$

*the Arnoldi process has generated an orthonormal basis $v_1, \ldots, v_k$ of the associated $k$-dimensional Krylov subspace and a Hessenberg matrix $\tilde{H}_k \in \mathbb{R}^{(k+1)\times k}$ with coefficients*

$$h_{j,k} := v_j^T A v_k, \quad 1 \le j \le k, \quad h_{k+1,k} := \left\| A v_k - \sum_{j=1}^{k} h_{j,k} v_j \right\|.$$

*If this Hessenberg matrix is brought to upper triangular form with the help of Givens rotations and the rotation for the $(i+1)$th row is given by $c_i$ and $s_i$, $i \leq k-1$, then the rotation for the $(k+1)$th row has Givens sine $s_k$ satisfying*

$$s_k^2 = \frac{h_{k+1,k}^2}{h_{k+1,k}^2 + (\sum_{j=1}^{k} c_{j-1}(\prod_{i=j}^{k-1}(-s_i))h_{j,k})^2},$$

*with* $c_0 := 1$.

Proof. In the GMRES method the Hessenberg matrix $\tilde{H}_k \in \mathbb{R}^{(k+1)\times k}$, $k \geq 1$, can be brought to upper triangular form by means of Givens rotations that zero out all non-zero lower subdiagonal elements. According to Lemma 3.2, after $(k-1)$ rotations the $k$th element of the last column has the form

$$h_{k,k}^* := \sum_{j=1}^{k} c_{j-1}\left(\prod_{i=j}^{k-1}(-s_i)\right)h_{j,k}.$$

To zero out the last entry $h_{k+1,k}$ of this column we define a Givens rotation for the $(k+1)$st row with

$$s_k^2 := \frac{h_{k+1,k}^2}{h_{k+1,k}^2 + (h_{k,k}^*)^2}. \qquad \square$$

By introducing the parameter-dependent matrix $A - by^T$, $y \in \mathbb{R}^n$, we create the opportunity to modify the values $h_{i,j}$ and hence the sines $s_k$ according to our own needs. In the case of an Arnoldi process applied to $A - by^T$ with vanishing initial guess, multiples of the first Arnoldi vector, $b$, are added to every Arnoldi vector. Krylov subspaces therefore remain the same as for the matrix $A$ and the influence of $y$ is simple to control.

**Lemma 3.4.** *The Arnoldi algorithm applied to the matrix $\hat{A} := A - by^T$, $y \in \mathbb{R}^n$, and to the right-hand side $b$ with initial approximation $x_0 = 0$ generates Arnoldi vectors $v_k$, $1 \leq k \leq n$, that are independent from the choice of $y$. Moreover, for $k \leq n-1$ the values*

$$\hat{h}_{j,k} := v_j^T \hat{A} v_k, \quad 2 \leq j \leq k,$$

*and*

$$\hat{h}_{k+1,k} := \left\| \hat{A}v_k - \sum_{j=1}^{k} \hat{h}_{j,k} v_j \right\|$$

*are independent from the choice of $y$.*

Proof. If the initial approximation is zero, then the first Arnoldi vector

$$v_1 = (b - \hat{A}x_0)/\|b - \hat{A}x_0\|$$

is equal to $b/\|b\|$. With the abbreviation $\alpha_1 := y^T v_1$ we obtain for the non-scaled second Arnoldi vector

$$\begin{aligned}\tilde{v_2} &= \hat{A}v_1 - \hat{h}_{1,1}v_1 = Av_1 - (v_1^T A v_1)v_1 - \alpha_1 b + \alpha_1 v_1^T b v_1 \\ &= Av_1 - (v_1^T A v_1)v_1 - \alpha_1(b - (b^T b)/\|b\| v_1) = Av_1 - (v_1^T A v_1)v_1,\end{aligned}$$

independent from $\alpha_1$ and hence from $y$. In general, with $\alpha_k := y^T v_k$ and because of the orthogonality of the Arnoldi vectors, the $(k+1)$st non-scaled Arnoldi vector becomes

$$\begin{aligned}\tilde{v_k} &= \hat{A}v_k - \sum_{j=1}^{k} \hat{h}_{j,k} v_j \\ &= Av_k - \sum_{j=1}^{k} (v_j^T A v_k) v_j - \alpha_k b + \alpha_k v_1^T b v_1 + \sum_{j=2}^{k} (v_j^T b \alpha_k) v_j \\ &= Av_k - \sum_{j=1}^{k} (v_j^T A v_k) v_j - \alpha_k (b - (b^T b)/\|b\| v_1) \\ &= Av_k - \sum_{j=1}^{k} (v_j^T A v_k) v_j,\end{aligned}$$

independent from $\alpha_k$ and $y$. With $\hat{h}_{k+1,k} = \|\tilde{v}_{k+1}\|$ we obtain the independence of $\hat{h}_{k+1,k}$, $k \leq n$, from $y$. As for the values $\hat{h}_{j,k}$, $2 \leq j \leq k$, we have

$$\hat{h}_{j,k} = v_j^T (A - b y^T) v_k = v_j^T A v_k,$$

because of the orthogonality of the Arnoldi vectors ($b = v_1 \|b\|$). □

The previous lemma states, in other words, that only the first row of the Hessenberg matrix involved in the GMRES process is dependent from the parameter vector $y$. Its influence on the Givens sines is given by Corollary 3.3. We thus have to our disposal an easy to handle tool for manipulating the associated Givens rotations. It is a well known fact, that in the GMRES method the residual norm at every iteration can be expressed as the product of all previously executed Givens rotation sines and the initial residual norm. In terms of geometry, the cosine of the angle between one residual vector and the next one decides about the decreasing of the residual norm. For our parameter-dependent matrix, this cosine appears to be sensible to changes of the parameter. In fact, we are able to turn the planes generated by $\hat{A}r_0, \ldots, \hat{A}^k r_0$ in such a way that the distance from $r_0$ to these planes becomes as small as possible, as long as it doesn't vanish.

**Theorem 3.5.** *Let $f_0 \geq f_1 \geq f_2 \ldots > f_n = 0$ be a non-increasing sequence of real values, let the system of linear equations*

$$\hat{A}x = b \tag{3.3}$$

*with $\|b\| = f_0$ be given by*

$$\hat{A} := A - by^T, \quad y \in \mathbb{R}^n,$$

*and let us choose the initial approximation $x_0 = 0$. If the Arnoldi algorithm applied to $A$ doesn't break down before the $n$th step, then there exists at least one $y \in \mathbb{R}^n$ such that the residual vectors $\hat{r}_k := b - \hat{A}\hat{x}_k$ obtained by application of the* GMRES *method to the system* (3.3) *satisfy*

$$\|\hat{r_k}\| = f_k, \quad 0 \leq k \leq n.$$

Proof. In the GMRES method we have the following recurrence formula for the residual norms (see for example [10], p. 167):

$$\|\hat{r}_k\| = |s_k \cdot \ldots \cdot s_1|\beta,$$

where $\beta := \|\hat{r}_0\| = \|b\|$. As a consequence,

$$\frac{\|\hat{r}_k\|}{\|\hat{r}_{k-1}\|} = |s_k|.$$

The theorem is proven if we show that we can find a vector $y \in \mathbb{R}^n$ such that all values $|s_k|$ satisfy

$$|s_k| = \frac{f_k}{f_{k-1}}, \quad 1 \leq k \leq n-1.$$

The last residuum vanishes automatically.

For $k = 1$, the Arnoldi process yields values $\hat{h}_{1,1}$ and $\hat{h}_{2,1}$ and the Givens sine that zeroes out (together with the Givens cosine $c_1$) $\hat{h}_{2,1}$ satisfies

$$s_1^2 = \frac{\hat{h}_{2,1}^2}{\hat{h}_{2,1}^2 + \hat{h}_{1,1}^2}.$$

Let us denote by $\alpha_1$ the value $y^T v_1$. Note that $\hat{h}_{2,1}$ is independent from $\alpha_1$ because of Lemma 3.4. As for $\hat{h}_{1,1}$, we have

$$\hat{h}_{1,1}^2 = (v_1^T(A - by^T)v_1)^2 = (v_1^T A v_1 - \beta\alpha_1)^2.$$

Hence $\hat{h}_{1,1}^2$ can have whatever non-negative value if we choose $\alpha_1$ accordingly. In other words, $|s_1|$ can assume, in dependency from $\alpha_1$, every positive value smaller than or equal to 1. In particular it can take the wanted value $f_1/f_0$ if we solve

$$\left(\frac{f_1}{f_0}\right)^2 = \frac{\hat{h}_{2,1}^2}{\hat{h}_{2,1}^2 + (v_1^T A v_1 - \beta\alpha_1)^2},$$

that is

$$\alpha_1 = \frac{\pm\sqrt{\frac{1-(f_1/f_0)^2}{(f_1/f_0)^2}}\,\hat{h}_{2,1} - v_1^T A v_1}{-\beta}.$$

This puts a first condition on the vector $y$, namely it fixes for the chosen $\alpha_1$ the value of $y^T v_1$. Now, let $k$ be greater than 1 and let us assume we have chosen all values $\alpha_i := y^T v_i$ such that $|s_i| = f_i/f_{i-1}$ for $i < k$. According to Corollary 3.3,

$$s_k^2 = \frac{\hat{h}_{k+1,k}^2}{\hat{h}_{k+1,k}^2 + (\sum_{j=1}^k c_{j-1}(\prod_{i=j}^{k-1}(-s_i))\hat{h}_{j,k})^2}.$$

In this expression, only $\hat{h}_{1,k}$ is dependent from the choice of $\alpha_k := y^T v_k$ because of Lemma 3.4. The second term of the denominator, including $\hat{h}_{1,k}$, can be written as

$$\begin{aligned}
&\left(\sum_{j=1}^{k} c_{j-1}\left(\prod_{i=j}^{k-1}(-s_i)\right)\hat{h}_{j,k}\right)^2 \\
&= \left(\hat{h}_{1,k}\prod_{i=1}^{k-1}(-s_i) + \sum_{j=2}^{k} c_{j-1}\left(\prod_{i=j}^{k-1}(-s_i)\right)\hat{h}_{j,k}\right)^2 \\
&= \left((-\beta\alpha_k + v_1^T A v_k)\prod_{i=1}^{k-1}(-s_i) + \sum_{j=2}^{k} c_{j-1}\left(\prod_{i=j}^{k-1}(-s_i)\right)\hat{h}_{j,k}\right)^2.
\end{aligned}$$

Again, this is a non-negative expression that is dependent from $\alpha_k$ and it can possibly assume the value 0. Therefore, an appropriate choice of $\alpha_k$ yields the desired equality $|s_k| = f_k/f_{k-1}$, $0 < \frac{f_k}{f_{k-1}} \le 1$. Possible choices are given by

$$\begin{aligned}
\alpha_k = &\left(\pm\sqrt{\frac{1-(f_k/f_{k-1})^2}{(f_k/f_{k-1})^2}}\,\hat{h}_{k+1,k} - v_1^T A v_k \prod_{i=1}^{k-1}(-s_i)\right. \\
&\left. - \sum_{j=2}^{k} c_{j-1}(\prod_{i=j}^{k-1}(-s_i))\hat{h}_{j,k}\right) \Big/ \left(-\beta\prod_{i=1}^{k-1}(-s_i)\right).
\end{aligned} \tag{3.4}$$

After $n-1$ steps we have defined all Arnoldi vectors $v_1, \ldots, v_n$ and have been putting $n-1$ conditions on the vector $y$:

$$y^T(v_1, \ldots, v_{n-1}) = (\alpha_1, \ldots, \alpha_{n-1}).$$

There exists at least one $y \in \mathbb{R}^n$ solving this under-determined linear system. $\square$

## 4 Substitution of the solution of the auxiliary system

We now turn our attention to various options of substituting the solution of the auxiliary system in the Sherman-Morrison formula in order to obtain the solution of the original system. In case the solution of the second system is exact, Sherman-Morrison also yields the exact solution of the first system. A first way to improve the GMRES method with the help of the Sherman-Morrison formula that comes to mind, is to try to define an auxiliary system that finds its exact solution earlier then the original system does. This, unfortunately, is not possible. As soon as the second system has found the exact solution, the corresponding Krylov subspace has maximal dimension. But at the same iteration number the Krylov subspace of the original system would have reached maximal dimension, because the Krylov subspaces of both systems are identical (see Lemma 3.4).

As a second option, one could substitute only approximate solutions of the second system in the Sherman-Morrison formula. As we were able in the previous section to construct auxiliary systems with arbitrary convergence speed, finding a "good" approximate solution $\hat{x}_k$, $k < n$, for such system should not be too difficult. Having done so, we could back-transform with the Sherman-Morrison formula as follows:

$$\begin{aligned} A^{-1}b = (\hat{A} + by^T)^{-1}b &= \hat{A}^{-1}b - \hat{A}^{-1}b(1 + y^T\hat{A}^{-1}b)^{-1}y^T\hat{A}^{-1}b \\ &\approx \hat{x}_k - (1 + y^T\hat{x}_k)^{-1}\hat{x}_k y^T\hat{x}_k \\ &= \left(1 - \frac{y^T\hat{x}_k}{1 + y^T\hat{x}_k}\right)\hat{x}_k = \frac{\hat{x}_k}{1 + y^T\hat{x}_k}. \end{aligned}$$

As far as $y^T\hat{x}_k \neq -1$, we can use

$$\bar{x}_k \equiv \frac{\hat{x}_k}{1 + y^T\hat{x}_k} \tag{4.1}$$

as an approximation for the original system. Due to the fact that $\bar{x}_k$ is only a scalar multiple of $\hat{x}_k$, we have $\bar{x}_k \in \mathcal{K}_k(\hat{A}, b)$. But again the equality $\mathcal{K}_k(\hat{A}, b) = \mathcal{K}_k(A, b)$ prevents any improvement with regards to the classical GMRES method, because the classical GMRES iterate $x_k$ for the original system already minimizes

$$\|b - Ax\| \quad \text{over all} \quad x \in \mathcal{K}_k(A, b) = \mathcal{K}_k(\hat{A}, b).$$

Summarizing, application of the Sherman-Morrison formula to the full GMRES method in the above proposed manner will not improve its convergence and this is essentially due to the identity of the involved Krylov subspaces. But initially we were interested in overcoming stagnation of the *restarted* GMRES method. When we use the restarted version, initial guesses at the beginning of every restart are nonzero and therefore Lemma 3.4 does not hold anymore and one can expect some improvement. In fact, we could use Theorem 3.5 to construct an auxiliary system who's say $k$ first iterations do not stagnate and apply GMRES($m$), $k < m$, to that specific auxiliary system. The philosophy behind this way of doing is that non-stagnation of the $k$ very first steps might cause non-stagnation during the $k$ first steps of restarts that follow too. Having found an approximation $\hat{x}$ of the second system, back-transformation with Sherman-Morrison according to (4.1) yields an approximation to the original system that is perhaps not as accurate as $\hat{x}$ is for the auxiliary system, but it is not unreasonable to expect $\bar{x}$ to be a better approximation then the approximations of the original stagnating GMRES($m$).

Let us demonstrate this with an example. It consists of a linear system that arises from the discretization of the differential equation

$$-\mathrm{e}^{-xy} \triangle u + (10 + y\mathrm{e}^{-xy})u_x + (10 + x\mathrm{e}^{-xy})u_y - 60u = 1 \tag{4.2}$$

on the unit square with Dirichlet boundary condition $u = 0$ on $\partial([0,1]^2)$. Finite difference approximation on a $20 \times 20$ grid yields the stiffness matrix $A \in \mathbb{R}^{400\times 400}$ and our right-hand side $b$ has only elements which have the value 0.5. So $\|b\| = 10$ and we choose $x_0 = 0$.

When we apply the GMRES method restarted after 30 steps to this system, the process stagnates. With Theorem 3.5 one could construct an auxiliary system with non-stagnating initial iterations. For example, let us ask the first 10 residual norms to fulfill

$$\|\hat{r}_0\| = 10, \quad \|\hat{r}_1\| = 9, \quad \|\hat{r}_2\| = 8, \quad \|\hat{r}_3\| = 7, \quad \|\hat{r}_4\| = 6, \quad \|\hat{r}_5\| = 5,$$
$$\|\hat{r}_6\| = 4, \quad \|\hat{r}_7\| = 3, \quad \|\hat{r}_8\| = 2, \quad \|\hat{r}_9\| = 1.5, \quad \|\hat{r}_{10}\| = 1.$$

Having defined a vector $y$ forcing such residual norms, we can apply GMRES(30) to the resulting system with matrix $A - by^T$. This process does not stagnate anymore. Moreover, when we use the approximations of this system to back-transform according to (4.1), these approximations to the original system do not yield stagnation either. This is shown in Fig. 1.

Graph GMRES(30) displays the restarted GMRES(30) applied to the first system and graph GMRES(30)^ concerns the auxiliary system. The gap between the curve for the first system after back-transformation and the curve for the second system results from the following equation:

$$b - A\bar{x} = b - A\left(\frac{\hat{x}}{1 + y^T\hat{x}}\right) = \frac{b + by^T\hat{x} - A\hat{x}}{1 + y^T\hat{x}} = \frac{b - (A - by^T)\hat{x}}{1 + y^T\hat{x}} = \frac{b - \hat{A}\hat{x}}{1 + y^T\hat{x}},$$

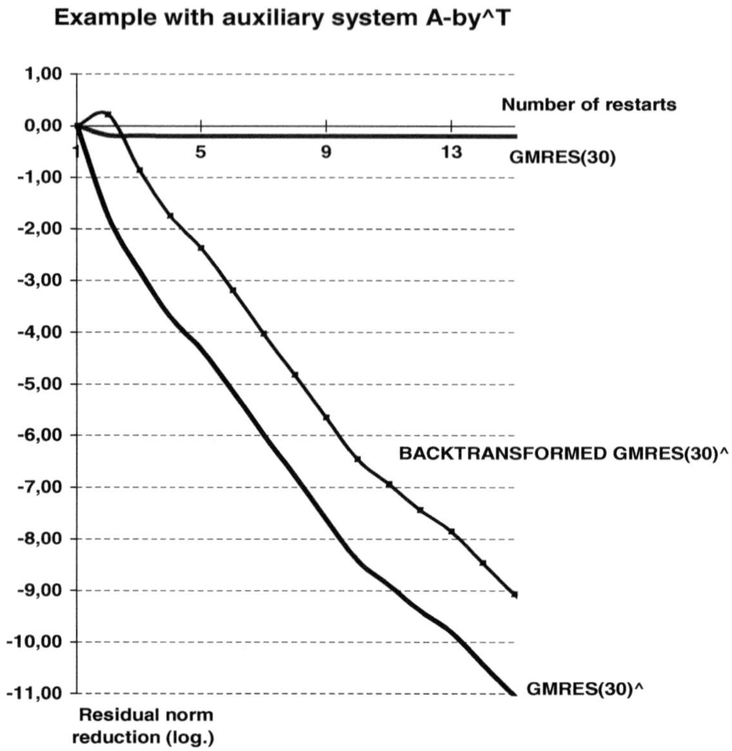


Figure 1. PDE-discretization stiffness matrix.

hence

$$\|b - A\bar{x}\| = \frac{\|b - \hat{A}\hat{x}\|}{|1 + y^T\hat{x}|}.$$

The loss of quality after back-transformation grows particularly large when the scalar $1 + y^T\hat{x}$ tends to 0. This appears to be the case whenever we force the auxiliary system to have too fast convergence speed. The proof is given in the following proposition. The result is not surprising since when we ask for fast convergence, our parameter vector $y$ will have comparatively large elements (see equation (3.4)). The matrix $A - by^T$ is then merely a perturbation of $by^T$, a singular rank-1 matrix.

**Proposition 4.1.** *With the notations of Theorem* 3.5, *let* $\hat{x}_i$ *denote the* $i$*th approximation calculated by the* GMRES *method applied to the system*

$$\hat{A}x = b,$$

*and let* $k < n$. *For every* $\varepsilon > 0$ *we can find a* $\delta > 0$ *such that if* $f_k/f_{k-1} < \delta$, *then*

$$|y^T\hat{x}_k + 1| < \varepsilon.$$

Proof. Let $V_k$ denote the matrix whose columns are the first $k$ Arnoldi vectors, $\tilde{H}_k$ denote the upper Hessenberg matrix of dimension $(k+1) \times k$ obtained with the computation of $V_k$ and $\beta = \|b\|$. In the $k$th step we have $k$ conditions for $y$:

$$y^T V_k = (\alpha_1, \ldots, \alpha_k), \quad i \leq k. \tag{4.3}$$

With $a := (\alpha_1, \ldots, \alpha_k)^T$, solutions for $y$ all satisfy

$$y \in V_k a + U,$$

where $U$ is the orthogonal complement in $\mathbb{R}^n$ to $\text{span}\{v_1, \ldots, v_k\}$. The $k$th approximation has because of the vanishing initial guess the form $V_k w_k$ with $w_k \in \mathbb{R}^k$ minimizing $\|\beta e_1 - \tilde{H}_k w\|$. Hence,

$$y^T \hat{x} = (V_k a + U)^T (V_k w_k) = a^T w_k. \tag{4.4}$$

The vector $w_k$ can be obtained by computing the QR-decomposition of $\tilde{H}_k$, $\tilde{H}_k = Q_k \tilde{R}_k$. With $R_k$ being the upper triangular matrix given by $\tilde{R}_k$ without its last row and $g$ being $Q_k^T \beta e_1$ without last element, we have

$$w_k = R_k^{-1} g.$$

According to Lemma 3.2, the last column of $R_k$ has elements

$$R_{i,k} = s_i \hat{h}_{i+1,k} + c_i \sum_{j=1}^{i} c_{j-1} \left( \prod_{l=j}^{i-1} (-s_l) \right) \hat{h}_{j,k}, \quad i \leq k,$$

and $g$ has elements

$$g_i = c_i \beta \prod_{l=1}^{i-1} (-s_l), \quad i \leq k.$$

For nonsingular upper triangular matrices the following recurrence formula holds:

$$R_k^{-1} = \begin{pmatrix} R_{k-1}^{-1} & -R_{k-1}^{-1} \cdot r_k / R_{k,k} \\ 0 & 1/R_{k,k} \end{pmatrix},$$

with $R_{k-1}$ being the left upper $(k-1) \times (k-1)$ part of $R_k$ and $r_k$ the last column of $R_k$ without its last element $R_{k,k}$. If $g = (g^*, g_k)^T$, then

$$\begin{aligned} w_k = R_k^{-1} g &= \begin{pmatrix} R_{k-1}^{-1} & -R_{k-1}^{-1} \cdot r_k / R_{k,k} \\ 0 & 1/R_{k,k} \end{pmatrix} \begin{pmatrix} g^* \\ g_k \end{pmatrix} \\ &= \begin{pmatrix} R_{k-1}^{-1} (g^* - g_k r_k / R_{k,k}) \\ g_k / R_{k,k} \end{pmatrix}, \end{aligned}$$

and hence

$$a^T w_k = (\alpha_1, \ldots, \alpha_{k-1}) R_{k-1}^{-1} (g^* - g_k r_k / R_{k,k}) + \alpha_k g_k / R_{k,k}. \tag{4.5}$$

Expressions that are dependent on $\alpha_k$ in (4.5) are $g_k r_k / R_{k,k}$ and $\alpha_k g_k / R_{k,k}$. We have

$$(r_k)_i / R_{k,k} = \frac{s_i \hat{h}_{i+1,k} + c_i \sum_{j=1}^{i} c_{j-1} (\prod_{l=j}^{i-1} (-s_l)) \hat{h}_{j,k}}{s_k \hat{h}_{k+1,k} + c_k \sum_{j=1}^{k} c_{j-1} (\prod_{l=j}^{k-1} (-s_l)) \hat{h}_{j,k}}, \quad i < k.$$

Bearing in mind that only $\hat{h}_{1,k} = v_1^T A v_k - \beta \alpha_k$ depends upon $\alpha_k$ and dividing by it yields

$$\lim_{\alpha_k \to \pm\infty} (r_k)_i / R_{k,k} = \frac{c_i \prod_{l=1}^{i-1} (-s_l)}{c_k \prod_{l=1}^{k-1} (-s_l)} = \frac{c_i}{c_k \prod_{l=i}^{k-1} (-s_l)}, \quad i < k.$$

Hence

$$\lim_{\alpha_k \to \pm\infty} (g^* - g_k r_k / R_{k,k})_i = c_i \beta \prod_{l=1}^{i-1} (-s_l) - \frac{c_i \beta \prod_{l=1}^{k-1} (-s_l)}{\prod_{l=i}^{k-1} (-s_l)} = 0, \quad i < k,$$

and

$$\lim_{\alpha_k \to \pm\infty} (\alpha_1, \ldots, \alpha_{k-1}) R_{k-1}^{-1} (g^* - g_k r_k / R_{k,k}) = 0. \tag{4.6}$$

The second term in (4.5) tends to

$$\begin{aligned} &\lim_{\alpha_k \to \pm\infty} (\alpha_k g_k) / R_{k,k} \qquad (4.7) \\ &= \lim_{\alpha_k \to \pm\infty} \frac{\alpha_k g_k}{s_k \hat{h}_{k+1,k} + c_k \sum_{j=1}^{k} c_{j-1} (\prod_{l=j}^{k-1} (-s_l)) \hat{h}_{j,k}} \\ &= \frac{g_k}{c_k (-\beta) \prod_{l=1}^{k-1} (-s_l)} = \frac{c_k \beta \prod_{l=1}^{k-1} (-s_l)}{c_k (-\beta) \prod_{l=1}^{k-1} (-s_l)} = -1 \end{aligned}$$

by dividing with $\alpha_k$.

Now let $\varepsilon > 0$. From (4.4), (4.5), (4.6) and (4.7) we obtain the existence of a positive integer $N$ such that

$$|y^T \hat{x}_k + 1| < \varepsilon, \quad \text{if } |\alpha_k| > N.$$

With (3.4) we can find a $\delta > 0$ satisfying

$$|\alpha_k| > N, \quad \text{if } f_k / f_{k-1} < \delta. \qquad \square$$

Thus attempts to overcome stagnation of the restarted GMRES method become ineffective when we ask for too dramatic residual norm decreasing in the auxiliary system. On the other hand, a reasonably defined, non-stagnating curve of the second system and back-transformation with Sherman-Morrison can successfully overcome stagnating, as was seen in the above example.

**Remark.** It is possible to use the above theory in case of an original system with non-vanishing first guess too. One would apply Theorem 3.5 with

$$b := v_1 = (b' - Ax_0)/\|b' - Ax_0\|,$$

$b'$ being the right-hand side of the original system. Then we obtain whatever convergence curve for

$$\hat{A} = A - v_1 y^T$$

with right-hand side $v_1$. With (2.2) one could find an approximate solution $\bar{x}$ of

$$Ax = v_1,$$

and hence,

$$A(\|b' - Ax_0\|\bar{x} + x_0) \approx b'.$$

# 5 An alternative way to apply the Sherman-Morrison formula

Because of the difficulties that occurred in the previous sections when the scalar $1 + y^T\hat{x}$ tends to 0, we propose a second way to substitute the approximate solution of the auxiliary system. It avoids the singularity at $y^T\hat{x} = -1$ and moreover, the decreasing of residual norms of auxiliary and original system will go hand in hand. In contrast with the preceding sections, let us define the auxiliary matrix $\hat{A}$ as follows:

$$\hat{A} := A - Ady^T, \quad y, d \in \mathbb{R}^n, \quad d \neq \mathbf{0}.$$

The first condition to put on $y$ is

$$y^T d = \gamma, \tag{5.1}$$

for some scalar $\gamma \in \mathbb{R}$, $\gamma \neq 1$. Then we have

$$\hat{A}d = (A - Ady^T)d = (1 - \gamma)Ad. \tag{5.2}$$

Denoting by $\hat{x}$ the calculated approximation of the auxiliary system, the Sherman-Morrison formula yields

$$\begin{aligned} A^{-1}b &= (\hat{A} + Ady^T)^{-1}b \qquad (5.3) \\ &= \hat{A}^{-1}b - \hat{A}^{-1}Ad(1 + y^T\hat{A}^{-1}Ad)^{-1}y^T\hat{A}^{-1}b \end{aligned}$$

$$\approx \hat{x} - \frac{d}{1-\gamma}\left(1+\frac{y^T d}{1-\gamma}\right)^{-1} y^T\hat{x}$$
$$= \hat{x} - \frac{d}{1-\gamma}\left(\frac{1-\gamma+\gamma}{1-\gamma}\right)^{-1} y^T\hat{x} = \hat{x} - (y^T\hat{x})d =: \bar{x}.$$

The singularity at $y^T d = 1$ is excluded because of (5.1). Also, in comparison with the previous section, the error of the back-transformed solution is far less sensible to the error at the auxiliary system, because the expression $\hat{A}^{-1}Ad$ is known to exactly equal $d/(1-\gamma)$. The residual vectors of original and auxiliary system even appear to be equal if we use (5.3) to compute the approximation of the first system:

$$b - \hat{A}\hat{x} = b - (A - Ady^T)\hat{x} = b - A\hat{x} + Ady^T\hat{x} = b - A(\hat{x} - (y^T\hat{x})d) = b - A\bar{x}.$$

In fact, this application of the Sherman-Morrison formula can be seen as right-preconditioning with the preconditioner

$$M := I_n - dy^T,$$

satisfying the restriction $y^T d \neq 1$. An approximate solution $\hat{z}$ of

$$AMz = b$$

yields an approximation $\bar{x} = M\hat{z} = \hat{z} - (y^T\hat{z})d$ of the unpreconditioned system with the same residual norm.

Our aim is now to formulate a process of minimization of residual norms that is analogous to the construction described in Theorem 3.5. An important difference with the situation in Theorem 3.5 is, that here Lemma 3.4 is not valid. Elements of the Hessenberg matrix are not anymore independent from the parameter vector $y$ and neither are the Krylov subspaces $\mathcal{K}_k(A, r_0)$ and $\mathcal{K}_k(\hat{A}, r_0)$ equal. Instead, we have

$$\hat{h}_{j,k} = v_j^T\hat{A}v_k = v_j^T(A - Ady^T)v_k = h_{j,k} - v_j^T A d\alpha_k, \tag{5.4}$$

where $h_{j,k} := v_j^T A v_k$ represents elements of the Hessenberg matrix of the original system and $\alpha_k := y^T v_k$. The subdiagonal elements change to

$$\hat{h}_{k+1,k} = \left\|\hat{A}v_k - \sum_{j=1}^{k}\hat{h}_{j,k}v_j\right\| = \left\|Av_k - \sum_{j=1}^{k} h_{j,k}v_j + \left(\sum_{j=1}^{k}(v_j^T Ad)v_j - Ad\right)\alpha_k\right\|.$$

But we can still attempt to influence the residual norms by changing the Givens sine $s_k$ which is given by the expression for $s_k^2$ from Corollary 3.3,

$$s_k^2 = \frac{\hat{h}_{k+1,k}^2}{\hat{h}_{k+1,k}^2 + (\sum_{j=1}^{k} c_{j-1}(\prod_{i=j}^{k-1}(-s_i))\hat{h}_{j,k})^2}. \tag{5.5}$$

Minimizing this Givens sine amounts to maximizing

$$\frac{(\sum_{j=1}^{k} c_{j-1}(\prod_{i=j}^{k-1}(-s_i))\hat{h}_{j,k})^2}{\hat{h}_{k+1,k}^2}.$$

With $w := \sum_{j=1}^{k}(v_j^T Ad)v_j - Ad$ and because of $\left(w, Av_k - \sum_{j=1}^{k} h_{j,k}v_j\right) = (w, Av_k)$, the denominator equals

$$\hat{h}_{k+1,k}^2 = \left\| Av_k - \sum_{j=1}^{k} h_{j,k}v_j + \alpha_k w \right\|^2 = h_{k+1,k}^2 + 2\alpha_k(w, Av_k) + \alpha_k^2\|w\|^2.$$

From (5.4) it can be seen that also the numerator consists of an expression that is quadratically dependent from the variable $\alpha_k$. In other words, we have to maximize a value that can be expressed as

$$\frac{a_1\alpha_k^2 + a_2\alpha_k + a_3}{b_1\alpha_k^2 + b_2\alpha_k + b_3},$$

for some $a_i, b_i \in \mathbb{R}$, $1 \le i \le 3$. Straightforward computations show that

$$\frac{a_1\alpha_k^2 + a_2\alpha_k + a_3}{b_1\alpha_k^2 + b_2\alpha_k + b_3} = \frac{a_1}{b_1}\left(1 + \frac{(a_2b_1 - a_1b_2)\alpha_k + a_3b_1 - a_1b_3}{a_1b_1\alpha_k^2 + a_1b_2\alpha_k + a_1b_3}\right).$$

Extrema of this expression are the roots of the first derivative of the involved quotient,

$$\frac{\partial}{\partial\alpha_k}\frac{(a_2b_1 - a_1b_2)\alpha_k + a_3b_1 - a_1b_3}{a_1b_1\alpha_k^2 + a_1b_2\alpha_k + a_1b_3} \tag{5.6}$$

$$= a_1\frac{(a_2b_1 - a_1b_2)(b_3 - b_1\alpha_k^2) - (a_3b_1 - a_1b_3)(b_2 + 2b_1\alpha_k)}{(a_1b_1\alpha_k^2 + a_1b_2\alpha_k + a_1b_3)^2} = 0.$$

As far as $\hat{h}_{k+1,k}$ is dependent from $\alpha_k$, it seems reasonable to minimize $s_k$ and hence the corresponding residual norm by taking for $\alpha_k$ the smaller root of equation (5.6) if $(a_2b_1 - b_2a_1)$ is negative and the larger root if $(a_2b_1 - a_1b_2)$ is positive. In the exceptional case that $\hat{h}_{k+1,k}$ is independent from $\alpha_k$ we can force whatever norm reducing (as in the previous section).

In theory one could successively minimize all $s_k$ with the help of (5.6) and define $y$ by solving

$$y^T(v_1, \ldots, v_{n-1}, d) = (\alpha_1, \ldots, \alpha_{n-1}, \gamma),$$

which is merely a question of orthogonalizing $d$ against $(v_1, \ldots, v_{n-1})$ and one matrix-vector multiplication. The resulting convergence curve cannot be drawn, as in Theorem 3.5, arbitrarily, but it will at every step find steepest possible descent (in dependency of the chosen vector $d$). In practice, as we aim to avoid stagnation of the restarted GMRES($m$) method, we will put only $k$, $k \leq m \ll n$, conditions on $y$. The corresponding algorithm could have the following form:

ALGORITHM GIVMIN($m,k$)-RESTARTED GMRES WITH MINIMIZATION OF GIVENS SINES

1. *Input:* $A \ldots$ *matrix;*
   $b \ldots$ *right-hand side;*
   $\varepsilon \ldots$ *tolerance for residual norm;*
   $x_0 \ldots$ *initial guess;*
   $m \ldots$ *number of steps after which* GMRES *is restarted;*
   $k \ldots$ *number of Givens sines to minimize at the beginning of every restart.*
2. *Initialization:* $y = \mathbf{0}$;
   $r_0 = b - Ax_0$; $\beta = \|r_0\|$; $r = r_0$.
3. **while** $\|r\|/\|r_0\| > \varepsilon$ **do**
   **do** $i = 1, m$
   $A = A - Ax_0 y^T$
   $\tilde{v} = Av_i$
   **do** $j = 1, i$
   $h_{j,i} = v_j^T \tilde{v}_i$
   $\tilde{v} = \tilde{v} - h_{j,i} v_j$
   **enddo**
   **if** $i \leq k$ **then**
   $h_i^* = v_i^T A x_0$
   *compute* $\alpha_i$, *minimizing root of* (5.6)
   $h_{j,i} = h_{j,i} - h_j^* \alpha_i, \ j \leq i$
   $\tilde{v} = \tilde{v} - \alpha_i (Ax_0) - \sum_{j=1}^{i} h_{j,i} v_j$
   **endif**
   $h_{i+1,i} = \|\tilde{v}\|, \quad v_{i+1} = \tilde{v}/h_{i+1,i}$
   **if** $i = k$ **then**
   $$\textit{compute } y \in \mathbb{R}^n \textit{ satisfying } \begin{pmatrix} v_1^T \\ \vdots \\ v_k^T \\ x_0^T \end{pmatrix} y = \begin{pmatrix} \alpha_1 \\ \vdots \\ \alpha_k \\ 0 \end{pmatrix}$$
   **endif**
   **enddo**

   *compute* $w_m \in \mathbb{R}^m$ *minimizing* $\|\beta e_1 - \tilde{H}_m w\|$
   *compute* $x_m := x_0 + V_m w_m$
   $x_0 = x_m - (y^T x_m) x_0$

$y = \mathbf{0}$
$r = b - Ax_0$.
**endwhile**

In this version we have chosen $\gamma = 0$ and the auxiliary vector $d$ to be equal to the actual approximation. In that way, with convergence, $\hat{A}$ will tend to have the form $A - by^T$ as was the case in the previous sections.

Note that computational costs of this algorithm are slightly higher than for GMRES($m$). One extra matrix-vector product is needed at every restart when the vector $d$ is not chosen to be equal to the actual approximation. During every restart we need to compute $k$ extra values $v_i^T Ad$ and to update the values $h_{j,i}$ and $\tilde{v}$ in dependency of the choice of $\alpha_i$, costing $n(k^2+3k)/2$ multiplications. To compute $y$ we need $2kn + n$ operations to orthogonalize $d$ against $v_1, \ldots, v_k$ and one matrix vector-product of dimension $n \times k$. The minimizer $\alpha_i$ can be computed with additional operations that are dependent from $n$ of order $2kn + n(k+1)k/2$. Finally, iteration numbers larger than $k$ require matrix vector-products with the auxiliary matrix, which costs $2n$ more multiplications then with the original matrix. As for storage costs, they will be the same as for GMRES($m$) at every restart, except for the storage of three more $n$-dimensional vectors, $y$, $d$ and $Ad$, and $k$ values $v_i^T Ad$. Note that it is not needed to compute $\hat{A}$ after the $k$th iteration, only matrix-vector products with $\hat{A}$ are required. While multiplying with $\hat{A} = A - Ady^T$, its special form can be exploited and the advantages of operations with sparse matrices are preserved.

## 6 Numerical experiments

At first let us consider an academic example. As stated earlier, the GIVMIN method avoids to eliminate unwanted eigenvalues from the spectrum of the involved matrix, but its aim is to take care of linear systems with 'harmless' eigenvalue distribution that are hard to solve for the GMRES method. An extreme case is the following matrix

$$A = \begin{pmatrix} 0 & & \ldots & 0 & 1 \\ 1 & 0 & \ldots & & 0 \\ 0 & 1 & \ddots & & \\ \vdots & 0 & \ddots & \ddots & \vdots \\ 0 & \ldots & 0 & 1 & 0 \end{pmatrix}$$

with right-hand side $b = e_1 = (1, 0 \ldots, 0)^T$ and the first guess

$$x_0 = (0, \ldots, 0, 2)^T.$$

(In order to be able to apply the above formulated GIVMIN algorithm we need a non-vanishing initial guess.) When GMRES is applied to this system, no improvement with regards to the initial residual norm $\|r_0\| = 1$ is made until the very last iteration, that is, $\|r_i\| = 1$, $1 \le i \le n-1$, $r_n = 0$. But all eigenvalues of $A$ have unit norm. As GMRES converges at the last step, no restarted GMRES($m$), $m < n$, will ever converge. Moreover, methods that seek to transform unwanted eigenvalues to eigenvalues with unit norm won't bring any improvement to GMRES($m$) either, they will stagnate for all possible initial guesses (except for the exact solution). The GIVMIN-algorithm however converges after the first iteration for whatever convergence criterion! Indeed, with the initial guess displayed above, we have

$$b - \hat{A}x_0 = b - (A - Ax_0y^T)x_0 = \begin{pmatrix} 1 \\ 0 \\ \vdots \\ 0 \end{pmatrix} - \begin{pmatrix} 2 \\ 0 \\ \vdots \\ 0 \end{pmatrix} = \begin{pmatrix} -1 \\ 0 \\ \vdots \\ 0 \end{pmatrix} =: v_1,$$

having put the condition $y^T x_0 = 0$. Then, with $\alpha_1 = y^T v_1$, the first step of the Arnoldi algorithm yields

$$\hat{h}_{1,1} = v_1^T \hat{A} v_1 = v_1^T (Av_1 - Ax_0\alpha_1) = 2\alpha_1, \quad \hat{h}_{2,1} = \|\hat{A}v_1 - \hat{h}_{1,1}v_1\| = 1.$$

Here we have the ideal situation from Theorem 3.5 where only $\hat{h}_{1,1}$ depends upon $y$. The first Givens sine can therefore assume whatever positive value smaller than 1:

$$s_1^2 = \frac{1}{1 + (2\alpha_1)^2}.$$

For example, with $\alpha_1 = 10^{10}$, the residual norm reduction $\|r_1\|/\|r_0\|$ of both auxiliary and original system are smaller then $\varepsilon = 10^{-10}$. In this situation, there is no local minimum for the Givens sine.

Let us come to a more realistic example. The matrix "Steam1", taken from the Matrix Market collection, has dimension $240 \times 240$, 3762 non-zero entries and its spectrum consists of 160 evenly distributed eigenvalues with norms ranging from 21711984.3 to 20918.0422 and the last 80 eigenvalues lie evenly distributed between 19.5655309 and $-0.768583444$. With right-hand side $b = e_1$, initial guess $x_0 = (0.01, \ldots, 0.01)^T$ and convergence criterion $\varepsilon = 10^{-8}$, the convergence curves of GMRES and GIVMIN are displayed in Figure 2. We took such a small $\varepsilon$ because the initial residual norm is relatively large, $\|r_0\| = 8407$. GIVMIN($m, k$) denotes the algorithm written out above with restart parameter $m$ and $k$ sine minimizations at the beginning of every restart. In this example the eigenvalue distribution of Steam1 certainly does have an influence on GMRES's behavior. Stagnation of the restarted GMRES($m$) starts to disappear for $m > 60$. But GIVMIN with only one minimization is about 5 times faster than full GMRES itself.

Next we will apply the above algorithm to the $400 \times 400$ stiffness matrix that we described earlier (see (4.2)). This matrix has evenly distributed eigenvalues but their norms are all small ($< 1$ with the exception of the first eigenvalue with norm 1.7), especially the last three ones ($\approx 0.08$). Taking as the

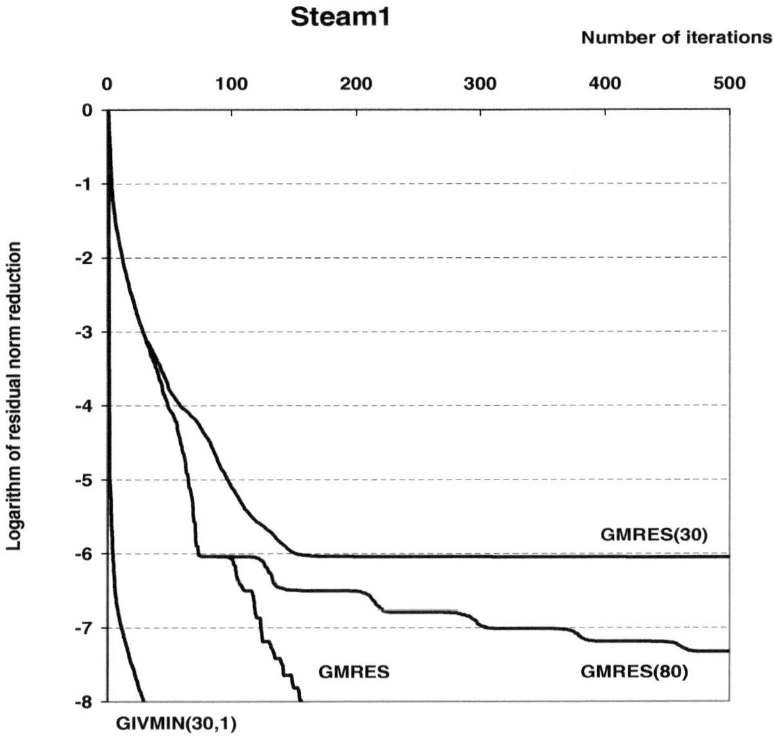


Figure 2. GMRES and GIVMIN applied to the matrix "Steam1".

right-hand side $e = (1/2, \ldots, 1/2)^T$ and setting $x_0 = (-1.685, \ldots, -1.685)^T$ (this initial guess minimizes the initial residual norm), the full GMRES method gives after 114 steps a residual norm reduction of $10^{-5}$. The restarted GMRES($m$) however, stagnates for all $m \leq 30$. Minimizing after every restart only the first Givens sine with respect to (5.6) sufficiently influences the process to overcome stagnation. With such a small parameter $k$ the amount of extra work in comparison with GMRES(30) is negligible. It is even possible to overcome stagnation for $m < 30$, though convergence is then very slow. Some results are displayed in Figure 3.

Our last matrix $A$ has the form $A = SDS^{-1}$ with $A, D, S \in \mathbb{R}^{100\times 100}$. $D$ is a diagonal matrix, $D = \text{diag}(1, \ldots, 100)$ and $S$ is bidiagonal with the element 1 on the diagonal and 1.1 on the upper subdiagonal. Our initial guess, minimizing the corresponding residual norm is $(3.44 \cdot 10^{-5}, \ldots, 3.44 \cdot 10^{-5})^T$, where $b = e$. With convergence criterion $\varepsilon = 10^{-5}$, full GMRES converges after 58 steps, taking 0.82 seconds. GMRES(25) stagnates and stagnation can be overcome with GIVMIN($25, k$), though minimization of only the first Givens sine is not sufficient to do so. This example demonstrates some results in dependency of the choice of the parameter $k$. The method GIVMIN is com-

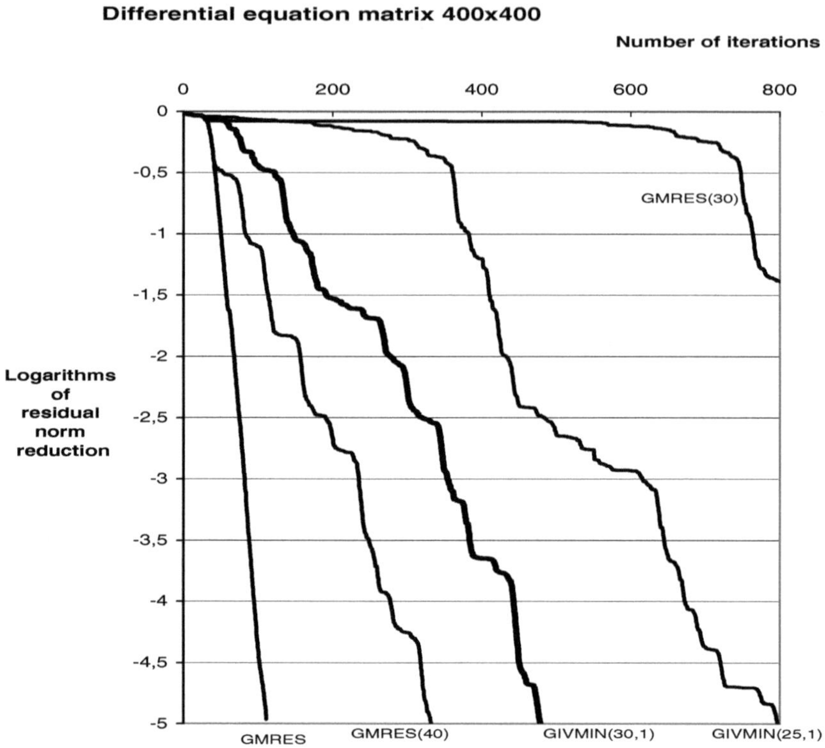


Figure 3. PDE-discretization matrix with GMRES and GIVMIN.

pared with, PGMRES($m, k_1, k_2, \ldots$), is described in [3]. At the $i$th restart of this method the system is preconditioned by right multiplication with a matrix modifying $k_i$ eigenvalues from the spectrum of the original matrix. The construction of the preconditioner is based on addition of Ritz vectors. This method starts to converge when we add at least 4 Ritz vectors: PGMRES(25,4) converges after 0.24 seconds. This is faster then GIVMIN(25,6) taking 0.39 seconds. On the other hand, adding very few Ritz vectors at every restart seems to be less advantageous. PGMRES(25,2,2,1) needs a few less iterations then GIVMIN(25,2) to convergence, but it is significantly slower (0.58 instead of 0.35 seconds). The curves are shown in Figure 4.

## 7 Conclusions and open questions

The foregoing examples were meant to only give a first impression of the possibilities and of the restrictions of Givens sine minimization in GMRES. From them it seems to be worth investigating options to improve the described basic GIVMIN algorithm. For example, it would be useful to know, for a given linear system, what vectors $d$ in $\hat{A} = A - Ady^T$ yield the best

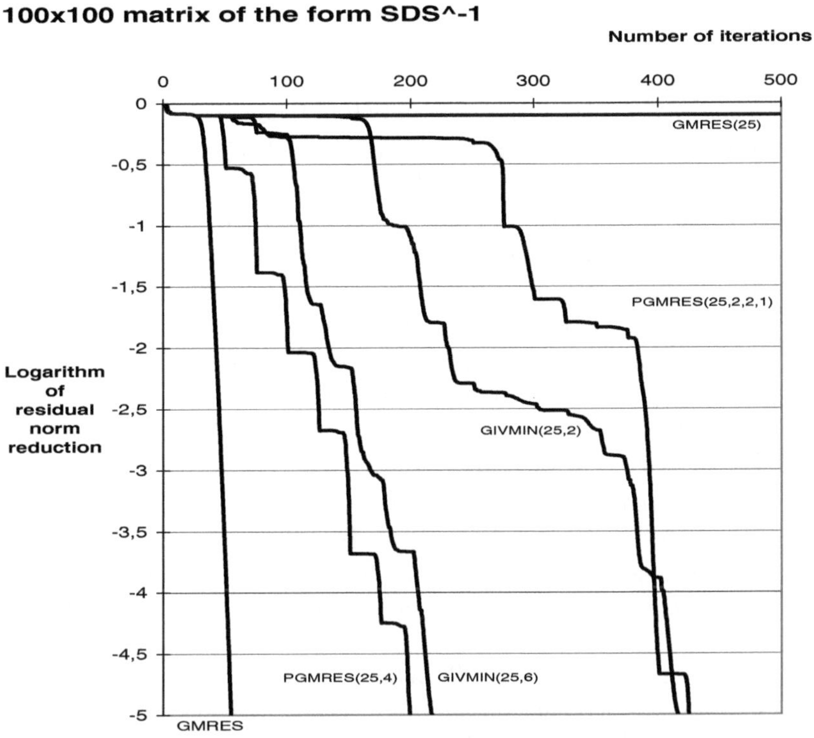


Figure 4. Comparison of GMRES, GIVMIN and PGMRES for matrix taken from Erhel [3].

minimization possibilities. Another choice we make is the choice of the concrete $y \in \mathbb{R}^n$ for whom we have only $k < n$ conditions. We do not have to our disposal a criterion yet to select this $y$ in such a way that it stimulates the convergence of the residual norms $\|r_i\|$, $i > k$. Also, one would expect the Sherman-Morrison formula gets more profitable when we make use of more than one parameter vector, that is $V \in \mathbb{R}^{n \times m}$, $m > 1$ in (2.1), but research in this direction hasn't yet been done. Finally, it might be possible to make use of the Sherman-Morrison formula in order to change the eigenvalue distribution of matrices too. As we are able to adapt the elements of the Hessenberg matrix $\hat{H} = V^T \hat{A} V$ according to our needs, it seems reasonable to try to change the eigenvalues of $\hat{H}$. They are approximate eigenvalues of $\hat{A}$.

## References


[1] M. Arioli, V. Pták, Z. Strakoš, *Krylov sequences of maximal length and convergence of GMRES*, BIT **38** (1998), 636–643.

[2] J. Baglama, D. Calvetti, G. H. Golub, L. Reichel, *Adaptively Preconditioned GMRES Algorithms*, SIAM J. Sci. Comput. **20** (1998), 243–269.

[3] J. Erhel, K. Burrage, B. Pohl, *Restarted GMRES preconditioned by deflation*, J. Comput. Appl. Math. **69** (1996), 303–318.

[4] M. Fiedler, *Special Matrices and Their Applications in Numerical Mathematics*, Nijhoff, Dordrecht, 1986.

[5] G. H. Golub, Ch. F. Van Loan, *Matrix Computation*, The John Hopkins University Press, Baltimore, MD, 1984.

[6] A. S. Householder, *The Theory of Matrices in Numerical Analysis*, Blaisdell, New York, 1964.

[7] R. B. Morgan, *A restarted GMRES method augmented with eigenvectors*, SIAM J. Matrix Anal. Appl. **16** (1995), 1154–1171.

[8] J. M. Ortega, W. C. Rheinboldt, *Iterative Solution of Nonlinear Equations in Severable Variables*, Academic Press, New York and London, 1970.

[9] Y. Saad, M. H. Schultz, *GMRES: A generalized minimal residual algorithm for solving nonsymmetric linear systems*, SIAM J. Sci. Stat. Comput. **7** (1986), 856–869.

[10] Y. Saad, *Iterative Methods for Sparse Linear Systems*, PWS Publishing Company, 1996.

[11] Y. Saad, *Further analysis of minimum residual iterations*, Numer. Linear Algebra Appl. **7** (2000), 67–93.

[12] Y. Saad, *Analysis of augmented Krylov subspace methods*, SIAM J. Matrix Anal. Appl. **18** (1997), 435-449.

[13] D. C. Sorensen, *Implicit application of polynomial filters in a k-step Arnoldi method*, SIAM J. Matrix Anal. Appl. **13** (1992), 357–385.

[14] Z. Strakoš, A. Greenbaum, and V. Pták, *Any nonincreasing convergence curve is possible for GMRES*, SIAM J. Matrix Anal. Appl. **17** (1996), 465–469.

[15] Z. Strakoš, A. Greenbaum, *Matrices that generate the same Krylov residual spaces*, IMA Volumes in Mathematics and its Applications, Vol. 50, Springer, 1992.

[16] R. Weiss, *Convergence behavior of generalized conjugate gradient methods*, PhD thesis, University of Karlsruhe, 1990.

[17] H. F. Walker, L. Zhou, *A simpler GMRES*, Numer. Linear Algebra Appl. **1** (1994), 571–581.

[18] J. Zítko, *Generalization of convergence conditions for a restarted GMRES*, Numer. Linear Algebra Appl. **7** (2000), 117–131.

# A Parallel CG Solver Based on Domain Decomposition and Non-Smooth Aggregation

Yuri V. Vassilevski

Institute of Numerical Mathematics, Russian Academy of Sciences, 8, Gubkina str., 119991 Moscow, Russia vasilevs@dodo.inm.ras.ru

## 1 Introduction

Methods based on overlapping Domain Decomposition (DD) are very convenient for the construction of black-box parallel solvers on unstructured meshes. Considerations of parallel efficiency require the minimal overlap of subdomains equal to element size. The minimal overlap implies a simple technology of automatic partitioning for unstructured meshes, minimal inter-processor exchanges and avoiding a double arithmetical work. Another advantage of the overlapping DD is that the interface between subdomains may be not matching with jumps in the diffusion coefficient. The basic overlapping DD preconditioner is the additive Schwarz preconditioner [10] which, in the case of minimal overlap, is just a block diagonal matrix whose blocks are preconditioners to respective diagonal blocks of the stiffness matrix. For problems with self adjoint second order elliptic operators, the condition number of thus preconditioned system matrix depends on the subdomain diameter ($\sim 1/H^2$), the width of the overlap ($\sim 1/\delta$) and the diffusion coefficient jump ($\sim \max \rho / \min \rho$). If the additive Schwarz preconditioner is equipped with a coarse subspace, the condition number estimate is reduced to $C(\rho)\left(1+\frac{H}{\delta}\right)^2$ [2] and in the case of a smooth coefficient $\rho$ it may be improved to $C\left(1+\frac{H}{\delta}\right)$ [3], [12].

In this paper, we consider a method based on the additive Schwarz algorithm which provides independence of the condition number of the diffusion coefficient in the region of the overlap and the estimate $C\left(1+\frac{H}{\delta}\right)$. The core of the method is the particular construction of the coarse subspace based on the aggregation in subdomains [1], [5]. The aggregation is non-smooth since the coarse subspace basis functions may be non-smooth. The number of degrees of freedom in the coarse subspace is far greater than the number of subdomains: the subspace incorporates fine mesh basis functions which are nontrivial in the overlap region. On the one hand, it makes the coarse subspace (aggregated) problem harder to solve. On the other hand, it allows us to eliminate the dependence of the coefficient. The approximate solution with the aggregated matrix is performed by a few BSOR sweeps which are rather efficient, since the aggregated matrix is not very stiff. Moreover, the iterative solution is easily parallelized. A practical feature of the proposed method is a simplicity of its implementation on unstructured meshes, both sequential and parallel [7]. Another peculiarity of the preconditioner is its hybrid form [8], [12], where the additive structure on the level of subdomains (for the sake of parallelism) is followed by the coarse subspace multiplicative correction (for the sake of better convergence).

The contents of the paper is as follows. A description of the model problem and assumptions on the domain decomposition in Section 2 are followed by a presentation of the method in Section 3. Numerical experiments illustrate the basic features of the method in Section 4.

## 2 Finite element problem

Let a polyhedral domain $\Omega \subset \mathbb{R}^3$, with a Lipschitz boundary $\partial\Omega$ be decomposed into $m$ overlapping regular shaped subdomains $\Omega_i$, with an overlap width $\delta$ and diameter $H$, i.e., $\bar{\Omega} = \bigcup_{i=1}^m \bar{\Omega}_i$. Each subdomain is partitioned into the interior part $\Omega_{i,\text{int}}$ and the overlapping part $\Omega_{i,\text{ovr}}$:

$$\Omega_{i,\text{ovr}} = \Omega_i \bigcap \Big( \bigcup_{i \neq j} \bar{\Omega}_j \Big), \quad \Omega_{i,\text{int}} \bigcap \Big( \bigcup_{i \neq j} \bar{\Omega}_j \Big) = \emptyset.$$

We denote by $J(i)$, $i = 1, \ldots, m$, the set of indices of $\Omega_j$ such that $\Omega_i \cap \Omega_j \neq \emptyset$.

Let $\Omega_i^h$ be a conformal simplicial regular triangulation of $\Omega_i$, $i = 1, \ldots, m$. We assume that in $\Omega_{i,\text{ovr}}$ the mesh $\Omega_i^h$ matches with $\Omega_j^h$, $j \in J(i)$. Therefore, $\bigcup_{i=1}^m \Omega_i^h$ defines a conformal triangulation $\Omega^h$ of $\Omega$. Furthermore, we assume that all the nodes of $\Omega_{i,\text{ovr}}^h$ belong to $\partial\Omega_{i,\text{ovr}}$. This implies that we shall use an overlapping domain decomposition with the minimal (one element) intersection of subdomains (see Fig. 1). We denote by $V^h$ the space of $P_1$ conforming finite elements on triangulation $\Omega^h$ with zero trace on $\partial\Omega$.

Let the bilinear form $a$ and the functional $l$ are given by

$$\begin{aligned}
&a(\cdot,\cdot)\colon H^1(\Omega) \times H^1(\Omega) \to \mathbb{R}, \\
&a(v,w) := \sum_{i=0}^{m} a_i(v,w), \\
&a_0(v,w) := \int_{\Omega \setminus \bigcup_{i=1}^m \Omega_{i,\text{int}}} \rho \nabla v \cdot \nabla w \,\mathrm{d}x, \\
&a_i(v,w) := \int_{\Omega_{i,\text{int}}} \rho \nabla v \cdot \nabla w \,\mathrm{d}x, \\
&l(\cdot)\colon H^1(\Omega) \to \mathbb{R}, \\
&l(v) := \int_\Omega f v \,\mathrm{d}x, \quad f \in L_2(\Omega),
\end{aligned}$$

where $\rho(x) > 0$ and $\rho(x)$ is a piece-wise constant function defined on $\Omega^h$.

The finite element problem: find $w \in V^h$ such that

$$a(w, v) = l(v), \quad v \in V^h \tag{1}$$

is reduced to the linear system

$$Au = f \tag{2}$$

with a sparse symmetric positive definite matrix $A \in \mathbb{R}^{n \times n}$. The solution entries are assumed to be associated with the mesh nodes.

## 3 Two-stage preconditioner

The preconditioner is based on a two-stage DD-method proposed in [7], [15]. At the first stage, the conventional additive Schwarz preconditioner is applied. We partition the entries of a vector $u$ into $m$ disjoint subsets associated with the meshes $\Omega^h_{i,\text{int}}$, $i = 1, \ldots, m$. The matrix $A$ admits a block representation associated with the above partitioning:

$$A = \begin{pmatrix} A_{11} & \ldots & A_{1m} \\ \vdots & \ddots & \vdots \\ A_{m1} & \ldots & A_{mm} \end{pmatrix}.$$

The diagonal blocks $A_{ii} \in \mathbb{R}^{n_i \times n_i}$, $i = 1, \ldots, m$, correspond to boundary value problems in the subdomains with the Dirichlet boundary conditions on interior boundaries.

Let $B_{ii} = B_{ii}^T > 0$ be a preconditioner for $A_{ii}$, $i = 1, \ldots, m$:

$$\alpha_1(B_{ii}^{-1}u, u) \leq (A_{ii}u, u) \leq \alpha_2(B_{ii}^{-1}u, u), \tag{3}$$

with positive constants $\alpha_1$, $\alpha_2$. Define a block diagonal matrix

$$B_1 = \begin{pmatrix} B_{11} & & \\ & \ddots & \\ & & B_{mm} \end{pmatrix}. \tag{4}$$

The matrix $B_1$ is the additive Schwarz preconditioner for $A$ with the minimal overlap of subdomains. It is a very simple preconditioner for $A$ and can be easily parallelized. However, its efficiency is affected by the number of subdomains $m$, the width of the overlap $\delta$ and the diffusion coefficient $\rho$. Both numerical evidence and theoretical considerations [4], [12] show that $\text{cond}(B_1A) \sim 1/\delta$, $\text{cond}(B_1A) \sim 1/H^2$, $\text{cond}(B_1A) \sim \max_x \rho(x)/\min_x \rho(x)$. In order to decrease the effect of the overlap and eliminate the dependence on the number of subdomains and variations of $\rho(x)$, we apply a correction step. The resulting two-stage preconditioner $B$ is implicitly described by its action on a vector $u \in \mathbb{R}^n$. Setting $v = Bu$, we have

$$w = B_1 u, \quad v = w + B_2(u - Aw), \tag{5}$$

where $B_2$ is a preconditioner described below. Formally, $B$ may be presented as follows:

$$B = B_1 + B_2(I - AB_1). \tag{6}$$

Clearly, it is the non-symmetric matrix even when $A = A^t$ and $B_i = B_i^t$, $i = 1, 2$. Therefore, for symmetric positive definite matrices the preconditioned conjugate gradient (PCG) method with preconditioner (6) may not be directly applied. However, if the initial vector $u_0$ in the PCG method is corrected as follows:

$$u_0 := u_0 + B_2(f - Au_0),$$

the method will converge provided $B_2$ has a general form (9) with $\tilde{B} = \tilde{A}^{-1}$. The reason is that the evaluation of the preconditioner $B$ on the PCG residuals is equivalent to a symmetric three-stage algorithm (see, for example, [8], [9]):

$$\bar{v} = B_2 u, \quad w = B_1(u - A\bar{v}), \quad v = w + B_2(u - Aw), \tag{7}$$

whose operator form is

$$B_h = B_1(I - AB_2) + B_2(I - AB_1(I - AB_2)). \tag{8}$$

We choose the preconditioner $B_2$ as follows. Let the number of nontrivial rows in the matrix $A_i = [A_{i1}, \dots, A_{i,i-1}, A_{i,i+1}, \dots, A_{i,m}]$ (without the diagonal block) be $\tilde{n}_i$. We define $\tilde{n} = \sum_{i=1}^m \tilde{n}_i + m$ and assume that rows of the matrix $A$ are ordered in such a way that in each matrix $A_i$ the nontrivial rows go first. Then the local *aggregation matrix* $T_{ii} \in \mathbb{R}^{n_i \times (\tilde{n}_i + 1)}$ is given by

$$T_{ii} = \begin{pmatrix} I_i & 0 \\ 0 & e_i \end{pmatrix}, \quad e_i = (1, \dots, 1)^t \in \mathbb{R}^{n_i - \tilde{n}_i},$$

where $I_i$ is the identity matrix. We define the global block diagonal aggregation matrix $T$ by

$$T = \begin{pmatrix} T_{11} & & \\ & \ddots & \\ & & T_{mm} \end{pmatrix},$$

and the aggregated stiffness matrix $\tilde{A}$ by

$$\tilde{A} = T^t A T, \quad \tilde{A} \in \mathbb{R}^{\tilde{n} \times \tilde{n}}.$$

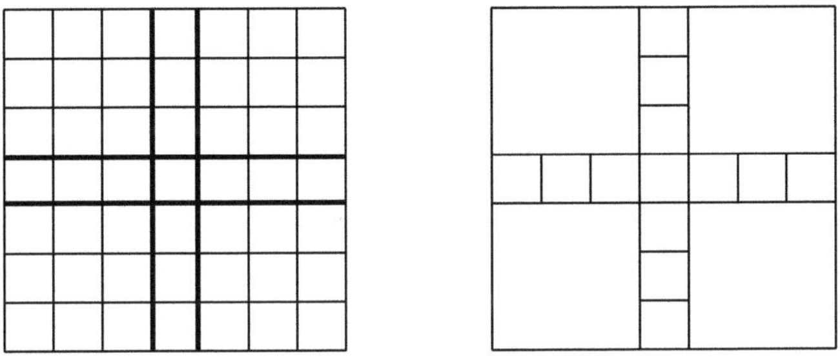

Figure 1. Mesh associated with the additive Schwarz (*left*) and the correction (*right*).

Let $\tilde{B}$ be the conventional BSOR smoother [11], [16] for $\tilde{A}$. Then, the preconditioner $B_2$ is defined implicitly by

$$B_2 = T\tilde{B}T^t. \tag{9}$$

Note that $B_2$ is a smoother in a subspace of aggregated vectors $\tilde{V} = \{v \in \mathbb{R}^n\colon v = T\tilde{v},\ \tilde{v} \in \mathbb{R}^{\tilde{n}}\}$. We motivate its construction as follows. The drawback of the additive Schwarz preconditioner $B_1$ is that it damps the error locally in subdomains but it does not control the error propagation on the global scale (dependence on $H$) and does not coordinate the error damping between neighboring subdomains (dependence on $\delta$). The approximate BSOR inversion of the aggregated matrix $\tilde{A}$ coordinates the mean subdomain values, matches the local errors in overlapping strips (see Fig. 1) and eliminates the sensitivity of $\mathrm{cond}(B_1A)$ to variations of $\rho$.

More precisely, if the diffusion coefficient $\rho(x)$ is smooth out of the region of the overlap: $\rho_{1,i} \le \rho(x) \le \rho_{2,i}$, $x \in \Omega_{i,\mathrm{int}}$, and both $\tilde{B}$ and $B_{ii}$, $i = 1, \dots, m$, precondition $\tilde{A}$ and $A_{ii}$ with constants of equivalence $\alpha_1$, $\alpha_2$, then for the method (8) the estimate holds [14]:

$$\mathrm{cond}(B_hA) \le 2\frac{\alpha_2}{\alpha_1}\left(1 + 2C \max_{1\le i\le m}\left(\frac{\rho_{2,i}}{\rho_{1,i}}\right)\frac{H}{\delta}\right). \tag{10}$$

Here $C$ is a generic positive constant. We notice that (10) holds true for a diffusion coefficient which is strongly varying and even heterogeneous in the region of the overlap $\bigcup_{i=1}^m \Omega_{i,\mathrm{ovr}}$. The numerical evidence shows that the requirement of the smoothness of $\rho(x)$ in $\Omega_{i,\mathrm{int}}$ is not necessary.

Typically, a few sweeps of BSOR are enough for preconditioning $\tilde{A}$ since the matrix $\tilde{A}$ is much less stiff than $A$. The number of iterations is not sensitive to variations of the coefficient $\rho(x)$. The techniques for searching the optimal relaxation parameter have been developed since 70's [6], [16]. Parallel realizations of BSOR are discussed, for example, in [11]. The constituents of the method make it very convenient for parallel implementation and for applications on unstructured meshes [7].

# 4 Numerical experiments

In the first part of this section we show that the presence of the smoother $\tilde{B}$ in (6) damps the sensitivity of the method to the number of subdomains. We set the mesh to be a uniform tetrahedralization of a cubic grid in $\Omega = (0,1)^3$, $h = 2^{-6}$ consisting of $N_T = 1572864$ elements. The ideal case is studied: $B_{ii} = A_{ii}^{-1}$, $\tilde{B} = \tilde{A}^{-1}$. Along with the problem (1) with $\rho(x) = f(x) = 1$ (Case 1), we consider the problem (Case 2)

$$-\Delta u + u = (1 + 3\pi^2) \cos \pi x_1 \cos \pi x_2 \cos \pi x_3 \quad \text{in } \Omega, \tag{11}$$
$$\frac{\partial u}{\partial n} = 0, \quad \text{on } \partial\Omega.$$

Table 1. Two-stage DD method versus additive Schwarz, number of PCG iterations for different decompositions.

| Preconditioner | $B_1$ | | $B$ | |
|---|---|---|---|---|
| decomposition \ problem | Case1 | Case2 | Case1 | Case 2 |
| $m = 4 \times 4 \times 4$ | 54 | 90 | 19 | 18 |
| $m = 6 \times 6 \times 6$ | 88 | 182 | 17 | 18 |
| $m = 16 \times 1 \times 1$ | 76 | 170 | 30 | 53 |

In Table 1 the number of PCG iterations required for the $10^6$-fold reduction of the initial residual is exhibited. Two methods are compared, the additive Schwarz preconditioner $B_1$ and the two-stage DD preconditioner $B$. We remark that for shape regular domain decompositions the method (6) is insensitive to the number of subdomains, whereas for irregular decompositions the smoother application damps the dependence.

The next example is the parallel solution of the diffusion equation (1) in the unit cube $\Omega = (0,1)^3$. We study three choices for $\rho(x)$. First, we take the uniform isotropic case with $\rho(x) \equiv 1$. Second, we consider the checkerboard-like jumps in the diffusion coefficient:

$$\rho(x) = \begin{cases} 1, & x \in \Omega_1 \cup \Omega_2 \cup \Omega_3 \cup \Omega_4, \\ 1000, & \text{otherwise,} \end{cases}$$

where $\Omega_1 = (0,0.5)^3$, $\Omega_2 = (0.5,1)^2 \times (0,0.5)$, $\Omega_3 = (0.5,1) \times (0,0.5) \times (0.5,1)$ and $\Omega_4 = (0,0.5) \times (0.5,1)^2$. Third, we consider the piecewise constant quasi-random diffusion coefficient given by

$$\rho(x)|_{e_i} = \begin{cases} 1, & \sin(1000x_{i1} + 3000x_{i2} + 5000x_{i3}) > 0, \\ 1000, & \text{otherwise,} \end{cases}$$

where $x_i = (x_{i1}, x_{i2}, x_{i3})$ is the barycenter of the tetrahedron $e_i$. We consider a uniform tetrahedralization of the cubic grid with $h = 2^{-5}$, $N_T = 196608$.

In addition, we artificially split the mesh elements between processors either by the plane $x_1 = 0.5$ (for $m = 2$), or by the planes $x_1 = 0.5$ and $x_2 = 0.5$ (for $m = 4$), or by the planes $x_1 = 0.5$, $x_2 = 0.5$ and $x_3 = 0.5$ (for $m = 8$). Therefore, the corresponding subdomains in the Schwarz method overlap along interfaces, where $\rho(x)$ has jumps. The subdomain preconditioner is chosen to be the algebraic multigrid method (one V-cycle of AMG1R5 [13]) and $\tilde{B}$ is the result of three BSOR iterations for the aggregated system. Table 2 shows that the convergence rate of the PCG method does not depend on the jumps in the diffusion coefficient and on the number of subdomains. Moreover, in all cases, the solver exhibits the good parallel properties. The time measurements have been performed on a Compaq TruCluster with processors cadenced to 667 MH.

Table 2. Number of PCG iterations and root's CPU time.

| | Case 1 | | Case 2 | | Case 3 | |
|---|---|---|---|---|---|---|
| $m$ | #iter PCG | sec | #iter PCG | sec | #iter PCG | sec |
| 2 | 14 | 1.22 | 12 | 1.02 | 13 | 1.13 |
| 4 | 14 | 0.58 | 12 | 0.58 | 14 | 0.62 |
| 8 | 14 | 0.38 | 13 | 0.35 | 13 | 0.33 |

Table 3. Number of PCG iterations and root's CPU time for $m = 4$ on adaptive unstructured meshes.

| $N_T$ | 9452 | 19802 | 28754 | 36810 | 52893 | 101344 | 164184 |
|---|---|---|---|---|---|---|---|
| #iter PCG | 8 | 11 | 12 | 13 | 14 | 17 | 20 |
| sec | 0.05 | 0.10 | 0.12 | 0.15 | 0.22 | 0.53 | 0.88 |
| $\frac{\text{sec}}{\#\text{iter CG} \;\; N_T} 10^7$ | 6.6 | 4.6 | 3.5 | 3.1 | 3.0 | 3.1 | 2.7 |

Finally, we consider the equation (1) in a domain with a reentrant corner, $\Omega = (0,1)^3 \setminus [0,0.5]^3$, and a singular right hand side $f(x) = \frac{1}{|x-x_0|}$ where $x_0 = (0.5, 0.5, 0.5)$. The solution possesses anisotropic edge singularities and a strong singularity at the reentrant corner point. We consider a sequence of adaptive unstructured meshes with local isotropic (towards the corner point) and anisotropic (towards the reentrant edges) refinements, see Fig. 2.

For this example, four BSOR sweeps are used for the approximate inversion of $\tilde{A}$ and one V-cycle of the AMG method for the evaluation of $B_{ii}$. In Table 3 the convergence rate and the arithmetical complexity of the PCG method is shown for 4-processor runs.

We note that the number of iterations grows as the number of elements $N_T$ is increased. However, the dependence on $N_T$ is moderate: 10-fold increase in $N_T$ only doubles the number of PCG iterations (#iter PCG). It is attributable to the two-fold feature of the method (6). On the one hand, the smaller

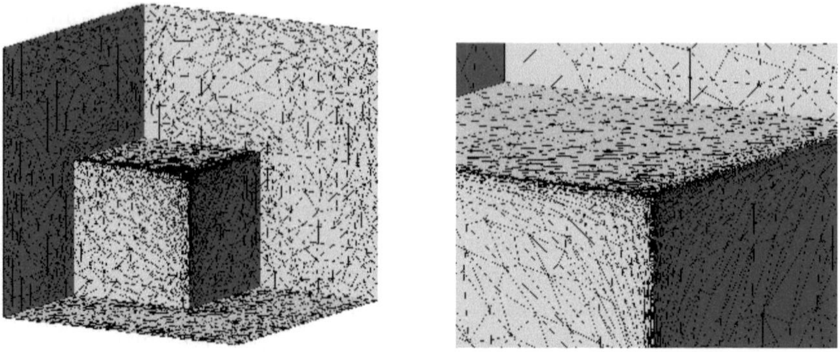

Figure 2. An example of the adaptive mesh in the domain with the reentrant corner.

order of $B_{ii}$, the better cond$(B_{ii}A_{ii})$, $i = 1, \ldots, m$, is. On the other hand, for uniform meshes cond$(BA)$ depends on the mesh size $h$ reciprocally (10), i.e., #iter PCG $\sim h^{-1/2}$. In our experiments, the meshes are not uniform but a weak dependence of #iter PCG on $N_T$ is observed as well. The arithmetical complexity per iteration per element is decreased and saturated as $N_T$ grows, resulting in a very good parallel scalability on the fine mesh. This is due to reduction of relative weight of interprocessor communications for large $N_T$.

In Table 4 we compare the convergence rate of the parallel solver for three types of adaptive meshes with $N_T \sim 10^5$: (1) quasi-uniform meshes, (2) roughly adapted (towards the corner point mostly), (3) well adapted to problem singularities (Fig. 2). The most important observation is that the number of PCG iterations is almost insensitive to both the structure of the mesh and the number of processors.

Table 4. Number of PCG iterations for $N_T \sim 10^5$.

| Mesh type \ $m$ | 2 | 4 | 6 | 8 |
|---|---|---|---|---|
| 1 | 15 | 16 | 17 | 17 |
| 2 | 17 | 19 | 19 | 19 |
| 3 | 18 | 20 | 20 | 19 |

## Conclusion

The iterative solution of the diffusion equation with highly varying diffusion coefficients was addressed. Being unsymmetric, the proposed two-stage preconditioner may be used in the PCG algorithm provided a proper initial guess is chosen. The preconditioner is based on the overlapping Domain Decomposition framework. Its coarse subspace is associated with the space of aggregated vectors. It has as many degrees of freedom in the region of the overlap as the original finite element space but only one degree of freedom for the remaining parts of each subdomain. Numerical robustness of the condition number cond($BA$) is shown with respect to the number of subdomains $m$ and variations of the diffusion coefficient $\rho$ in the region of the overlap. The weak dependence of the convergence rate on mesh unstructureness and irregularity is shown numerically. The method is easily parallelized and exploits the nonzero structure of the stiffness matrix only. This gives the framework for a parallel black-box solution algorithm.

**Acknowledgement.** The author is grateful to Professors M. Garbey and D. Tromeur-Dervoux for fruitfull discussions, and to Dr. K. Lipnikov for the assistance in performing numerical experiments.

## References

[1] M. Bresina, *Robust iterative solvers on unstructured meshes*, PhD thesis, University of Colorado at Denver, Denver, CO, 1997.

[2] T. Chan, B. Smith, and J. Zou, *Overlapping Schwarz methods on unstructured meshes using non-matching coarse grids*, Numer. Math. **73** (1996), 149–167.

[3] M. Dryjam, O. Widlund, *Domain decomposition algorithms with small overlap*, SIAM J. Sci. Comput. **15** (1994), 604–620.

[4] I. Graham, M. Hagger, *Unstructured additive Schwarz-CG method for elliptic problems with highly discontinuous coefficients*, SIAM J. Sci. Comp. **20** (1999), 2041–2066.

[5] E. Jenkins, C. Kees, C. Kelley, and C. Miller, *An aggregation-based domain decomposition preconditioner for groundwater flow*, SIAM J. Sci. Comp. **25** (2001), 430–441.

[6] D. Kincaid, J. Respess, D. Young, and R. Grimes, *Algorithm 586 ITPACK 2C: A fortran package for solving large sparse linear systems by adaptive accelerated iterative methods*, ACM Transactions on Mathematical Software **8** (1982), 302–322.

[7] K. Lipnikov, Yu. Vassilevski, *Parallel adaptive solution of 3D boundary value problems by Hessian recovery*, Comput. Methods Appl. Mech. Engrg. **192** (2003), 1495–1513.

[8] J. Mandel, *Hybrid domain decomposition with unstructured subdomains*, Contemporary Mathematics, AMS, **157** (1994), 103–112.

[9] J. Mandel, *Balancing domain decomposition*, Comm. Numer. Methods Engrg. **9** (1993), 233–241.

[10] A. Matsokin, S. Nepomnyaschikh, *A Schwarz alternating method in a subspace*, Soviet Mathematics **29** (1985), 78–84.

[11] Y. Saad, *Iterative Methods for Sparse Linear Systems*, PWS Publishing Co., Boston, 1996.

[12] B. Smith, P. Bjorstad, and W. Gropp, *Domain Decomposition: Parallel Multilevel Methods for Elliptic Partial Differential Equations*, Cambridge University Press, 1996.

[13] K. Stüben, *Algebraic multigrid (AMG): experiences and comparisons*, Appl. Math. Comput. **13** (1983), 419–452.

[14] Yu. Vassilevski, *A hybrid domain decomposition method based on aggregation*, in preparation.

[15] Yu. Vassilevski, *Iterative Solvers for the Implicit Parallel Accurate Reservoir Simulator (IPARS), II: parallelization issues*, TICAM report 00-33, The University of Texas at Austin, 2000, http:// www.ticam.utexas.edu/reports/2000/index.html.

[16] D. Young, *Iterative Solution of Large Linear Systems*, Academic Press, New York, 1971.

# Deflation in Preconditioned Conjugate Gradient Methods for Finite Element Problems

Fred Vermolen,[1] Kees Vuik,[2] Guus Segal[3]

[1]Department of Applied Mathematical Analysis, Delft University of Technology, Mekelweg 4, 2628 CD Delft, Netherlands F.J.Vermolen@its.tudelft.nl
[2]Department of Applied Mathematical Analysis, Delft University of Technology, Mekelweg 4, 2628 CD Delft, Netherlands C.Vuik@its.tudelft.nl
[3]Department of Applied Mathematical Analysis, Delft University of Technology, Mekelweg 4, 2628 CD Delft, Netherlands g.segal@math.tudelft.nl

## 1 Introduction

Large linear systems are solved for modelling many scientific and engineering applications. Often these systems result from a discretization of model equations using Finite Elements, Finite Volumes or Finite Differences. The systems tend to become very large for three dimensional problems. Some models involve both time and space as independent parameters and therefore it is necessary to solve such a linear system efficiently at all time-steps.

In this paper we only consider symmetric positive definite (SPD) matrices. Presently, direct methods (such as an LU-decomposition) and iterative methods are available to solve such a linear system. However, for large sparse coefficient matrices fill-in causes a loss of efficiency (in computer memory and number of floating point operations). For such a case iterative methods are a better alternative. Furthermore, if a time integration is necessary, then the solution of the preceding time-step can be used as a starting vector for the algorithm to get the result on the next time-step. This too supports the use of iterative methods.

Iterative methods such as Gauss-Seidel, Jacobi, SOR, and Chebyshev-methods can be used, however, convergence is in general slow and it is often very expensive to determine good estimates of parameters on which they depend. To avoid these problems, the conjugate gradient method is used. We deal with an application from transport in porous media where we encounter extreme contrasts in the coefficients of the partial differential equation. The large contrasts are caused by the layered domain where neighbouring layers in Fig. 1 extreme contrasts in mobility. Here a preconditioning is necessary and we use a standard incomplete Cholesky factorization as a preconditioner

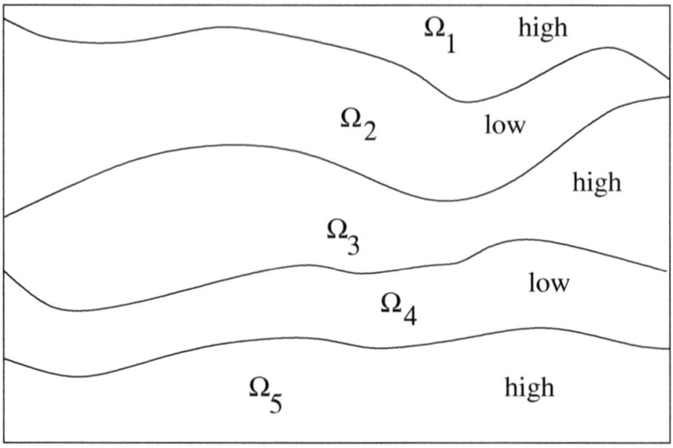


Figure 1. An example of the geometry of the domain with high and low permeability regions.

for the conjugate gradient method (ICCG) to improve convergence behavior. Furthermore, deflation is applied to get rid of remaining extremely small eigenvalues that delay convergence. Vuik et al. [17] proposed a scheme based on physical deflation in which the deflation vectors are continuous and satisfy the original partial differential equation on a subdomain. A different variant involves the so-called algebraic deflation with discontinuous deflation vectors. Here convergence is speeded up. This was also subject of [16]. In that paper a comparison is given between physical deflation vectors and algebraic deflation vectors. Two choices of algebraic deflation vectors are applied:

- algebraic deflation vectors restricted to high permeability layers,
- algebraic deflation vectors for each layer.

From numerical experiments it follows that the second choice gives faster convergence. Furthermore, this option turned out to be more efficient for many applications than the use of physical deflation vectors. Therefore, we limit ourselves to the use of algebraic projection vectors for *each* layer.

For references related to the Deflated ICCG method we refer to the overview given in [17] and [16]. The DICCG method has already been successfully used for complicated magnetic field simulations [2]. A related method is recently presented in [11]. In [4] deflation is used to accelerate block-IC preconditioners combined with Krylov subspace methods in a parallel computing environment.

The DICCG method is related to coarse grid correction, which is used in domain decomposition methods [6], [11]. Therefore insight in a good choice of the deflation vectors can probably be used to devise comparable strategies for coarse grid correction approaches.

We assume that the domain $\Omega$ consists of a number of disjoint sets $\Omega_j$, $j = 1, \ldots, m$, such that $\bar{\Omega} = \bigcup_{j=1}^{m} \bar{\Omega}_j$. The division in subdomains is motivated by jumps in the coefficients and/or the data distribution used to parallelize the solver. For the construction of the deflation vectors it is important which type of discretization is used: cell centered or vertex centered.

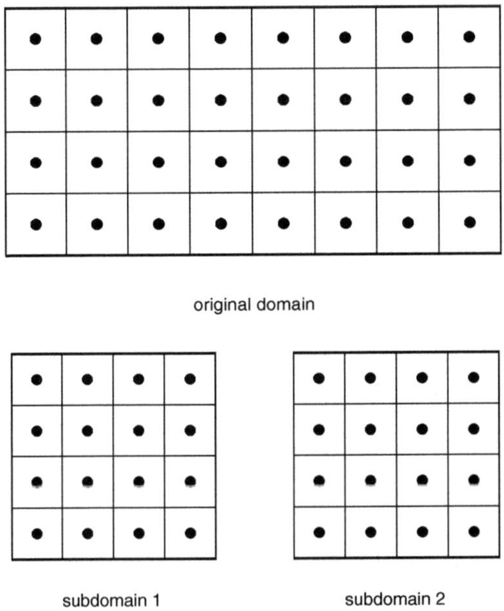


Figure 2. Domain decomposition for a cell centered discretization.

**Cell centered**

For this discretization the unknowns are located in the interior of the finite volume. The domain decomposition is straightforward as can be seen in Fig. 2. The algebraic deflation vectors $z_j$ are uniquely defined as:

$$z_j(x_i, y_i) = \begin{cases} 1, & \text{for } (x_i, y_i) \in \Omega_j, \\ 0, & \text{for } (x_i, y_i) \in \Omega \setminus \Omega_j. \end{cases}$$

**Vertex centered**

If a vertex centered discretization is used the unknowns are located at the boundary of the finite volume or finite element. Two different ways for the data distribution are known [12]:

- Element oriented decomposition: each finite element (volume) of the mesh is contained in a unique subdomain. In this case interface nodes occur.
- Vertex oriented decomposition: each node of the mesh is an element of a unique subdomain. Now some finite elements are part of two or more subdomains.

For Finite Elements only the last option is commonly used. Note that the vertex oriented decomposition is not well suited to combine with a finite element method. Therefore, we restrict ourselves to the element oriented decomposition, see Fig. 3. As a consequence of this the deflation vectors can overlap at interfaces.

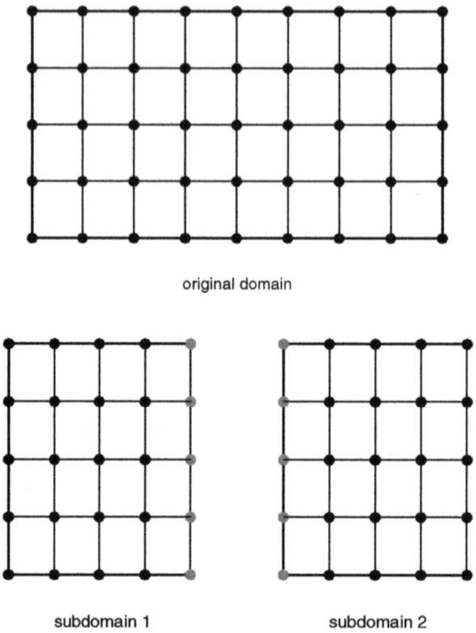


Figure 3. Domain decomposition for a vertex centered discretization. The grey nides are the interface nodes.

In our previous work we always use non-overlapping deflation vectors. In [17], [18] the interface vertices are only elements of the high permeability subdomains, whereas in [4] no interface vertices occur due to a cell centered discretization. The topic of this paper is: how to choose the value of the deflation vectors at interface points in order to obtain an efficient, robust and parallelizable black-box deflation method.

First we briefly present the mathematical model that we use to compare the various deflation vectors. Subsequently we give the algorithm and describe different versions of deflation. This is followed by a description of the numerical experiments.

## 2 The mathematical model

We denote the horizontal and vertically downward pointing coordinates by $x$ and $y$. Flow in porous media is often modelled by the following coupled scaled problem:

$$\left.\begin{aligned} &\frac{\partial S}{\partial t} + \nabla \cdot (\boldsymbol{q}S) = \nabla \cdot (D(S)\nabla S), \\ &\nabla \cdot \boldsymbol{q} = 0, \\ &\frac{1}{\sigma}\boldsymbol{q} + \nabla p - gS\, \boldsymbol{e}_y = \boldsymbol{0}. \end{aligned}\right\} \qquad (\mathrm{P}_0)$$

The above equations are supplemented with appropriate initial and boundary conditions. In above equations $S$ (-), $\boldsymbol{q}$ (m/s) and $p$ (Pa) are the unknown saturation, discharge and pressure respectively. The unit vector in the $y$-direction is represented by $\boldsymbol{e}_y$ and the contstant of gravity is denoted by $g$. The time is denoted by $t$ (s) and $\sigma$ is the (known) mobility. Porous media mostly consist of several layers where the mobility varies between several orders of magnitude. In this work we take $\sigma$ as a (piecewise) constant function. For an overview of the equations that occur in modeling flow in porous media we refer to the books of among others Bear [1] and Lake [9].

For the 2-dimensional case it is favourable [3], [13] to introduce the stream function $\boldsymbol{\psi}$ such that $\boldsymbol{q} = \nabla \times \boldsymbol{\psi}$. Since here $\nabla$ and $\boldsymbol{q}$ respectively work and are given for the $(x, y)$ plane only, it follows that $\boldsymbol{\psi}$ only has a non-constant $z$-component, i.e. $\boldsymbol{\psi} = \langle 0, 0, \psi \rangle$. For more mathematical details on the existence of such a stream function, we refer to the book of Temam [14]. Further, $\partial_z(\cdot) = 0$ and hence after taking the curl over the third equation of ($\mathrm{P}_0$) we are faced with

$$-\nabla \cdot \left(\frac{1}{\sigma} \nabla \psi\right) = g \frac{\partial S}{\partial y}.$$

If one imposes no-flow conditions over the boundary of $\Omega$, then it follows that

$$\psi = 0 \text{ on } \Gamma,$$

where $\Gamma$ represents the boundary of $\Omega$. This implies that the equations in ($\mathrm{P}_0$) change into

$$\left.\begin{aligned} \frac{\partial S}{\partial t} + \left\langle -\frac{\partial \psi}{\partial y}, \frac{\partial \psi}{\partial x} \right\rangle \cdot \nabla S &= \nabla \cdot (D(S) \nabla S), \\ -\nabla \cdot \left(\frac{1}{\sigma} \nabla \psi\right) &= g \frac{\partial S}{\partial y}, \end{aligned}\right\} \qquad (\mathrm{P}_1)$$

for $(x, y) \in \Omega$. We focus on the solution of the second equation of ($\mathrm{P}_1$) by a Finite Element Method, and hence consider a variational formulation:

$$\left.\begin{aligned} &\text{Find } \psi \in H_0^1(\Omega) \ \ (\psi|_\Gamma = 0) \text{ such that} \\ &\int_\Omega \nabla v \cdot \frac{1}{\sigma} \nabla \psi \, \mathrm{d}A = \int_\Omega g \frac{\partial S}{\partial y} v \, \mathrm{d}A \ \text{ for all } v \in H_0^1(\Omega). \end{aligned}\right\} \qquad (\mathrm{P}_2)$$

In problem ($\mathrm{P}_2$) $\sigma$ is allowed to be piecewise continuous and hence, the solution $\psi$ is only piecewise smooth. This problem is solved by the use of a standard Galerkin Finite Element Method, with $\psi = \sum_{i=1}^n \psi_i v_i$ $(v_i|_\Gamma = 0)$, with piecewise linear element functions $v_i$. In our examples we take a layered domain with

$$\sigma(x, y) = \begin{cases} \sigma_{\max} = 10^7, & (x, y) \in \overline{\Omega}_{2j+1}, \\ \sigma_{\min} = 1, & (x, y) \in \Omega_{2j}, \end{cases}$$

where we suppose that the closed domain $\overline{\Omega}$ consists of the union of $m$ closed subdomains $\overline{\Omega}_1, \ldots, \overline{\Omega}_m$ (see Fig. 1). In some applications, the high and low mobility ($\sigma$) respectively correspond to sand and shale layers. We will also use this terminology to refer to the high and low permeability layers. From the Galerkin discretization it follows inmediately that accross an interface the coefficients in the discrete equation varies several orders of magnitude.

Discretization by the use of Galerkin's method results into a matrix-vector equation of type

$$A\,\underline{x} = \underline{b},$$

where $A \in \mathbb{R}^{n\times n}$, $\underline{x} \in \mathbb{R}^n$, and $\underline{b} \in \mathbb{R}^n$, respectively, represent the discretization (or stiffness) matrix, solution vector, and right-hand side vector. Using a FEM approach the discretization matrix is sparse, symmetric and positive definite (SPD). Furthermore, the discretization is chosen such that the interfaces between consecutive layers coincide with gridpoints. For the case of large jumps in the coefficient $\sigma$ the condition of the discretization-matrix is very large. The remainder of the paper is devoted to the efficient solution of the above matrix-vector equation when $n$ is large.

## 3 Solution of the matrix equation

Since $A$ is symmetric and positive definite, the conjugate gradient method is a natural candidate to solve the matrix equation. After $k$ iterations the $\|\cdot\|_A$-norm ($\|\cdot\|_A := \sqrt{(\cdot, A.)}$) of the error is bounded from above by the well-known result of Luenberger, which can also be found in [5]

$$\|\underline{x} - \underline{x}_k\|_A \leq 2\Big(\frac{\sqrt{\kappa}-1}{\sqrt{\kappa}+1}\Big)^k \|\underline{x} - \underline{x}_0\|_A, \tag{1}$$

where $\kappa$ denotes the condition number of the matrix $A$ and $\underline{x}$ is the exact solution of the system. Further, in the above expression, $\underline{x}_0$ and $\underline{x}_k$, respectively, represent the initial estimate of the solution $\underline{x}$ and the result after $k$ conjugate gradient iterations. For the $\|\cdot\|_2$-norm ($\|\cdot\|_2 := \sqrt{(\cdot,\cdot)}$), it can be proven that

$$\sqrt{\lambda_{\min}}\|\cdot\|_2 \leq \|\cdot\|_A \leq \sqrt{\lambda_{\max}}\|\cdot\|_2. \tag{2}$$

The above inequalities are combined with expression (1) to obtain

$$\|\underline{x} - \underline{x}_k\|_2 \leq 2\sqrt{\kappa}\Big(\frac{\sqrt{\kappa}-1}{\sqrt{\kappa}+1}\Big)^k \|\underline{x} - \underline{x}_0\|_2. \tag{3}$$

This estimate is standard and can be found in the book of Golub and van Loan [5]. It gives an upper bound for the error for the $\|\cdot\|_2$-norm.

Vuik et al. [17] observe that the number of small eigenvalues (order $10^{-7}$) of $A$ is equal to the number of gridnodes in the low mobility layer ($\sigma = 10^{-7}$)

plus the number of high permeability layers that have a low permeability layer on top. The conjugate gradient method converges to the exact solution only once all small eigenvalues have been 'discovered'. The number of eigenvalues is reduced by the use of a preconditioner (Incomplete Cholesky decomposition or even a diagonal scaling). However, still a number of small eigenvalues remain for the preconditioned matrix. These small eigenvalues persist due to the fact that at each interface between low-and high permeability layers a homogeneous Neumann condition is effectively adopted by the sand-layer. This makes the blocks of the discretization matrix that correspond to the sandwiched sandlayers almost singular. This observation is formulated in terms of the following theorem, which is proven by Vuik et al. [17]:

**Theorem 1.** *Let* $\varepsilon := \dfrac{\sigma_{\min}}{\sigma_{\max}}$ *be small enough,* $D = \operatorname{diag}(A)$ *and let* $r$ *be the number of layers with* $\sigma$ *of order one between low* $\sigma$ *layers. Then the diagonally scaled matrix* $D^{-1/2} A\, D^{-1/2}$ *has exactly* $r$ *eigenvalues of order* $\varepsilon$.

We remark here that Theorem 1 is extended to an incomplete Cholesky preconditioning, see Vuik et al [18]. The preconditioning aims at improving the condition of the matrix. However, for this case $r$ small eigenvalues persist. In experiments we use an incomplete Cholesky decomposition as a preconditioner for the symmetric positive definite discretization matrix.

In the next section we consider the DICCG-method as proposed by Vuik et al. [17] for the Laplace problem with extreme contrasts of the coefficients. The aim is to get rid of the remaining very small eigenvalues of the preconditioned matrix $\tilde{A} = L^{-T}L^{-1}A$, where $L^{-T}L^{-1} \approx A^{-1}$ is the IC-preconditioner.

### 3.1 Deflation

In this subsection we analyze the elimination of the small eigenvalues of $A$ by deflation. Therefore, we first prove that, if the deflation matrix is constructed by eigenvectors of $A$, then its corresponding eigenvalues are transformed into zero for the product of the deflation matrix and the discretization matrix $A$. Here we need properties like symmetry and that the deflation matrix is a projection. These properties are proven first. Suppose that $\lambda_i$ and $\vec{z}_i$, $i \in \{1, \ldots, n\}$, respectively represent eigenvalues and orthonormal eigenvectors of the symmetric discretization matrix $A \in \mathbb{R}^{n \times n}$ (such that $\vec{z}_i^T \vec{z}_j = \delta_{ij}$), and define the matrix $\overline{P} \in \mathbb{R}^{n \times n}$ by

$$\overline{P} := I - \sum_{j=1}^{m} \vec{z}_j \vec{z}_j^T, \qquad m \leq n.$$

Then

$$\overline{P}A\, \vec{z}_i = A\, \vec{z}_i - \sum_{j=1}^{m} \vec{z}_j \vec{z}_j^T A\, \vec{z}_i = A\, \vec{z}_i - \lambda_i \sum_{j=1}^{m} \vec{z}_j \vec{z}_j^T \vec{z}_i = \vec{0}$$
$$\forall i \in \{1, \ldots, m\},$$

$$\overline{P}A\,\vec{z}_k = A\,\vec{z}_k - \sum_{j=1}^{m} \vec{z}_j\vec{z}_j^T A\,\vec{z}_k = A\,\vec{z}_k - \lambda_k \sum_{j=1}^{m} \vec{z}_j\vec{z}_j^T \vec{z}_k = \lambda_k \vec{z}_k$$
$$\forall k \in \{m+1, \dots, n\}.$$

Hence, we state the following result:

**Theorem 2.** *Let* $\overline{P} := I - \sum_{i=1}^{m} \vec{z}_i\vec{z}_i^T$, *where* $\vec{z}_i$ *and* $\lambda_i$ *are respectively orthogonal eigenvectors and eigenvalues of the matrix* $A$. *Then*

1. $\overline{P}A$, *for* $j > m$, *and* $A$ *have the same eigenvalues* $\lambda_j$, *and the corresponding eigenvalues of* $A$ *given by* $\lambda_j$ *for* $j \leq m$ *are all zero for the matrix* $\overline{P}A$;
2. *the matrix* $\overline{P}$ *is a projection.*

Proof. The first statement is proven by the argument above Theorem 2. To prove the second statement, we compute

$$\begin{aligned}\overline{P}^2 &= \left(I - \sum_{i=1}^{m} \vec{z}_i\vec{z}_i^T\right)\left(I - \sum_{i=1}^{m} \vec{z}_i\vec{z}_i^T\right)\\ &= I - 2\sum_{i=1}^{m} \vec{z}_i\vec{z}_i^T + \left(\sum_{i=1}^{m} \vec{z}_i\vec{z}_i^T\right)\left(\sum_{i=1}^{m} \vec{z}_i\vec{z}_i^T\right)\\ &= I - \sum_{i=1}^{m} \vec{z}_i\vec{z}_i^T = \overline{P}.\end{aligned}$$

The second last equality results from orthonormality of the eigenvectors. Hence, the matrix $\overline{P}$ is a projection. □

From Theorem 2 it follows that $\overline{P}$ and $\overline{P}A$ are singular. Now we introduce the matrix $P$ for a general choice of deflation vectors:

**Definition 1.** The deflation matrix $P$ is defined by

$$P := I - AZ(Z^TAZ)^{-1}Z^T.$$

We show that $P$ can be written as $\overline{P}$ when the columns of $Z$ are eigenvectors of $A$.

**Theorem 3.** *Let* $Z \in \mathbb{R}^{n\times m}$: $Z = (\vec{z}_1, \dots, \vec{z}_m)$ *where* $\vec{z}_i$ *are the orthonormal eigenvectors with eigenvectors* $\lambda_i$ *of the matrix* $A$ *and let* $P$ *be defined as in Definition* 1. *Then* $P = \overline{P}$. *Moreover,* $P$ *is a projection for all choices of* $Z \in \mathbb{R}^{n\times m}$.

Proof. If $P = I - A\,Z\,(Z^T A\,Z)^{-1} Z^T$, then

$$
\begin{aligned}
P &= I - A\,(\vec{z}_1, \dots, \vec{z}_m) \left( \begin{pmatrix} \vec{z}_1^T \\ \vdots \\ \vec{z}_m^T \end{pmatrix} A\,(\vec{z}_1, \dots, \vec{z}_m) \right)^{-1} \begin{pmatrix} \vec{z}_1^T \\ \vdots \\ \vec{z}_m^T \end{pmatrix} \\
&= I - (A\,\vec{z}_1, \dots, A\,\vec{z}_m) \begin{pmatrix} \vec{z}_1^T A\,\vec{z}_1 & \dots & \vec{z}_1^T A\,\vec{z}_m \\ \vdots & \vdots & \vdots \\ \vec{z}_m^T A\,\vec{z}_1 & \dots & \vec{z}_m^T A\,\vec{z}_m \end{pmatrix}^{-1} \begin{pmatrix} \vec{z}_1^T \\ \vdots \\ \vec{z}_m^T \end{pmatrix} \\
&= I - (\lambda_1 \vec{z}_1, \dots, \lambda_m \vec{z}_m)\, \mathrm{diag}\Big(\frac{1}{\lambda_1}, \dots, \frac{1}{\lambda_m}\Big) \begin{pmatrix} \vec{z}_1^T \\ \vdots \\ \vec{z}_m^T \end{pmatrix} \\
&= I - \sum_{i=1}^{m} \vec{z}_i \vec{z}_i^T = \overline{P}.
\end{aligned}
$$

This proves the first statement. The second statement is proven by direct multiplication:

$$P^2 = I - 2AZ(Z^T A\,Z)^{-1} Z^T + AZ(Z^T AZ)^{-1} Z^T AZ(Z^T AZ)^{-1} Z^T = P.$$

Hence $P$ is a projection. □

**Corollary 1.** *The matrix $PA$ is symmetric positive semi-definite.*

Proof. Since $A^T = A$, the symmetry of $PA$ is established by

$$(PA)^T = (A - AZ(Z^T A\,Z)^{-1} Z^T A)^T = A - AZ(Z^T AZ)^{-1} Z^T A = PA.$$

Further, $P$ is a projection and $A$ is symmetric positive definite; hence from Lemma 2.1 by Frank and Vuik [4] it follows that $PA$ is positive semi-definite. □

Furthermore, the matrix $PA$, where $P := I - AZ(Z^T AZ)^{-1} Z^T$, is singular, since

$$PA\,Z = A\,Z - A\,Z\,(Z^T A\,Z)^{-1} Z^T A\,Z = A\,Z - A\,Z = 0.$$

This is also shown in Theorem 4.1 in Vuik et al [17]. Consider a matrix $A$ with eigenvalues $\{\lambda_1, \dots, \lambda_m, \lambda_{m+1}, \dots, \lambda_n\}$ and let $Z = (\vec{z}_1, \dots, \vec{z}_m)$ where $A\vec{z}_i = \lambda_i \vec{z}_i$, $i \in \{1, \dots, m\}$ are eigenvectors of $A$ that correspond to eigenvalues $\{\lambda_1, \dots, \lambda_m\}$. Then with $P$, according to Definition 1, it follows that $PA$ has eigenvalues $\{0, \dots, 0, \lambda_{m+1}, \dots, \lambda_n\}$.

**Theorem 4.** *Let $P \in \mathbb{R}^{n\times n}$ be defined as in Definition* 1 *with* $Z = (\vec{z}_1, \dots, \vec{z}_m)$. *Then the null-space of $P$ is spanned by the independent set* $\{A\,\vec{z}_1, \dots, A\,\vec{z}_m\}$, *i.e.* $\text{null}(P) = \text{Span}\{A\vec{z}_1, \dots, A\,\vec{z}_m\}$.

Proof. Since $PAZ = 0$, it follows that $A\vec{z}_i \in \text{null}\, P$ and hence $\dim \text{null}\, P \geq m$. If $V := \text{Span}\{\vec{z}_1, \dots, \vec{z}_m\}$, then from the Direct Sum Theorem (see for instance [8]) it follows $\mathbb{R}^n = V \oplus V^{\perp}$, where $V^{\perp} = \{\vec{y} \in \mathbb{R}^n : \vec{y} \perp V\}$. Hence, $\dim V^{\perp} = n - m$. Suppose $\vec{y} \in V^{\perp}$, then

$$P\vec{y} = \vec{y} - AZ(Z^T A\, Z)^{-1} Z^T \vec{y} = \vec{y}.$$

Hence, $\dim \text{col}\, P \geq n - m$. Since $\dim \text{null}\, P + \dim \text{col}\, P = n$, this implies

$$\dim \text{null}\, P \leq m.$$

Consequently, using $\dim \text{null}\, P \geq m$, we have $\dim \text{null}\, P = m$. Since $\{A\,\vec{z}_1, \dots, A\vec{z}_m\}$ represents a linearly independent set of $m$ vectors in $\text{null}\, P$ and $\dim \text{null}\, P = n$, it follows from the Basis-Theorem (see for instance [10]) that

$$\text{null}\, P = \text{Span}\{A\vec{z}_1, \dots, A\vec{z}_m\}.$$

This proves the theorem. □

The matrix $P$ is referred to as the deflation matrix. The vectors $\vec{z}_1, \dots, \vec{z}_m$ are referred to as the projection vectors and they are chosen such that their span approaches the span of the small eigenvectors of $\tilde{A}$. The advantage of working with the matrix $P\tilde{A}$ rather than with $\tilde{A}$ is that the smaller eigenvalues of $\tilde{A}$ are transferred to zero eigenvalues of $PA$ which do not influence the convergence of the CG-method.

### 3.2 Deflated incomplete Cholesky preconditioned conjugate gradients

The elimination of the small eigenvalues of $A$ takes place by using the projection matrix $P$. We then solve

$$PA\,\vec{\tilde{x}} = P\vec{b}, \tag{4}$$

with the ICCG-method, where $PA$ is singular. The solution of the above equation is not unique. We obtain the solution of the matrix-equation $A\vec{x} = \vec{b}$ by

$$\vec{x} = (I\ - P^T)\vec{x} + P^T\vec{x}. \tag{5}$$

Note that $\vec{\tilde{x}}$ is the solution of equation (4). In this paragraph we first establish uniqueness of the above $P^T\vec{\tilde{x}}$, given any $\vec{\tilde{x}}$ that satisfies equation (4). Therefore, we first need to establish that for the projected solution we have $P^T\vec{x} = P^T\vec{\tilde{x}}$ and that hence $P^T\vec{\tilde{x}}$ is unique. Equation (4) is written as

$P(A\,\vec{\bar{x}} - \vec{b}) = \vec{0}$, where $A\vec{\bar{x}} - \vec{b} \in \operatorname{null} P$. Since $PA\ Z\ = 0$, the vectors $A\vec{z}_i$ are in the null-space of $P$, i.e. $A\vec{z}_i \in \operatorname{null} P$. Since, we know from Theorem 4 that $\operatorname{null} P = \operatorname{Span}\{A\vec{z}_1, \dots, A\ \vec{z}_m\}$, the vectors in the null-space of $P$ can be written as linear combinations $\vec{w} = \sum_{i=1}^{m} \alpha_i A\ \vec{z}_i$. Therefore, it can be seen that one can write for the vector $\vec{\bar{x}}$:

$$\vec{\bar{x}} = A^{-1}\vec{b} + A^{-1}\vec{w} = A^{-1}\vec{b} + \sum_{i=1}^{m} \alpha_i \vec{z}_i. \tag{6}$$

Since $A$ is not singular, the first term in the right-hand side is uniquely determined. We investigate the result obtained from multiplication of the second term in the right-hand side with the matrix $P^T$. For convenience we look at the product $P^T Z$:

$$P^T Z\ = (I\ - Z\ (Z^T A\ Z)^{-T} Z^T A) Z\ = Z\ - Z\ (Z^T A\ Z)^{-1} Z^T A\ Z\ = 0.$$

Hence, this product is zero and the second term of the right-hand side of equation (6) vanishes after multiplication with $P^T$, since $Z = (\vec{z}_1, \dots, \vec{z}_m)$. Hence, the vector $P^T\vec{\bar{x}}$ is unique and we have

$$P^T\vec{\bar{x}} = P^T A^{-1}\vec{b} = P^T\vec{x}.$$

For the first term of the right-hand side of equation (5) we note that

$$(I\ - P^T)\vec{x} = Z(Z^T AZ)^{-1} Z A\vec{x} = Z(Z^T AZ)^{-1} Z\vec{b}.$$

Hence, this term is also uniquely determined and since the dimension of the matrix $Z^T AZ$ is small, the computation of this term is relatively cheap. This is summarized in the next theorem:

**Theorem 5.** *Let $P$ be defined as in Definition* 1 *and $\vec{\bar{x}}$ as in equation* (4). *Then*

1. *$P^T\vec{\bar{x}}$ is unique and $P^T\vec{\bar{x}} = P^T\vec{x}$,*
2. *the unique solution of $A\vec{x} = \vec{b}$ can be written by*

$$\vec{x} = Z(Z^T AZ)^{-1} Z^T\vec{b} + P^T\vec{\bar{x}},$$

*where $\vec{\bar{x}}$ is a solution of equation* (4).

We further note that $(I - P^T)Z = Z(Z^T AZ)^{-1} Z^T AZ = Z$. Hence,

$$(I - P^T)(\vec{z}_1, \dots, \vec{z}_m) = (\vec{z}_1, \dots, \vec{z}_m).$$

From this it follows that $\operatorname{Span}\{\vec{z}_1, \dots, \vec{z}_m\} =: V\ \subset \mathbb{R}^n$ is in the eigenspace of eigenvalue $\lambda = 1$ of the matrix $(I - P^T)$.

For completeness, we present the algorithm of the deflated ICCG.

**Algorithm 1** (DICCG [17]):

$k = 0$, $\vec{\tilde{r}}_0 = P\vec{r}_0$, $\vec{p}_1 = \vec{z}_0 = L^{-T}L^{-1}\vec{\tilde{r}}_0$

**while** $\|\vec{\tilde{r}}_k\|_2 > \varepsilon$

$$k = k + 1$$

$$\alpha_k = \frac{\vec{\tilde{r}}_{k-1}^{\,T}\vec{z}_{k-1}}{\vec{p}_k^{\,T} PA\,\vec{p}_k}$$

$$\vec{x}_k = \vec{x}_{k-1} + \alpha\vec{p}_k$$

$$\vec{\tilde{r}}_k = \vec{\tilde{r}}_{k-1} - \alpha_k PA\,\vec{p}_k$$

$$\vec{z}_k = L^{-T}L^{-1}\vec{\tilde{r}}_k$$

$$\beta_k = \frac{\vec{\tilde{r}}_k^{\,T}\,\vec{z}_k}{\vec{\tilde{r}}_{k-1}^{\,T}\,\vec{z}_{k-1}}$$

$$\vec{p}_{k-1} = \vec{z}_k + \beta_k\vec{p}_k$$

**end while**

The conjugate gradient method is reported to converge for symmetric positive definite matrices. However, Kaasschieter ([7], Section 2) notes that eigenvalues of a symmetric positive semi-definite matrix that are zero do not contribute to the convergence of the CG-method. Furthermore, he concludes that the singular system can be solved by conjugate gradients as long as system (4) is consistent ($P\vec{b} \in \mathrm{Col}(PA)$). Van der Sluis and van der Vorst [15] note that the convergence may be much faster than bounds (1) and (3) predict when eigenvectors are clustered.

### 3.3 Choice of deflation vectors

We apply several choices for the deflation vectors to solve the following problem:

$$\left.\begin{aligned} -\nabla\cdot(k\ \nabla\psi) &= 0, \quad (x,y)\in\Omega, \\ \psi &= 1, \quad (x,y)\in\Gamma_D, \\ \frac{\partial\psi}{\partial n} &= 0, \quad (x,y)\in\Gamma_N. \end{aligned}\right\} \tag{P$_3$}$$

The domain $\Omega$ is divided into subdomains $\Omega_j$ such that $\overline{\Omega} = \bigcup_{i=1}^m \overline{\Omega}_i$ and $\aleph \subseteq \{1,\ldots,m\}$ denotes the set of indices that correspond to subdomains with high permeability, i.e.,

$$k(x,y) = \begin{cases} k_{\max} = 1, & (x,y)\in\Omega_j,\ j\in\aleph, \\ k_{\min} = \varepsilon, & (x,y)\in\Omega_j,\ j\in\{1,\ldots,m\}\setminus\aleph. \end{cases}$$

Further, we assume that if $\overline{\Omega}_i \cap \overline{\Omega}_j \neq \emptyset$ then the value of $k$ in $\Omega_i$ is not equal to the value of $k$ in $\Omega_j$. We take $\varepsilon = 10^{-7}$. For each subdomain $\Omega_i$ we introduce a deflation vector $\vec{z}_j$ as follows:

$$z_j(x,y) = \begin{cases} 0 & \text{for } (x,y) \in \Omega \setminus \overline{\Omega}_j, \\ \in [0,1] & \text{for } (x,y) \in \overline{\Omega}_j \setminus (\Omega_j \cup (\Gamma_N \cap \overline{\Omega}_j)), \\ 1 & \text{for } (x,y) \in \Omega_j. \end{cases}$$

Note that in the finite element formulation the Dirichlet boundary points do not participate in the solution of the matrix-vector equation. An example of the geometry is shown in Fig. 1. We investigate the following choices where $\vec{z}_j$ is varied for $(x,y) \in \overline{\Omega}_j \setminus (\Omega_j \cup (\Gamma_N \cap \overline{\Omega}_j))$.

1. **non overlapping** projection vectors:

$$z_j(x,y) = \begin{cases} 1 & \text{for } (x,y) \in \overline{\Omega}_j \setminus (\Omega_j \cup \Gamma), \quad j \in \aleph, \\ 0 & \text{for } (x,y) \in \overline{\Omega}_j \setminus (\Omega_j \cup \Gamma), \quad j \in \{1,\dots,m\} \setminus \aleph, \end{cases}$$

2. **complete overlapping** projection vectors:

$$z_j(\vec{x}) = 1 \quad \text{for } \overline{\Omega}_j \setminus (\Omega_j \cup \Gamma), \quad j \in \{1,\dots,m\},$$

3. **average overlapping** projection vectors of the subdomains:

$$z_j(\vec{x}) = \frac{1}{2} \quad \text{for } \overline{\Omega}_j \setminus (\Omega_j \cup \Gamma), \quad j \in \{1,\dots,m\},$$

4. **weighted overlapping** projection vectors of the subdomains:

$$z_j(x,y) = \begin{cases} \dfrac{k_{\max}}{k_{\max} + k_{\min}} & \text{for } \overline{\Omega}_j \setminus (\Omega_j \cup \Gamma), \quad j \in \aleph, \\ \dfrac{k_{\min}}{k_{\max} + k_{\min}} & \text{for } \overline{\Omega}_j \setminus (\Omega_j \cup \Gamma), \quad j \in \{1,\dots,m\} \setminus \aleph. \end{cases}$$

Note that weighted overlap is approximated by no overlapping when $k_{\min} \ll k_{\max}$ and by average overlapping whenever $k_{\min} = k_{\max}$. After some theoretical results we investigate the four choices by numerical experiment for both the case that $k_{\min} \ll k_{\max}$ and $k_{\min} = k_{\max}$. Subsequently, we apply the deflation principle to parallel computation.

We consider the case that constrasts are large, i.e. $k_{\min} \ll k_{\max}$, $\varepsilon = 10^{-7}$. We now show that the choice of average overlap is not suitable for this case where $\text{cond}(A) = O(\frac{1}{\varepsilon})$. In the next theorem we refer to areas with $k_{\min}$ and $k_{\max}$ as low mobility and high mobility layers, respectively. Now we will show that the average overlapping projection vectors do not approximate the span of the eigenvectors that belong to eigenvalues that are of the order $O(\varepsilon)$.

**Assumption 1.** *We assume that the finite element discretization is consistent, which means that discretization error is zero for a constant function.*

*We further assume that the off-diagonal entries of the discretization matrix A are non-positive.*

Let $\boldsymbol{x}_m$ denote the position of grid point $m$. A consequence of the above assumption is

$$\sum_{j=1}^{n} a_{mj} = 0 \quad \text{for} \quad \boldsymbol{x}_m \in \Omega \setminus \Gamma_D. \tag{7}$$

Before we state the theorem we introduce the index set $\Pi_i \subset \{1, \ldots, n\}$ as the set of indices of $\vec{v}_i$ (which is the projection vector that corresponds to subdomain $\Omega_i$) that correspond to grid points, which are on the boundary between two consecutive subdomains with $k = k_{\max} = 1$ and $k = k_{\min} = \varepsilon$. Further, we denote the neighbouring grid points of index set $\Pi_i$ (located in $\Omega_i$) by $\tilde{\Pi}_i$.

**Theorem 6.** *If the finite element discretization satisfies Assumption* 1 *and if the discretization matrix is irreducible, then for* $D = \operatorname{diag}(a_{11} \ldots a_{nn})$ *we have*

1. $\|D^{-1}A\, \vec{v}_i\|_\infty \sim 1$ *for all* $i$*, regardless the value of* $k$ *in the subdomain, for average overlapping delation vectors,*
2. $\|D^{-1}A\, \vec{v}_i\|_\infty = O(\varepsilon)$ *for all* $i$ *corresponding to the subdomains* $\overline{\Omega}_i \cap \Gamma_D = \emptyset$ *with* $k = k_{\max} = 1$*, for the cases of non-overlapping, completely overlapping and weighted overlapping deflation vectors,*
3. $\|D^{-1}A\, \vec{v}_i\|_\infty \sim 1$ *for all values of* $i$ *that are not incorporated in part* 2*, i.e.,* $k = k_{\min} = \varepsilon$ *or* $\overline{\Omega}_i \cap \Gamma_D \neq \emptyset$ *and* $k = k_{\max} = 1$.

Proof. We start with $\boldsymbol{x}_m \in \Omega_i \cup (\overline{\Omega}_i \cap \Gamma_N)$ and $m \in \{1, \ldots, n\} \setminus (\Pi_i \cup \tilde{\Pi}_i)$, where $\boldsymbol{x}_m$ is not a neighbouring point of a Dirichlet point, then $(v_i)_m$ is equal to its neighbours. From consistency of the discretization it follows

$$(A\, v_i)_m = 0 \quad m \in \{1, \ldots, n\} \setminus (\Pi_i \cup \tilde{\Pi}_i),$$

regardless the value of $k$ in the subdomain $i$. For $m \in \Pi_i$ we split its neighbours into the sets $J_m^H$ and $J_m^H$, respectively denoting the neighbouring gridnodes in the high and low permeability layers. Further, we introduce the set $J_m^I$ representing the set of indices corresponding to neighbours of point $m$ with index in $\Pi_i$. We proceed with $m \in \Pi_i$, then for the discretization matrix we have

$$\left.\begin{aligned} a_{mj} &\sim 1 \quad \text{for } j \in J_m^H, \\ a_{mj} &= O(\varepsilon) \quad \text{for } j \in J_m^L. \end{aligned}\right\}$$

Multiplication of $A$ with vector $v_i$ gives for component $m$

$$(A\, v_i)_m = \sum_{j=1}^{n} a_{mj}(v_i)_j = a_{mm}(v_i)_m + \sum_{j \in J_m^L \cup J_m^H \cup J_m^I} a_{mj}(v_i)_j.$$

Since $a_{mm} = -\sum_{j=1,\, j\neq m}^{n} a_{mj} \sim 1$ and $(v_i)_m = (v_i)_j$ for $j \in J_m^I$ when the subdomains are layered, the above relation implies

$$(Av_i)_m = \sum_{j \in J_m^L \cup J_m^H} a_{mj}((v_i)_j - (v_i)_m). \tag{8}$$

Now, we estimate $(Av_i)_m$ by the use of equation (8) for the three cases in the theorem.

1. Consider average overlapping deflation vectors. Then

$$\sum_{j \in J_m^L} a_{mj}((v_i)_j - (v_i)_m) = O(\varepsilon).$$

Further, note that from irreducibility of the matrix $A$ it follows that $a_{mj} \sim 1$ for at least one $j \in J_m^H \neq \emptyset$. Hence, we have

$$\sum_{j \in J_m^L} a_{mj}((v_i)_j - (v_i)_m) \sim 1.$$

Therefore, with $D_{mm} = a_{mm} \sim 1$, it follows after using equation (8) that

$$\|D^{-1}A\, \vec{v}_i\|_\infty \sim 1, \quad \text{for all subdomains.}$$

This proves part 1 of the theorem.

2. Consider $i$ corresponding to subdomain $\overline{\Omega_i} \cap \Gamma_D = \emptyset$ with $k = k_{\max} = 1$. Then for $m \in \Pi_i$ we have for all cases of overlap

$$\sum_{j \in J_m^L} a_{mj}((v_i)_j - (v_i)_m) = O(\varepsilon),$$

and

$$\sum_{j \in J_m^H} a_{mj}((v_i)_j - (v_i)_m) = \begin{cases} O(\varepsilon) & \text{for weighted overlap,} \\ 0 & \text{for complete and no overlap.} \end{cases}$$

Substitution of the above relations into equation (8) gives

$$(Av_i)_m = O(\varepsilon).$$

Since $D_{mm} \sim 1$, it follows for non-overlapping, complete overlap and weighted overlap that

$$(D^{-1}A\, v_i)_m = O(\varepsilon) \quad \text{for subdomains where } k = k_{\max} = 1.$$

For subdomains $\Omega_i$ where $\overline{\Omega}_i \cap \Gamma_D = \emptyset$ and $k = k_{\max} = 1$, it follows, with $(Av_i)_m = 0$ for $m \in \{1, \ldots, n\} \setminus \Pi_i$, that

$$\|D^{-1}A\ v_i\|_\infty = O(\varepsilon).$$

This proves part 2 of the theorem.

3. Now we consider $\Omega_i$ where $k = k_{\max} = 1$ and with $\overline{\Omega}_i \cap \Gamma_D \neq \emptyset$, then for a certain grid point $m$ that neighbours a Dirichlet grid point, we do not satisfy equation (8), but

$$\sum_{j=1}^{n} a_{mj} \sim 1,$$

since $k = k_{\max} = 1$. This implies that

$$\sum_{j=1}^{n} a_{mj}(v_i)_j \sim 1,$$

and hence

$$\|D^{-1}A\ v_i\|_\infty \sim 1, \quad \text{for } \overline{\Omega}_i \cap \Gamma_D \neq \emptyset \text{ and } k\ = k_{\max} = 1.$$

For $m \in \tilde{\Pi}_i$ we have for a subdomain, where $k = \varepsilon$,

$$\sum_{j=1}^{n} a_{mj}((v_i)_j - (v_i)_m) = O(\varepsilon) \quad \text{for no overlap and weigthed overlap.}$$

Since $a_{mm} = D_{mm} \sim \varepsilon$ for $m \in \tilde{\Pi}_i$ when $k = \varepsilon$, we obtain

$$\|D^{-1}A\ v_i\|_\infty \sim 1, \quad \text{for no overlap or weighted overlap.}$$

For $m \in \Pi_i$ we have for a subdomain, where $k = \varepsilon$,

$$\sum_{j \in J_m^H} a_{mj}((v_i)_j - (v_i)_m) \sim 1 \quad \text{for complete overlap.}$$

For complete overlap, we also have $D_{mm} \sim 1$ for $m \in \Pi_i$ and hence, one obtains

$$\|D_{-1}A\ v_i\|_\infty \sim 1 \quad \text{for complete overlap,}$$

whenever $i$ corresponds to a subdomain where $k = k_{\min} = \varepsilon$. This proves the theorem. □

An important conclusion here is that the average overlapping deflation vectors with $\varepsilon \ll 1$ do not approximate the span of eigenvectors corresponding to the small eigenvalues of $D^{-1}A$. The above theorem is proven for diagonal scaling. We extend the result to an incomplete Cholesky decomposition for algebraic projection vectors with weighted and complete overlap and no overlap. A similar theorem is proven by Vuik et al [18] for physical deflation vectors. This is formulated in the following theorem:

**Theorem 7.** *If the finite element discretization is consistent under Assumption 1, then*

$$\|L^{-T}L^{-1}A\,\vec{v}_i\|_2 = O(\varepsilon)$$

*for non-overlapping, completely and weighted overlapping projection vectors.*

Proof. The proof is based on Theorem 6.

$$\|L^{-T}L^{-1}A\,\vec{v}_i\|_2 = \|L^{-T}L^{-1}DD^{-1}A\,\vec{v}_i\|_2 \leq \lambda_{\max}(L^{-T}L^{-1}D)\|D^{-1}A\,\vec{v}_i\|_2.$$

Vuik et al [18] (Theorem 2.2) prove that $\lambda(L^{-T}L^{-1}D)$ is bounded also for $\varepsilon \to 0$. Further, since $\|D^{-1}A\vec{v}_i\|_2 \leq \sqrt{n}\|D^{-1}A\vec{v}_i\|_\infty$, we obtain

$$\|L^{-T}L^{-1}A\,\vec{v}_i\|_2 \leq \lambda_{\max}(L^{-T}L^{-1}D)\sqrt{n}\|D^{-1}A\vec{v}_i\|_\infty.$$

Since we know from Theorem 6

$$\|D^{-1}A\vec{v}_i\|_\infty \leq O(\varepsilon)$$

for non-, completely and weighted overlapping projection vectors, we obtain

$$\|L^{-T}L^{-1}A\,\vec{v}_i\|_2 \leq \lambda_{\max}(L^{-T}L^{-1}D)\sqrt{n}\,O(\varepsilon).$$

This proves the assertion. □

Note that the above theorem does not say anything for the case of incomplete Cholesky preconditioning and average overlapping projection vectors. The result in Theorem 6 is expected to hold for average overlapping with incomplete Cholesky preconditioning as well. This is also observed by numerical experiments.

## 4 Numerical experiments

For the experiments we consider test-problem ($P_3$) for a rectangular domain $\Omega$. It easily follows that the solution of this test-problem is $\psi = 1$. The problem is solved by the use of a Finite Element method. We have done experiments with the different overlapping between subsequent subdomains. The parameter $k$ is allowed to have large jumps. To illustrate the need for deflation, when $k$ has large jumps, we start with the setting where we have horizontal layers of alternating permeability where $\varepsilon = 10^{-7}$. Here we take seven layers and 3200 ($80 \times 40$) elements per layer and show a convergence result obtained from a Cholesky preconditioned conjugate gradient method when no deflation is applied in Fig. 4. It can be seen in the graph of the residual $\|\vec{b} - A\vec{x}_k\|_2$ that convergence is fast at the early stages. Subsequently, a non-monotonic behaviour is observed which is due to the presence of three eigenvalues that are of the order of $10^{-7}$ in accordance to Theorem 1.

These eigenvalues are due to the three sand layers that are sandwiched between low permeability shale layers. The convergence speed of the exact error $\|\vec{x}_k - \vec{x}_{\text{true}}\|_2$, however, is slow at the early stages and hence the solution has a poor quality then. Further for the sake of illustration, we plot the smallest eigenvalue of the preconditioned discretization matrix as a function of the iteration number. Only after 'discovery' of all the (small) eigenvalues convergence sets in. We see that although the problem is small, convergence is poor. To illustrate the increase of speed of convergence by the use of deflation, we present the results from deflation for the same problem in Fig. 5. Here physical deflation has been used, where only for each high permeability layer, i.e. sand layer, that is surrounded by shale-layers a deflation vector is used. From the results it can be seen that the smallest eigenvalue is of the order of 0.01 and that convergence of the residual is monotonous. Further, the exact error converges fast from the start. The computation has been finished in about 99 iterations instead 290 iterations in Fig. 4. So it is clear that deflation increases the convergence speed. Similar results can be found in the paper of Vuik et al [17].

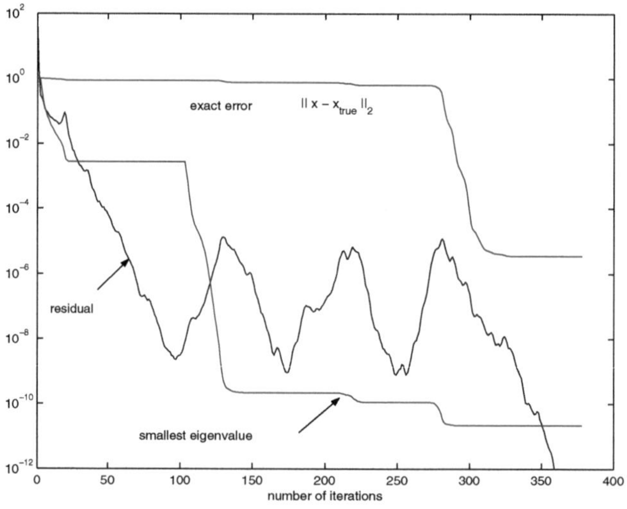


Figure 4. Convergence behavior of the incomplete Cholesky preconditioned CG-method for 7 subdomains with large jumps of the coefficient $k$ in ($P_3$). The $\|\cdot\|_2$-norm of the residual and error have been presented. The number of elements is 3200 per layer.

### 4.1 Algebraic deflation vectors

We consider a comparison between deflation with algebraic projection vectors without overlap and deflation with physical projection vectors when contrasts in the permeability are very high $\varepsilon = 10^{-7}$. The results have been plotted in Figs. 5 and 6. From both figures it is clear that algebraic deflation without

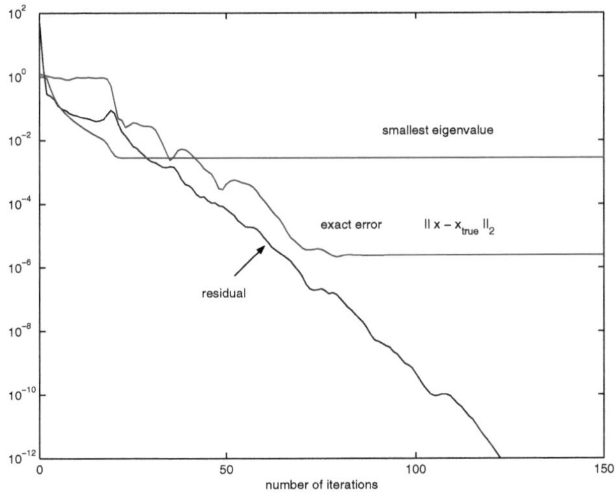


Figure 5. Convergence behavior of the deflated incomplete Cholesky preconditioned CG-method for 7 subdomains with large jumps of the coefficient $k$ in ($P_3$). The $\|\cdot\|_2$-norm of the residual and error have been presented. Here results obtained by the use of physical deflation are shown. The number of elements is 3200 per layer.

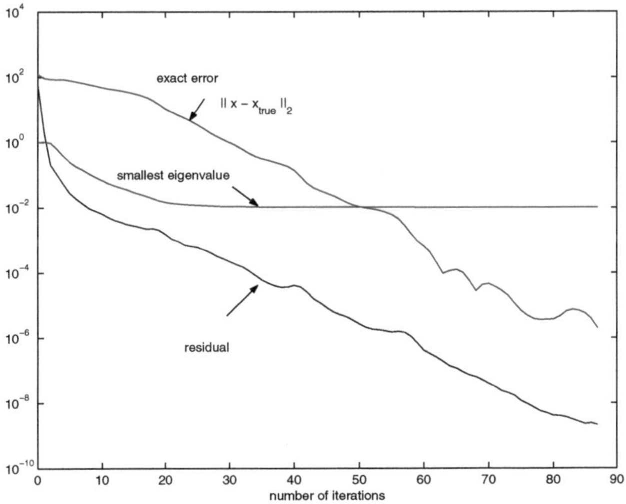


Figure 6. Convergence behavior of the deflated incomplete Cholesky preconditioned CG-method for 7 subdomains with large jumps of the coefficient $k$ in ($P_3$). The $\|\cdot\|_2$-norm of the residual and error have been presented. Here results obtained by the use of algebraic deflation are shown. Algebraic deflation is here by no overlap for the projection vectors. The number of elements is 3200 per layer.

overlap, where for each layer there is a projection vector, requires fewer iterations than physical deflation. In Table 1 we present the number of iterations needed and computing time to obtain convergence for different number of elements per subdomain.

Table 1. Computation time and speed of convergence for physical and algebraic deflation vectors for several mesh-sizes.

| Number of elements | Physical deflation Time (s) | Algebraic deflation Time (s) | $N_{\text{phys}}$ | $N_{\text{alg}}$ |
|---|---|---|---|---|
| 50 | 0.02 | 0.02 | 16 | 15 |
| 200 | 0.07 | 0.07 | 28 | 26 |
| 800 | 0.46 | 0.39 | 52 | 44 |
| 3200 | 3.49 | 2.92 | 99 | 84 |
| 12800 | 24.20 | 18.72 | 173 | 140 |
| 41200 | 171.18 | 129.61 | 298 | 255 |

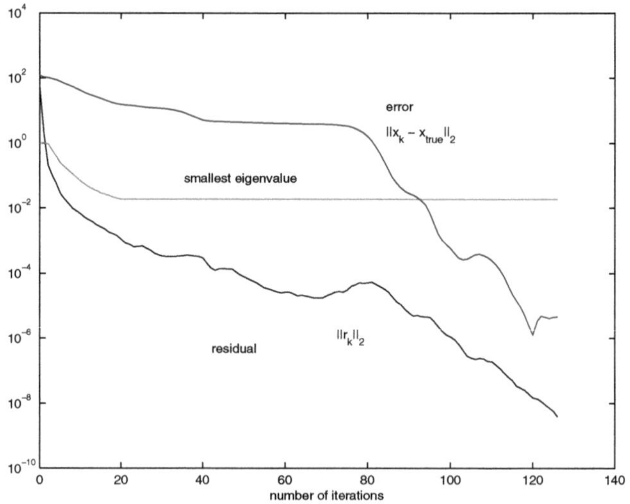


Figure 7. Convergence behavior of the deflated incomplete Cholesky preconditioned CG-method for 7 subdomains with large jumps of the coefficient $k$ in ($P_3$). The $\|\cdot\|_2$-norm of the residual and error have been presented. Here results obtained by the use of algebraic deflation are shown. Algebraic deflation is here by complete overlap for the projection vectors. The number of elements is 3200 per layer.

From Table 1 it follows that algebraic deflation vectors without overlap gives a better convergence than physical deflation vectors, especially when the number of elements becomes large. Furthermore, the CPU-time is larger for physical deflation vectors, which is due to solving the homogeneous partial differential equation when the projection vector is determined. Note, however, that the number of projection vectors is larger for algebraic deflation.

From Fig. 5 and 6 it can also be seen that the smallest absolute value of the non-zero eigenvalues is less small when algebraic deflation is used. This is assumed to be the cause for the increase of speed of convergence when algebraic projections vectors are used.

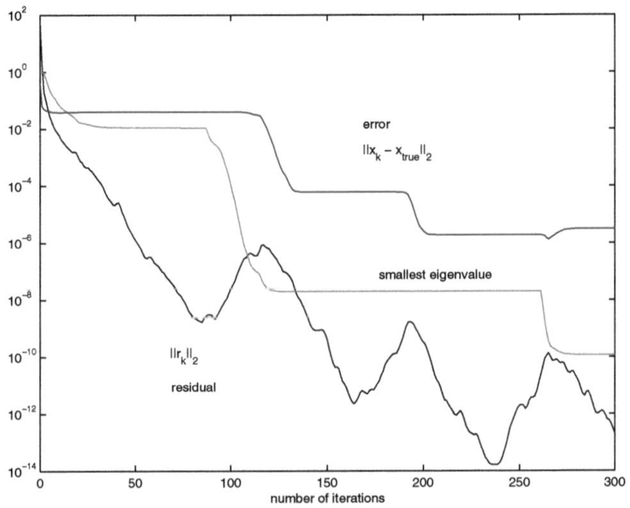


Figure 8. Convergence behavior of the deflated incomplete Cholesky preconditioned CG-method for 7 subdomains with large jumps of the coefficient $k$ in ($P_3$). The $\|\cdot\|_2$-norm of the residual and error have been presented. Here results obtained by the use of algebraic deflation are shown. Algebraic deflation is here by average overlap for the projection vectors. The number of elements is 3200 per layer.

For the sake of illustration we show the evolution of the residual, smallest eigenvalue and error during the conjugate gradient iterations for the various implementations of deflation. In Fig. 7 we show the convergence for a system of seven subdomains with 3200 triangular elements. Here the projection vectors are chosen with complete overlap at the interface points. From Figs. 6 and 7 we see that the choice of complete overlapping projection vectors gives a slower convergence than non overlapping projection vectors. Subsequently, we show convergence with deflation vectors chosen with average overlap at the interfaces in Fig. 10. It can be seen that the smallest eigenvalue for average overlap is in the order of $10^{-8}$ by which convergence is deteriorated. From Figs. 6, 7 and 8 it is concluded that no overlapping deflation vectors give the best choice when contrasts are huge. Furthermore, average overlap gives the worst convergence behaviour. This is further illustrated by Fig. 9, where the exact error is plotted for the different options of deflation during the iterations. Computations with weighted overlapping projection vectors give the same results as for no-overlap in the case of sharp contrasts in the permeability, which follows from the definition of the several overlappings.

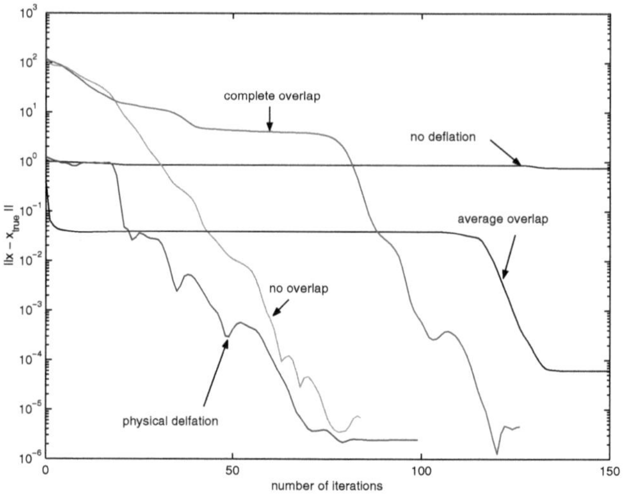


Figure 9. Convergence behavior of the deflated incomplete Cholesky preconditioned CG-method for 7 subdomains *with* large jumps of the coefficient $k$ in ($P_3$). The $\|\cdot\|_2$-norm of error is presented for the various choices of deflation. The number of elements is 3200 per layer.

Similar computations, without sharp contrasts, $\varepsilon = 1$, are presented in Fig. 10. This figure indicates that average overlap gives the best results, although the differences are not as striking as for the case where there are large contrasts. Note that weighted overlap and average overlap are equivalent here. This difference is small due to the absence of extremely small eigenvalues. Further, the computations with complete overlap gives the poorest convergence. From the figures and computations with weighted overlap, whose results are omitted in this paper, it is concluded that weighted overlap always gives a good convergence, since it mimics no overlap when constrasts are high and average overlap when constrasts do not exist. This is an important insight for future parallelization of the deflated preconditioned conjugate gradient method. We further note that computations with weighted overlap give the same results as for average overlapping for the case of no contrasts of the permeability.

We conclude that weighted overlap is most robust and this will always give the best choice for use in a 'blackbox' algorithm.

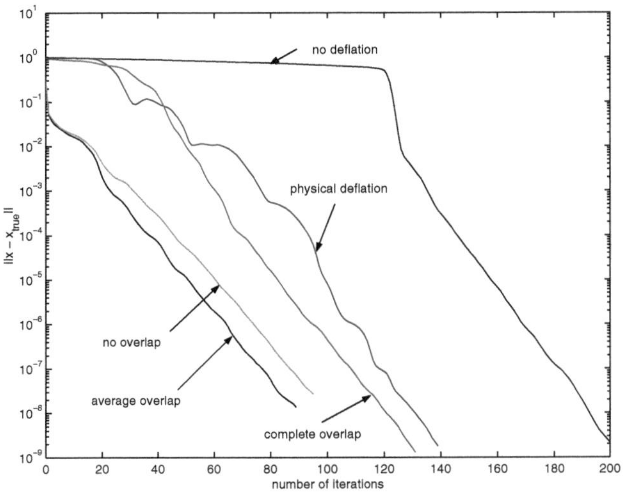


Figure 10. Convergence behavior of the deflated incomplete Cholesky preconditioned CG-method for 7 subdomains *without* large jumps of the coefficient $k$ in ($P_3$). The $\| \cdot \|_2$-norm of the error is presented for the various choices of deflation. The number of elements is 3200 per layer.

## 4.2 Parallelization

The deflated preconditioned conjugate gradient method is very suitable for parallelization. Parallelization is still a topic of research. The first results are given here, which have been computed for a layered domain as in Fig. 1. In the first series of numerical experiments, we show the convergence behaviour when the number of subdomains is increased, using a constant number of grid points, so the total number of gridnodes increases. Subsequently, we show the convergence behaviour for the case that the number of blocks increases such that the total number of gridnodes remains constant.

### 4.2.1 Increase of the size of the domain of computation

Here we consider a rectangular domain $\Omega^n$ that consists of the equisized subdomains $\Omega_1, \ldots, \Omega_n$, $\overline{\Omega}^n = \overline{\Omega}_1 \cup \ldots \cup \overline{\Omega}_n$, $\overline{\Omega}^{n+1} = \overline{\Omega}^n \cup \overline{\Omega}_{n+1}$. In this section we take the permeability constant over the whole domain. Further we use algebraic projection vectors with average overlap. We compare the results from the following computation methods:

1. Sequential ICCG without deflation (SICCG);
2. Sequential ICCG with deflation (SDICCG);
3. Parallel ICCG without deflation (PICCG);
4. Parallel ICCG with deflation (PDICCG).

The number of iterations needed for convergence is plotted as a function of the number of subdomains for all four approaches in Fig. 11. The number of grid points per subdomain is $80 \times 80$. It is seen that the parallel ICCG

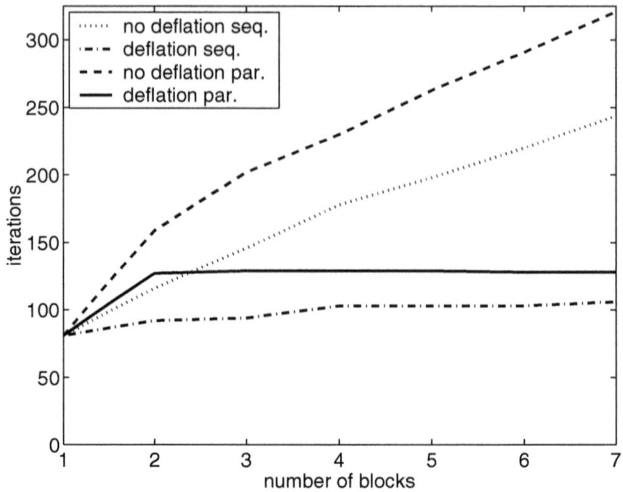


Figure 11. The number of iterations as the number of subdomains increases for the various methods of computation.

(PICCG) requires more iterations than the sequential ICCG (SICCG). This is a common observation. For both cases the number of iterations needed for convergence increases rapidly as the size of the domain of computation increases. However, for both deflated methods the number of iterations is lower than for the non-deflated methods. Further, as the number of subdomains, i.e. the size of the domain of computation, increases the number of needed iterations becomes independent of the number of subdomains.

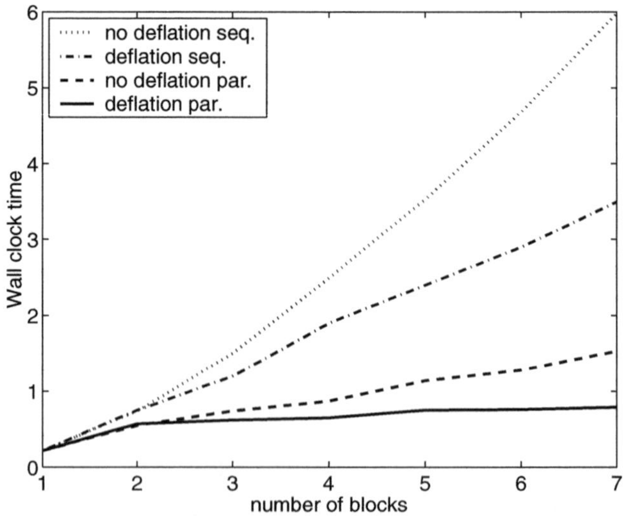


Figure 12. The wall-clock time as the number of subdomains increases for the various methods of computation.

As a further illustration we show the wall-clock time for the four different approaches in Fig. 12. The wall-clock time is measured on a Beowulf cluster. It can be seen from this figure that the wall-clock time is significantly smaller for the deflated ICCG (SDICCG) for the sequential computation. So, for sequential computations deflation is attractive to increase the speed of computation. Further, parallelization gives a significant speed-up, but as the number of subdomains increases, the wall-clock time continues to increase when no deflation is used (PICCG). However, if deflation is used in the parallel computations, the wall-clock time decreases and even becomes almost independent of the number of added subdomains. It is therefore concluded in this section that deflation accellerates computation in both a sequential and parallel computer environment, if the solution of an elliptic problem as in ($P_3$) is computed.

### 4.2.2 Increase of the number of deflation vectors in a given domain

Given a square domain, we increase the number of deflation vectors. Further, we assume that the permeability is constant over the whole domain. We use $150 \times 150$ elements over the domain.

We present the results of the computations in Table 2 where we use the methods SDICCG, DICCG and PDICCG (sequential with deflation, parallel without deflation and parallel with deflation) for the cases of 3 and 6 blocks. The number of iterations for the convergence of the SICCG method (sequential, without deflation) is 142 iterations.

Table 2. Number of iterations for a square domain with $150 \times 150$ elements.

| Number of blocks | SDICCG | PICCG | PDICCG |
|---|---|---|---|
| 3 | 106 | 198 | 141 |
| 6 | 101 | 198 | 138 |

We remark that we used average overlap for the projection vectors in the deflated method. From the above table it is seen that deflation gives a reduction of the number of required iterations for the sequential computations (compare column 2 with the 142 iterations for the SICCG method). Further, the number of iterations increases when a parallel method is applied. However, deflation applied in a parallelized method reduces the number of iterations again. Hence, deflation is again recommended to use in both sequential and parallelized computations.

# 5 Conclusions

We investigated various choices of deflation vectors, which are used in the Deflated ICCG method. It is found that the choice of the deflation vectors at the interfaces plays a crucial role in the convergence rate. Summarized, the following is concluded so far:

- As the domain is divided into more subdomains, the number of iterations needed for convergence decreases. This effect has been observed for $\sigma = 1$. Furthermore, this observation does not depend on the choice of the values of the deflation vectors at the boundaries. Further, deflation makes the parallel preconditioned conjugate gradient method scalable: the wall-clock time becomes invariant with respect to the number of blocks if the number of blocks is increased and the number of gridnodes per block is constant.
- For the case of no contrasts of the permeability between subsequent layers, it is observed that average overlap between subsequent deflation vectors is superior to no overlapping. Here the use of complete overlapping projection vectors is unsuitable. Whereas, for cases with large contrasts of permeability the use of average overlapping projection vectors is not suitable. This is caused by the fact that the span of the projection vectors approximates the span of the eigenvectors that belong to the small eigenvectors badly.
- We introduce the method of 'weighted overlap', which mimics average and no overlap for respectively the cases of no contrasts and very large contrasts of the permeability. It is observed that this choice gives the best convergence behavior for both the presence and absence of sharp contrasts until now.

## References

[1] J. Bear, *Dynamics of Fluids in Porous Meida*, Elsevier, New York, 1972.

[2] H. De Gersem, K. Hameyer, *A deflated iterative solver for magnetostatic finite element models with large differences in permeability*, Eur. Phys. J. Appl. Phys. **13** (2000), 45–49.

[3] G. De Josselin de Jong, *Singularity distributions for the analysis of multiple flyuid flow in porous media*, Journal of Geothermal Research **65** (1960), 3739–3758.

[4] J. Frank, C. Vuik, *On the construction of deflation-based preconditiones*, SIAM J. Sci. Comput. **23** (2001), 442–562.

[5] G. H. Golub, C. F. van Loan, *Matrix Computations*, Third edition, The Johns Hopkins University Press, Baltimore, 1996.

[6] C. B. Jenssen, P.Å. Weinerfelt, *Coarse grid correction scheme for implicit multiblock Euler calculations*, AIAA Journal **33** (1995), 1816–1821.

[7] E. F. Kaasschieter, *Preconditioned conjugate gradients for solving singular systems*, J. Comput. Appl. Math. **24** (1988), 265–275.

[8] E. Kreyszig, *Introductory Functional Analysis with Applications*, Willey, New York, 1989.

[9] L. W. Lake, *Enhanced Oil Recovery*, Pretience-Hall, Englewood Cliffs, 1989.

[10] D. C. Lay, *Linear Algebra and Its Applications*, Addison-Wesley, Longman Scientific, Reading, Massachusetts, 1996.

[11] A. Padiy, O. Axelsson, and B. Polman, *Generalized augmented matrix preconditioning approach and its application to iterqative solution of ill-conditioned algebraic systems*, SIAM J. Matrix Anal. Appl. **22** (2000), 793–818.

[12] E. Perchat, L. Fourment and T. Coupez, *Parallel incomplete factorisations for generalised Stokes problems: application to hot metal forging simulation*, Report, EPFL, Lausanne, 2001.

[13] G. J. M. Pieters, *Stability analysis for a saline layer formed by uniform using finite elements*, Report RANA 01-07, Eindhoven University of Technology, Eindhoven, 2001.

[14] R. Temam, *Navier-Stokes Equations, Theory and Numerical Analysis*, Elsevier Science Publishers, Amsterdam, 1984.

[15] A. van der Sluis, H. van der Vorst, *The rate of convergence of conjugate gradients*, Numer. Math. **48** (1986), 543–560.

[16] C. Vuik, A. Segal, L. el Yaakoubi and E. Dufour, *A comparison of various deflation vectors applied to elliptic problems with discontinuous coefficients*, Appl. Numer. Math. **41** (2002), 219–233.

[17] C. Vuik, A. Segal, and J. A. Meijerink, *An efficient preconditioned DG method for the solution of a class of layered problems with extreme contasts in the coefficients*, J. Comput. Phys. **152** (1999), 385–403.

[18] C. Vuik, A. Segal, J. A. Meijerink, and G. T. Wijma, *The construction of projection vectors for a Deflated ICCG method applied to problems with extreme contrasts in the coefficients*, J. Comput. Phys. **172** (2001), 426–450.

# Nonsmooth Equation Method for Nonlinear Nonconvex Optimization

Ladislav Lukšan,[1] Ctirad Matonoha,[2] Jan Vlček[3]

[1]Institute of Computer Science, Academy of Sciences of the Czech Republic, Pod vodárenskou věží 2, CZ–182 07 Praha 8, Czech Republic luksan@cs.cas.cz and Technical University of Liberec, Hálkova 6, CZ–461 17 Liberec, Czech Republic

[2]Institute of Computer Science, Academy of Sciences of the Czech Republic, Pod vodárenskou věží 2, CZ–182 07 Praha 8, Czech Republic luksan@cs.cas.cz and Technical University of Liberec, Hálkova 6, CZ–461 17 Liberec, matonoha@cs.cas.cz and Technical University of Liberec, Hálkova 6, CZ–461 17 Liberec, Czech Republic

[3]Institute of Computer Science, Academy of Sciences of the Czech Republic, Pod vodárenskou věží 2, CZ–182 07 Praha 8, Czech Republic luksan@cs.cas.cz and Technical University of Liberec, Hálkova 6, CZ–461 17 Liberec, vlcek@cs.cas.cz and Technical University of Liberec, Hálkova 6, CZ–461 17 Liberec, Czech Republic

## 1 Introduction

In this contribution, we are concerned with a general nonlinear programming problem:

(NP): Find the minimum of function $f(x)$ on the set given by constraints $c_I(x) \le 0$ and $c_E(x) = 0$, where $f\colon R^n \to R$, $c_I\colon R^n \to R^{m_I}$ and $c_E\colon R^n \to R^{m_E}$ are twice continuously differentiable mappings ($c_I \le 0$ is considered by elements), $I = \{1, \ldots, m_I\}$, and $E = \{m_I + 1, \ldots, m_I + m_E\}$.

Necessary conditions (the KKT-conditions) for the solution of problem (NP) (if gradients of active constraints are linearly independent) have the following form

$$\begin{aligned} g(x,u) &= 0, \\ c_I(x) &\le 0, \quad u_I \ge 0, \quad U_I C_I(x) e = 0, \\ c_E(x) &= 0, \end{aligned} \tag{1.1}$$

where

$$\begin{aligned} g(x,u) &= \nabla f(x) + \sum_{i \in I} u_i \nabla c_i(x) + \sum_{i \in E} u_i \nabla c_i(x) \\ &= \nabla f(x) + A_I(x) u_I + A_E(x) u_E \end{aligned}$$

and $A_I(x) = [\nabla c_i(x): i \in I]$, $A_E(x) = [\nabla c_i(x): i \in E]$. Here $u_I \in R^{m_I}$, $u_E \in R^{m_E}$ are vectors of Lagrange multipliers, $U_I = \mathrm{diag}(u_i: i \in I)$, $C_I(x) = \mathrm{diag}(c_i(x): i \in I)$, and $e$ is the vector with unit elements.

The second set of conditions in (1.1), the so-called complementarity conditions, can be transformed to equations using the Fischer-Burmeister function

$$\psi(a,b) = \sqrt{a^2+b^2} - (a+b),$$

which is zero if and only if $a \geq 0$, $b \geq 0$ and $ab = 0$. Therefore, complementarity conditions are satisfied if and only if $\psi(u_i, -c_i(x)) = 0$, $i \in I$, and (1.1) can be replaced by the nonlinear system

$$\begin{aligned} g(x,u) &= 0, \\ \psi_I(u_I, -c_I(x)) &= 0, \\ c_E(x) &= 0, \end{aligned} \tag{1.2}$$

where $\psi_I(u_I, -c_I(x)) = [\psi(u_i, -c_i(x)): i \in I]^T$.

The Fischer-Burmeister function $\psi(a,b)$ is continuously differentiable at a point where $|a| + |b| \neq 0$ and is semismooth if $|a| + |b| = 0$. Moreover, the function $\psi^2(a,b)$ is continuously differentiable everywhere [2]. Therefore, (1.2) is a system of semismooth equations, which can be solved by the Newton method [10], [2], [8]. The gradient and the Clarke subdifferential of the Fischer-Burmeister function are given by formulas

$$\nabla\psi(a,b) = \begin{bmatrix} \dfrac{a}{\sqrt{a^2+b^2}} - 1 \\ \dfrac{b}{\sqrt{a^2+b^2}} - 1 \end{bmatrix}, \quad |a| + |b| \neq 0, \tag{1.3}$$

$$\partial\psi(0,0) = \mathrm{conv} \bigcup_{\phi\in[0,2\pi]} \begin{bmatrix} \cos\phi - 1 \\ \sin\phi - 1 \end{bmatrix}. \tag{1.4}$$

Formula (1.4) implies that $[-1,-1]^T \in \partial\psi(0,0)$. Therefore, setting $r(a,b) = \sqrt{a^2+b^2}$ for $|a| + |b| \neq 0$ and $r(a,b) = 1$ for $|a| + |b| = 0$, we obtain

$$\begin{bmatrix} \dfrac{a}{r(a,b)} - 1 \\ \dfrac{b}{r(a,b)} - 1 \end{bmatrix} \in \partial\psi(a,b). \tag{1.5}$$

Linearizing system (1.2) by using (1.5), we obtain a step of the Newton method

$$\begin{bmatrix} G & A_I & A_E \\ (R_I + C_I)R_I^{-1}A_I^T & -(R_I - U_I)R_I^{-1} & 0 \\ A_E^T & 0 & 0 \end{bmatrix} \begin{bmatrix} \Delta x \\ \Delta u_I \\ \Delta u_E \end{bmatrix} = - \begin{bmatrix} g \\ \psi_I \\ c_E \end{bmatrix}, \tag{1.6}$$

where

$$G = G(x,u) = \nabla^2 f(x) + \sum_{i \in I} u_i \nabla^2 c_i(x) + \sum_{i \in E} u_i \nabla^2 c_i(x),$$

$A_I = A_I(x)$, $A_E = A_E(x)$, $C_I = C_I(x)$, $U_I = U_I(x)$, $R_I = \mathrm{diag}(r(u_i, -c_i(x)) : i \in I)$, $g = g(x,u)$, $\psi_I = \psi_I(u_I, -c_I(x))$, and $c_E = c_E(x)$. Note that linear system (1.6) is not suitable for iterative solvers in general since it is nonsymmetric and can have unsuitable diagonal elements. Therefore, an equivalent system is constructed and solved as is shown in Section 2.

The algorithm of a nonsmooth equation method can be roughly described in the following form. For given vectors $x \in R^n$, $u_I \in R^{m_I}$, $u_E \in R^{m_E}$ we determine direction vectors $\Delta x$, $\Delta u_I$, $\Delta u_E$ by solving linear system equivalent to (1.6). Furthermore, we choose step-length $\alpha > 0$ and set $x := x + \alpha \Delta x$, $u_I := u_I + \alpha \Delta u_I$, $u_E := u_E + \alpha \Delta u_E$. Individual parts of the algorithm require more detailed explanation.

The use of the Fischer-Burmeister function for solving dense medium-size nonlinear programming problems is studied theoretically in [3] and [4]. Unfortunately, no computational experiments are presented. The aim of this paper is an application of the nonsmooth equation approach to the solution of large sparse problems and, especially, the use of the indefinitely preconditioned conjugate gradient method in this case.

## 2 Direction determination

System (1.6) is nonsymmetric with the dimension $n + m_I + m_E$. This system can be symmetrized by the multiplication of a diagonal matrix and reduced by the elimination of inactive constraints. In the subsequent text, we use the simplified notation $c_i = c_i(x)$, $r_i = r(u_i, -c_i(x))$, $\psi_i = \psi(u_i, -c_i(x)) = r_i - (u_i - c_i)$ so $\partial \psi_i / \partial u_i = (u_i - r_i)/r_i$, $\partial \psi_i / \partial c_i = (c_i + r_i)/r_i$.

**Definition 1.** Let $0 < \varepsilon_0 < 1/2$. We say that the constraint with index $i \in I$ is active, if

$$-\partial \psi_i / \partial u_i \le \varepsilon_0 \partial \psi_i / \partial c_i, \tag{2.1}$$

which can be written in the form

$$r_i - u_i \le \varepsilon_0 (r_i + c_i). \tag{2.2}$$

**Lemma 1.** *Let the constraint with index $i \in I$ be active. Then $u_i > 0$ and $(4/5)\varepsilon_0 < (r_i + c_i)/r_i \le 2$. Furthermore, there exist positive numbers $\varepsilon_1$ and $\varepsilon_2$ such that the constraint with index $i \in I$ is active if $u_i > 0$ and $|c_i| \le \varepsilon_1 u_i$ and inactive if $u_i \le 0$ or $|c_i| > \varepsilon_2 u_i$.*

Proof. (a) Obviously, $0 \le (r_i - u_i)/r_i \le 2$ and $0 \le (r_i + c_i)/r_i \le 2$. Assume that $u_i \le 0$. Then $(r_i - u_i)/r_i \ge 1$ which together with the previous inequalities implies $(r_i - u_i)/(r_i + c_i) = ((r_i - u_i)/r_i)(r_i/(r_i + c_i)) \geq 1/2$, which is in contradiction with (2.2), since $\varepsilon_0 < 1/2$. Using (1.5), we obtain

$$(1 - \partial\psi_i/\partial c_i)^2 + (1 + \partial\psi_i/\partial u_i)^2 = \frac{c_i^2 + u_i^2}{r_i^2} = 1 \tag{2.3}$$

so $\partial\psi_i/\partial c_i$ and $-\partial\psi_i/\partial u_i$ lie on the unit circle centred at point $[1, 1]$. If the constraint with index $i \in I$ is active, then (2.2) implies that $-\partial\psi_i/\partial u_i$ lies under the line defined by (2.1) (with equality), see Fig. 1.

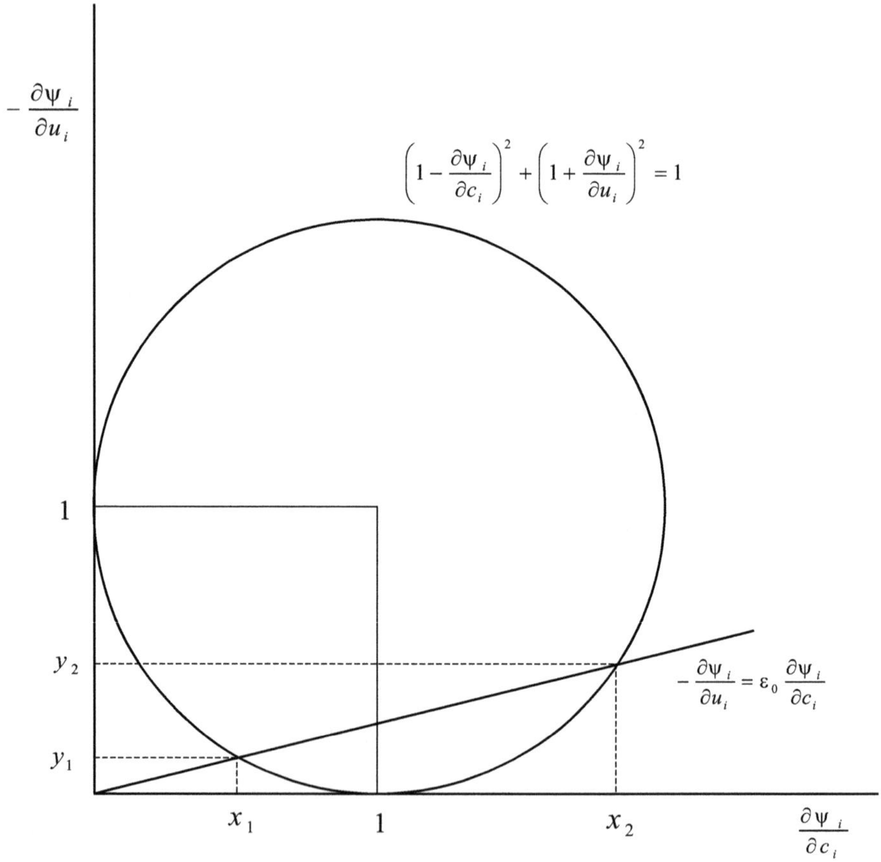


Figure 1

Denote by $(x_1, y_1)$ and $(x_2, y_2)$ intersections of the circle (2.3) and the line defined by (2.1) (with $x_1 \leq x_2$). Then $y = \varepsilon_0 x$ and

$$(1 + \varepsilon_0^2)x^2 - 2(1 + \varepsilon_0)x + 1 = 0.$$

If the constraint with index $i \in I$ is active, then

$$\frac{\partial \psi_i}{\partial c_i} \geq x_1 = \frac{1 + \varepsilon_0 - \sqrt{2\varepsilon_0}}{1 + \varepsilon_0^2} > \frac{4}{5}\varepsilon_0$$

(we used the inequality $\varepsilon_0 < 1/2$).

(b) Obviously, $y_1 = \varepsilon_0 x_1 > 0$. Furthermore, (2.3) implies $x_2 \leq 2$, which gives $y_2 = \varepsilon_0 x_2 < 1$. If $-\partial\psi_i/\partial u_i \leq y_1$, then (2.1) holds and the constraint with index $i \in I$ is active and if $-\partial\psi_i/\partial u_i > y_2$, then (2.1) cannot hold and the constraint with index $i \in I$ is inactive. We have proved above that the constraint with index $i \in I$ is inactive when $u_i \leq 0$. Assuming $u_i > 0$, we can write

$$-\frac{\partial \psi_i}{\partial u_i} = 1 - \frac{u_i}{\sqrt{u_i^2 + c_i^2}} = 1 - \frac{1}{\sqrt{1 + \lambda_i^2}}, \tag{2.4}$$

where $\lambda_i = |c_i|/u_i$, $i \in I$. Since the function on the right-hand side of (2.4) is increasing (with the value 0 for $\lambda_i = 0$ and the limit 1 for $\lambda_i \to \infty$), there exist positive numbers $\varepsilon_1$ and $\varepsilon_2$ such that $-\partial\psi_i/\partial u_i \leq y_1$ if $|c_i|/u_i \leq \varepsilon_1$ and $-\partial\psi_i/\partial u_i > y_2$ if $|c_i|/u_i > \varepsilon_2$. □

**Theorem 1.** *Let the pair $x^* \in R^n$ and $u_I^* \in R^{m_I}$ solve (1.1) and the strict complementarity conditions $u_i^* + |c_i(x^*)| > 0$ for all $i \in I$ hold. Then there exist neighbourhoods $\mathcal{N} \subset R^n$ and $\mathcal{M}_I \subset R^{m_I}$ of $x^*$ and $u_I^*$ such that the constraint with index $i \in I$ is active when $c_i(x^*) = 0$ and inactive when $u_i^* = 0$ whenever $x \in \mathcal{N}$ and $u_I \in \mathcal{M}_I$. Moreover, if $x \to x^*$ and $u_I \to u_I^*$, then $(r_i - u_i)/(r_i + c_i) \to 0$ for all active constraints and $(r_i - u_i)/(r_i + c_i) \to \infty$ for all inactive constraints.*

Proof. Let $u_i^* > 0$. Since $c_i(x^*) = 0$ and function $c_i(x)$ is continuous, there exist neighbourhoods $\mathcal{N}_i \subset R^n$ of $x^*$ and $\mathcal{M}_i \subset R$ of $u_i^*$ such that $u_i > u_i^*/2$ if $u_i \in \mathcal{M}_i$ and $|c_i(x)| < \varepsilon_1 u_i^*/2$ if $x \in \mathcal{N}_i$, so the constraint with index $i \in I$ is active by Lemma 1. Let $u_i^* = 0$. Since $c_i(x^*) < 0$ and the function $c_i(x)$ is continuous, there exist neighbourhoods $\mathcal{N}_i \subset R^n$ of $x^*$ and $\mathcal{M}_i \subset R$ of $u_i^*$ such that $|c_i(x)| > |c_i(x^*)|/2$ if $x \in \mathcal{N}_i$ and $u_i < |c_i(x^*)|/(2\varepsilon_2)$ if $u_i \in \mathcal{M}_i$ so the constraint with index $i \in I$ is inactive by Lemma 1. If we set $\mathcal{N} = \bigcap_{i\in I} \mathcal{N}_i$ and put $\mathcal{M}_I$ equal to the Cartesian product of $\mathcal{M}_i$, $i \in I$, then the constraint with index $i \in I$ is active if $c_i(x^*) = 0$ and inactive if $u_i^* = 0$ whenever $x \in \mathcal{N}$ and $u_I \in \mathcal{M}_I$. Moreover, if $x \to x^*$ and $u_I \to u_I^*$, then $u_i \to u_i^* > 0$ and $c_i \to 0$ for all active constraints, implying $r_i \to u_i^*$ and $(r_i - u_i)/(r_i + c_i) \to 0$. Similarly, $u_i \to 0$ and $c_i \to c_i(x^*) < 0$ for all inactive constraints, implying $r_i \to |c_i(x^*)|$ and $(r_i - u_i)/(r_i + c_i) \to \infty$. □

Let us denote active quantities by $\hat{u}_i$, $\hat{c}_i$, $\hat{r}_i$, $\hat{\psi}_i$ and inactive quantities by $\check{u}_i$, $\check{c}_i$, $\check{r}_i$, $\check{\psi}_i$. Eliminating equations corresponding to inactive quantities from (1.6), we obtain

$$\Delta\check{u}_I = \check{M}_I^{-1}[(\check{R}_I + \check{C}_I)^{-1}\check{R}_I\check{\psi}_I + \check{A}_I^T\Delta x], \tag{2.5}$$

where $\check{M}_I = (\check{R}_I + \check{C}_I)^{-1}(\check{R}_I - \check{U}_I)$ and

$$\begin{bmatrix} \hat{G} & \hat{A}_I & A_E \\ \hat{A}_I^T & -\hat{M}_I & 0 \\ A_E^T & 0 & 0 \end{bmatrix} \begin{bmatrix} \Delta x \\ \Delta\hat{u}_I \\ \Delta u_E \end{bmatrix} = - \begin{bmatrix} \hat{g} \\ (\hat{R}_I + \hat{C}_I)^{-1}\hat{R}_I\hat{\psi}_I \\ c_E \end{bmatrix}, \tag{2.6}$$

where $\hat{M}_I = (\hat{R}_I + \hat{C}_I)^{-1}(\hat{R}_I - \hat{U}_I)$ and

$$\hat{G} = G + \check{A}_I\check{M}_I^{-1}\check{A}_I^T,$$
$$\hat{g} = g + \check{A}_I\check{M}_I^{-1}(\check{R}_I + \check{C}_I)^{-1}\check{R}_I\check{\psi}_I.$$

Condition (2.2) guarantees that the matrices $\hat{M}_I$ and $\check{M}_I^{-1}$ are bounded. Therefore, also $\hat{G}$ is bounded (if $G$ and $A$ are bounded). Moreover, Theorem 1 insures that $\hat{M}_I \to 0$ if $x \to x^*$, $u_I \to u_I^*$ and the strict complementarity conditions hold at the solution point. The property $\hat{M}_I \to 0$ is very useful, since system (2.6) can be efficiently preconditioned by the way described in [5] if $\hat{M}_I$ has small elements. The second set of equations of (2.6) was obtained after multiplying corresponding set of equations of (1.6) by the diagonal matrix $(\hat{R}_I + \hat{C}_I)^{-1}\hat{R}_I$ whose elements lie in the bounded interval $[1/2, 5/(4\varepsilon_0)]$ by Lemma 1. If $\varepsilon_0$ has a reasonable value (usually $\varepsilon_0 = 1/10$), then this scaling does not influence the final inexact solution of (2.6).

Vector $\Delta\hat{u}_I$ is determined by solving system (2.6) and vector $\Delta\check{u}_I$ is computed from (2.5). System (2.6), which is symmetric and indefinite of order $n + \hat{m}_I + m_E$, can be solved either directly by using the sparse Bunch-Parlett decomposition or iteratively by using preconditioned Krylov-subspace methods for symmetric indefinite systems. Motivated by [5] and [7], we will investigate the preconditioner

$$C = \begin{bmatrix} \hat{D} & \hat{A}_I & A_E \\ \hat{A}_I^T & -\hat{M}_I & 0 \\ A_E^T & 0 & 0 \end{bmatrix}, \tag{2.7}$$

where $\hat{D}$ is a positive definite diagonal matrix derived from a diagonal of $\hat{G}$. Let us denote the matrix of system (2.6) by $K$. Then the following theorems are proved in [9].

**Theorem 2.** *Consider preconditioner* (2.7) *applied to system* (2.6). *Then matrix* $KC^{-1}$ *has at least* $\hat{m}_I + 2m_E$ *unit eigenvalues but, if* $\hat{G} - \hat{D}$ *is nonsingular (the usual case), then at most* $\hat{m}_I + m_E$ *linearly independent eigenvectors corresponding to these eigenvalues exist. The other eigenvalues of the matrix* $KC^{-1}$ *are exactly eigenvalues of matrix* $Z_E^T\tilde{G}Z_E(Z_E^T\tilde{D}Z_E)^{-1}$, *where* $[Z_E, A_E]$ *is a nonsingular square matrix,* $Z_E^T A_E = 0$, $Z_E^T Z_E = I$ *and where* $\tilde{G} = \hat{G} + \hat{A}_I\hat{M}_I^{-1}\hat{A}_I^T$, $\tilde{D} = \hat{D} + \hat{A}_I\hat{M}_I^{-1}\hat{A}_I^T$. *If* $Z_E^T\tilde{G}Z_E$ *is positive definite then all eigenvalues are positive.*

**Theorem 3.** *Consider preconditioner* (2.7) *applied to system* (2.6) *and assume that $\hat{G} - \hat{D}$ is nonsingular. Then Krylov subspace*

$$\mathcal{K} = \text{span}\{\bar{r}, KC^{-1}\bar{r}, (KC^{-1})^2\bar{r}, \dots\},$$

*where $\bar{r} \in R^{n+\hat{m}_I+m_E}$, has a dimension of at most* $\min(n+1, n-m_E+2)$.

**Theorem 4.** *Consider conjugate gradient method with preconditioner* (2.7) *applied to system* (2.6). *Assume that initial estimation of the solution is chosen in such a way that $\hat{h}_I = 0$ and $h_E = 0$ at the start of the algorithm ($h_x$, $\hat{h}_I$ and $h_E$ are parts of the residual vector). Let matrix $Z_E^T\tilde{G}Z_E$ be positive definite. Then:*

(a) *Vector $\Delta^* x$ (the first part of the solution) is found after $n - m_E$ iterations at most.*
(b) *The algorithm cannot break down before $\Delta^* x$ is found.*
(c) *Error $\|\Delta x - \Delta^* x\|$ converges to zero at least R-linearly with quotient*

$$\frac{\sqrt{\kappa}-1}{\sqrt{\kappa}+1},$$

*where $\kappa$ is the spectral condition number of matrix $Z_E^T\tilde{G}Z_E(Z_E^T\tilde{D}Z_E)^{-1}$.*
(d) *If $\Delta x = \Delta^* x$, then also $\Delta\hat{u}_I = \Delta^*\hat{u}_I$ and $\Delta^* u_E$ can be determined by the formula*

$$\Delta^* u_E = \Delta u_E + (A_E^T\tilde{D}^{-1}A_E)^{-1}A_E^T\tilde{D}^{-1}h_x.$$

Theorem 2 shows that the efficiency of a Krylov-subspace method preconditioned by (2.7) does not depend strongly on the choice of active and inactive variables (matrix $\tilde{G}$ is the same for every splitting). Nevertheless, the splitting significantly influences the resulting solution since the equations corresponding to inactive variables are solved precisely and only active variables are obtained by the inexact iterative process.

Although the Fischer-Burmeister function has good theoretical properties, we finally used the modified Fischer-Burmeister function in our algorithm, since it was more efficient in computational experiments. The modified Fischer-Burmeister function $\psi^\varepsilon(a,b) = \sqrt{a^2+b^2+\varepsilon^2} - (a+b)$, where $\varepsilon$ is a sufficiently small positive number, is differentiable everywhere and $r = \sqrt{a^2+b^2+\varepsilon^2}$ is continuous at the origin. The following theorem characterizes inaccuracy, which appears in the complementarity conditions if the modified Fischer-Burmeister function is used.

**Theorem 5.** *Let $\psi^\varepsilon(a,b) = \delta$ and $|a| \le |b|$. Then*

$$|a| \le 2|\delta| + \sqrt{|\varepsilon^2 - \delta^2|/2},$$
$$|a||b| \le \frac{1}{2}(\varepsilon + 3|\delta|)\max(|\varepsilon - \delta|, |a| + |b|).$$

Proof. Let $\psi^\varepsilon(a,b) = \delta$. Then

$$\sqrt{a^2+b^2+\varepsilon^2} - a - b = \delta$$

so

$$a^2+b^2+\varepsilon^2 = (a+b+\delta)^2 = a^2+b^2+\delta^2+2ab+2(a+b)\delta$$

or

$$2|a||b| \le |\varepsilon^2-\delta^2| + 2(|a|+|b|)|\delta|. \tag{2.8}$$

This implies

$$2|a| \le \frac{|\varepsilon^2-\delta^2|}{|b|} + 2\left(\frac{|a|}{|b|}+1\right)|\delta| \le \frac{|\varepsilon^2-\delta^2|}{|a|} + 4|\delta|,$$

so $|a|$ lies between roots of the quadratic equation $2|a|^2 - 4|a||\delta| - |\varepsilon^2-\delta^2| = 0$ which have opposite signs. Therefore,

$$\begin{aligned} |a| &\le |\delta| + \frac{1}{2}\sqrt{4\delta^2 + 2|\varepsilon^2-\delta^2|} \\ &\le |\delta| + \frac{1}{2}(2|\delta| + \sqrt{2|\varepsilon^2-\delta^2|}) = 2|\delta| + \sqrt{|\varepsilon^2-\delta^2|/2}. \end{aligned}$$

Using (2.8) again, we obtain

$$2|a||b| \le |\varepsilon-\delta|(|\varepsilon|+|\delta|) + 2(|a|+|b|)|\delta|.$$

Substituting $\max(|\varepsilon-\delta|, |a|+|b|)$ for both $|\varepsilon-\delta|$ and $|a|+|b|$, we obtain the second assertion of the theorem. □

There are two important cases. First, if $\psi^\varepsilon(a,b) = 0$, then $\delta = 0$ and, therefore, $|a| \le \varepsilon/\sqrt{2}$, $|a||b| \le (\varepsilon/2)\max(\varepsilon, |a|+|b|)$. On the other hand, if $|\delta| = \varepsilon$ then $|a| \le 2\varepsilon$, $|a||b| \le 2\varepsilon(|a|+|b|)$.

The use of the modified Fischer-Burmeister function is advantageous, since the case when $|a|+|b| = 0$ is not exceptional. Moreover, the use of the basic Fischer-Burmeister function can lead to difficulties caused by computations in the finite precision arithmetic. We have tested both these possibilities and obtained better results with the modified Fischer-Burmeister function. Note that all equations used in this section hold for the modified Fischer-Burmeister function if we set $r_i = \sqrt{c_i^2+u_i^2+\varepsilon^2}$.

## 3 Step-length selection

Step-length $\alpha > 0$ can be determined in many ways. Basic requirement for the selection of step-length is the satisfying of a suitable goal criterion. This criterion is usually a merit function which is a combination of the objective function and a measure of constraint violation. Motivated by [5] and [7], we use the following merit function

$$\begin{aligned} P(\alpha) = {} & f(x+\alpha\Delta x) + (u_I + \Delta u_I)^T c_I(x+\alpha\Delta x) \\ & + (u_E + \Delta u_E)^T c_E(x+\alpha\Delta x) \\ & + \frac{\sigma}{2}\|\psi_I(u_I + \alpha\Delta u_I, -c_I(x+\alpha\Delta x))\|^2 \\ & + \frac{\sigma}{2}\|c_E(x+\alpha\Delta x)\|^2, \end{aligned} \tag{3.1}$$

where $\sigma \geq 0$, which is a generalization of the merit function introduced in [5]. The following theorem holds.

**Theorem 6.** *Let $\Delta x$, $\hat{\Delta} u_I$, $\Delta u_E$ be obtained by an inexact solution of* (2.6) *such that*

$$\begin{bmatrix} \hat{G} & \hat{A}_I & A_E \\ \hat{A}_I^T & -\hat{M}_I & 0 \\ A_E^T & 0 & 0 \end{bmatrix} \begin{bmatrix} \Delta x \\ \Delta \hat{u}_I \\ \Delta u_E \end{bmatrix} + \begin{bmatrix} \hat{g} \\ (\hat{R}_I + \hat{C}_I)^{-1}\hat{R}_I\hat{\psi}_I \\ c_E \end{bmatrix} = \begin{bmatrix} h_x \\ \hat{h}_I \\ h_E \end{bmatrix},$$

*where $h_x$, $\hat{h}_I$, $h_E$ are parts of the residual vector and let $\check{\Delta} u_I$ is given by* (2.5). *Then*

$$\begin{aligned} P'(0) = {} & -(\Delta x)^T G\Delta x - \sigma(\|\psi_I\|^2 + \|c_E\|^2) \\ & + (\Delta x)^T h_x + \sigma(\hat{\psi}_I^T(\hat{R}_I + \hat{C}_I)\hat{R}_I^{-1}\hat{h}_I + c_E^T h_E). \end{aligned} \tag{3.2}$$

*If*

$$\sigma > -\frac{(\Delta x)^T G\Delta x}{\|\psi_I\|^2 + \|c_E\|^2}, \tag{3.3}$$

*and if* (2.6) *is solved with a sufficient precision, namely, if*

$$(\Delta x)^T h_x + \sigma(\hat{\psi}_I^T(\hat{R}_I + \hat{C}_I)\hat{R}_I^{-1}\hat{h}_I + c_E^T h_E) < (\Delta x)^T G\Delta x + \sigma(\|\psi_I\|^2 + \|c_E\|^2), \tag{3.4}$$

*then $P'(0) < 0$.*

Proof. Differentiating (3.1) by $\alpha$, we find that

$$\begin{aligned} P'(0) = {} & (\Delta x)^T(\nabla f + A_I(u_I + \Delta u_I) + A_E(u_E + \Delta u_E)) \\ & + \sigma\psi_I^T((R_I + C_I)R_I^{-1}A_I^T\Delta x - (R_I - U_I)R_I^{-1}\Delta u_I) \\ & + \sigma c_E^T A_E^T \Delta x. \end{aligned} \tag{3.5}$$

Using (2.5) and (2.6), we obtain equalities

$$\begin{aligned} & (\Delta x)^T(\nabla f + A_I(u_I + \Delta u_I) + A_E(u_E + \Delta u_E)) \\ & \qquad = -(\Delta x)^T G\Delta x + (\Delta x)^T h_x, \\ & \psi_I^T((R_I + C_I)R_I^{-1}A_I^T\Delta x - (R_I - U_I)R_I^{-1}\Delta u_I) \\ & \qquad = -\|\psi_I\|^2 + \hat{\psi}_I^T(\hat{R}_I + \hat{C}_I)\hat{R}_I^{-1}\hat{h}_I, \\ & c_E^T A_E^T \Delta x = -\|c_E\|^2 + c_E^T h_E, \end{aligned}$$

which after substituting into (3.5) give (3.2). If (3.3) holds, then the right-hand side of (3.4) is positive so if (2.6) is solved with a sufficient precision, then (3.4) holds and $P'(0) < 0$ by (3.2). □

Condition (3.3) restricts the choice of parameter $\sigma$ weakly. If the matrix $G$ is positive semidefinite, any value $\sigma \geq 0$ satisfies this condition. Inequality (3.4) gives one possibility for the computation of parameter $\sigma$, which implies the inequality $P'(0) < 0$. But it is usually more efficient for practical computation to choose parameter $\sigma$ as a constant and replace the matrix $G$ by a positive definite diagonal matrix if condition $P'(0) < 0$ does not hold.

To improve local convergence properties of the method, we can use another merit function

$$\begin{aligned} Q(\alpha) = & \frac{1}{2}\|g(x+\alpha\Delta x, u_I+\alpha\Delta u_I, u_E+\alpha\Delta u_E)\|^2 \\ & + \frac{1}{2}\|\psi_I(u_I+\alpha\Delta u_I, -c_I(x+\alpha\Delta x))\|^2 \\ & + \frac{1}{2}\|c_E(x+\alpha\Delta x)\|^2. \end{aligned} \tag{3.6}$$

This function is a natural criterion for nonlinear equation solvers. Since the square of the Fischer-Burmeister function is continuously differentiable, the same property has the function $Q(\alpha)$. The following theorem characterizes properties of this function.

**Theorem 7.** *Let the assumptions of Theorem 6 be satisfied. Then*

$$Q'(0) = -\|g\|^2 - \|\psi_I\|^2 - \|c_E\|^2 + g^T h_x + \hat{\psi}_I^T(\hat{R}_I + \hat{C}_I)\hat{R}_I^{-1}\hat{h}_I + c_E^T h_E. \tag{3.7}$$

*If* (2.6) *is solved with a sufficient precision, namely if*

$$g^T h_x + \hat{\psi}_I^T(\hat{R}_I + \hat{C}_I)\hat{R}_I^{-1}\hat{h}_I + c_E^T h_E < \|g\|^2 + \|\psi_I\|^2 + \|c_E\|^2, \tag{3.8}$$

*then* $Q'(0) < 0$.

Proof. Differentiating (3.6) by $\alpha$, we obtain

$$\begin{aligned} Q'(0) = & g^T(G\Delta x + A_I\Delta u_I + A_E\Delta u_E) \\ & + \psi_I^T((R_I + C_I)R_I^{-1}A_I^T\Delta x - (R_I - U_I)R_I^{-1}\Delta u_I) + c_E^T A_E^T \Delta x. \end{aligned} \tag{3.9}$$

Using (2.5) and (2.6), we obtain equalities

$$\begin{aligned} g^T(G\Delta x + A_I\Delta u_I + A_E\Delta u_E) &= -g^T g + g^T h_x, \\ \psi_I^T((R_I + C_I)R_I^{-1}A_I^T\Delta x - (R_I - U_I)R_I^{-1}\Delta u_I) & \\ = -\|\psi_I\|^2 + \hat{\psi}_I^T(\hat{R}_I + \hat{C}_I)\hat{R}_I^{-1}\hat{h}_I&, \\ c_E^T A_E^T \Delta x = -\|c_E\|^2 + c_E^T h_E&, \end{aligned}$$

which after substituting into (3.9) give (3.7). The right-hand side of (3.8) is always positive so if (2.6) is solved with a sufficient precision, then (3.8) holds and $Q'(0) < 0$ by (3.7). □

Both functions $P(\alpha)$ and $Q(\alpha)$ have some advantages and disadvantages. Function $P(\alpha)$ is more closely related to the optimization problem. Using restarts, this function decreases the probability to find the saddle point of the objective function on the feasible set. Unfortunately, function $P(\alpha)$ is not sufficient to prove the global convergence. On the other hand, function $Q(\alpha)$ can be used in the globally and superlinearly convergent Newton method [2], [8]. Unfortunately, this method converges to a KKT point satisfying (1.1), which may not be the required minimum if the second order sufficient conditions do not hold. We have implemented and tested four different strategies for step-length selection. All of these strategies use function $P'(0)$ for restarts. Using a sufficiently small value of $\sigma$, we compute $P'(0)$ and if $P'(0) \geq 0$, then we replace matrix $G$ by some positive definite diagonal matrix and repeat the direction determination. In the first strategy, using the Armijo procedure with initial estimate $\alpha = \min(1, \overline{\Delta}/\|\Delta x\|)$, we accept the first value for which $P(\alpha) < P(0)$ (since function $Q(\alpha)$ is not used, the first derivatives need not be computed during the step-length selection). In the second strategy, the value $\alpha$ is accepted if at least one of the conditions $P(\alpha) < P(0)$ and $Q(\alpha) < Q(0)$ is satisfied. In the third strategy, the condition $P(\alpha) < P(0)$ is used after restart and the condition $Q(\alpha) < Q(0)$ is used otherwise. In the fourth strategy no condition for step-length selection is used, i.e., the initial estimate $\alpha = \min(1, \overline{\Delta}/\|\Delta x\|)$ is always accepted.

## 4 Description of the algorithm

The above considerations can be summarized in the algorithmic form.

**Algorithm 1.**

**Data:** Parameter for the active constraint definition $\varepsilon_0$. Minimum precision for the direction determination $0 < \overline{\omega} < 1$. Line-search parameter $0 < \beta < 1$. Step bound $\overline{\Delta} > 0$.

**Input:** Sparsity pattern of matrices $\nabla^2 F$ and $A$. Initial choice of the vector $x$.

**Step 1:** *Initiation.* Choose $\sigma > 0$ (e.g. $\sigma = 1$). For $i \in I \cup E$ set $u_i := 0$. Compute the value $f(x)$ and vectors $c_I(x)$, $c_E(x)$. Set $k := 0$.

**Step 2:** *Termination.* Compute the matrix $A := A(x)$ and vector $g := g(x, u)$. If conditions (1.1) are satisfied with a sufficient precision, then terminate the computation. Otherwise set $k := k + 1$.

**Step 3:** *Approximation of the Hessian matrix.* Compute approximation $G$ of the Hessian matrix $G(x, u)$ by using the differences of gradient $g(x, u)$ as in [1].

**Step 4:** *Direction determination.* Split constraints into active and inactive and build linear system (2.6). Determine positive definite diagonal matrix $\hat{D}$ as an approximation of the diagonal of $\hat{G}$ and determine a representation of preconditioner (2.7) (Bunch-Parlett decomposition or Schur complement based representation, see [7]). Writing system (2.6) in the form $Kd = b$, set $\omega = \min(\|b\|, 1/k, \overline{\omega})$ and determine direction vectors $\Delta x$, $\Delta \hat{u}_I$ and $\Delta u_E$ as an inexact solution of (2.6) (with the precision $\|Kd - b\| \le \omega \|b\|$) by the preconditioned Krylov-subspace method. Compute vector $\Delta \check{u}_I$ by (2.5). Compute the directional derivative $P'(0)$ of the merit function $P(\alpha)$ by (3.5).

**Step 5:** *Restart.* If $P'(0) \ge 0$, determine a positive definite diagonal matrix $D$ by the procedure given in [7], set $G = D$ and go to Step 4.

**Step 6:** *Step-length selection.* Find the minimum integer $l \ge 0$ using one of the following strategies. Strategy 1: $P(\beta^l) < P(0)$. Strategy 2: $P(\beta^l) < P(0)$ or $Q(\beta^l) < Q(0)$. Strategy 3: $P(\beta^l) < P(0)$ after restart and $Q(\beta^l) < Q(0)$ otherwise. Strategy 4: $l = 0$, (i.e., $\alpha = 1$). Finally, set $\alpha = \beta^l$, $x := x + \alpha \Delta x$, $u_I := u_I + \alpha \Delta u_I$, $u_E := u_E + \alpha \Delta u_E$, compute value $f(x)$ and vectors $c_I(x)$ and $c_E(x)$ and go to Step 2.

## 5 Numerical experiments

Algorithm 1 was tested by using five sets each containing 18 test problems. These sets were obtained as a modification of test problems for equality constrained minimization given in [6], which can be downloaded (together with report [6]) from the page `http://www.cs.cas.cz/~luksan/test.html`. In the first set, equalities $c(x) = 0$ are replaced by inequalities $c(x) \ge 0$. In the second set, equalities $c(x) = 0$ are replaced by inequalities $c(x) \le 0$. The third set was generated from the first set by adding box constraints $x \ge 0$. The fourth set was generated from the second set by adding box constraints $x \le 0$. The fifth set contains inequalities $-1 \le x \le 1$ and $-1 \le c(x) \le 1$. All problems used contain 1000 variables (vector $x$ has the dimension 1000). The results of the tests are listed in Table 1–Table 4, where NIT is the total number of iterations, NFV is the total number of function evaluations, NFG is the total number of gradient evaluations (NFG is much greater than NIT, since second order derivatives are computed by using gradient differences), NCG is the total number of CG iterations, NRS is the total number of restarts and NFAIL gives the number of failures for a given set (the number of problems which has not been solved).

Table 1. Nonsmooth equation method – Strategy 1.

| Set | NIT | NFV | NFG | NCG | NRS | TIME | NFAIL |
|---|---|---|---|---|---|---|---|
| 1 | 1091 | 1444 | 10176 | 23958 | 89 | 21.78 | – |
| 2 | 723 | 1657 | 5142 | 17790 | 188 | 20.91 | 1 |
| 3 | 691 | 1216 | 5220 | 56643 | 91 | 22.12 | – |
| 4 | 341 | 931 | 2571 | 1483 | 72 | 7.99 | 2 |
| 5 | 580 | 897 | 4387 | 7196 | 137 | 12.59 | – |

Table 2. Nonsmooth equation method – Strategy 2.

| Set | NIT | NFV | NFG | NCG | NRS | TIME | NFAIL |
|---|---|---|---|---|---|---|---|
| 1 | 1323 | 1691 | 13858 | 29288 | 182 | 30.84 | – |
| 2 | 764 | 1595 | 6996 | 18113 | 196 | 22.19 | 1 |
| 3 | 735 | 1249 | 6660 | 68421 | 90 | 25.92 | – |
| 4 | 399 | 1261 | 4217 | 1938 | 84 | 10.29 | 2 |
| 5 | 500 | 705 | 4445 | 6149 | 86 | 10.53 | – |

Table 3. Nonsmooth equation method – Strategy 3.

| Set | NIT | NFV | NFG | NCG | NRS | TIME | NFAIL |
|---|---|---|---|---|---|---|---|
| 1 | 863 | 3843 | 9408 | 61887 | 78 | 23.78 | 2 |
| 2 | 630 | 2957 | 6822 | 19748 | 121 | 15.63 | 3 |
| 3 | 613 | 2533 | 6842 | 38612 | 114 | 20.21 | 1 |
| 4 | 480 | 2137 | 5340 | 5182 | 92 | 13.70 | 3 |
| 5 | 526 | 2163 | 5605 | 4467 | 82 | 12.42 | 1 |

Table 4. Nonsmooth equation method – Strategy 4.

| Set | NIT | NFV | NFG | NCG | NRS | TIME | NFAIL |
|---|---|---|---|---|---|---|---|
| 1 | 1111 | 1112 | 10295 | 38876 | 111 | 23.96 | – |
| 2 | 1386 | 1386 | 9794 | 31914 | 306 | 49.71 | – |
| 3 | 1621 | 1622 | 16566 | 31784 | 175 | 37.28 | 1 |
| 4 | 709 | 791 | 5311 | 35419 | 232 | 36.10 | 3 |
| 5 | 1395 | 1396 | 10170 | 11515 | 292 | 29.13 | 1 |

For a comparison, Table 5 and Table 6 contain results obtained by using the interior-point method described in [9].

Table 5. Interior-point method – Strategy 1.

| Set | NIT | NFV | NFG | NCG | NRS | TIME | NFAIL |
|---|---|---|---|---|---|---|---|
| 1 | 740 | 1059 | 5477 | 45935 | 98 | 27.24 | – |
| 2 | 736 | 1895 | 5321 | 58664 | 215 | 41.26 | 1 |
| 3 | 720 | 1378 | 5260 | 31350 | 82 | 28.33 | – |
| 4 | 378 | 538 | 2788 | 12666 | 20 | 19.50 | 3 |
| 5 | 570 | 723 | 4276 | 5876 | 55 | 14.48 | – |

Table 6. Interior point method – Strategy 4.

| Set | NIT | NFV | NFG | NCG | NRS | TIME | NFAIL |
|---|---|---|---|---|---|---|---|
| 1 | 786 | 786 | 5691 | 25602 | 55 | 15.08 | – |
| 2 | 717 | 717 | 5092 | 20289 | 119 | 24.07 | – |
| 3 | 907 | 907 | 6816 | 25639 | 84 | 22.66 | – |
| 4 | 458 | 458 | 3405 | 20793 | 49 | 22.76 | 3 |
| 5 | 609 | 612 | 4292 | 4183 | 21 | 12.24 | 1 |

## 6 Conclusions

In this contribution, we describe an implementation of the nonsmooth equation method for solving large and sparse general nonlinear programming problems. Computational experiments show that this approach is very reasonable. Nevertheless, KKT equations belong to the class of problems which are difficult for application of the Newton method (this observation is also known from the equality constrained optimization, see [5]). For this reason, the use of natural merit function $Q(\alpha)$, which has the best theoretical properties [2], is not efficient for practical computations, see Table 3. The simple merit function $P(\alpha)$, which was chosen to give negative directional derivative for a broad class of directions obtained as inexact solutions of system (2.6), has better practical properties. Unfortunately, this function does not allow us to prove the global convergence in general. Surprisingly good results were obtained without using a merit function, see Table 4. However, parameters $\rho$ and $\overline{\Delta}$ have to be chosen more carefully in this case. Another conclusion, which follows from our computational experiments, is that the nonsmooth equation method based on the Fischer-Burmeister function (with Strategy 1) is competitive with the interior-point method described in [9], see Table 5. The main advantage of the nonsmooth equation approach is that no slack variables are used and the matrix $\hat{M}_I$ is positive definite independently of the signs of Lagrange multipliers. Therefore, a special strategy for step-length selection ensuring positivity of slack variables and Lagrange multipliers need not be used. The nonsmooth equation method is relatively sensitive to the choice of parameters $\rho$ and $\overline{\Delta}$ and the system (2.6) is frequently ill-conditioned due to the linear dependence of gradients of the active constraints (the number of active constraints can be greater than the number of variables).

**Acknowledgement**. This work was supported by grant No. 201/00/0080 of the Grant Agency od the Czech Republic.

## References

[1] A.R. Curtis, M.J.D. Powell, and J.K. Reid, *On the estimation of sparse Jacobian matrices*, IMA J. Apl. Math. **3** (1974), 117–119.

[2] T. De Luca, F. Facchinei, and C. Kanzow, *A semismooth equation approach to the solution of nonlinear complementarity problems*, Math. Programming **75** (1996), 407–439.

[3] H. Qi, L. Qi, *A new QP-free globally convergent, locally superlinearly convergent algorithm for inequality constrained optimization*, SIAM J. Optim. **11** (2000), 112–132.

[4] D. Li, N. Yamashita, and M. Fukushima, *A nonsmooth equation based BFGS method for solving KKT systems in mathematical programming*, J. Optim. Theory Appl. **109** (2001), 123–167.

[5] L. Lukšan, J. Vlček, *Indefinitely preconditioned inexact Newton method for large sparse equality constrained nonlinear programming problems*, Numer. Linear Algebra Appl. **5** (1998), 219–247.

[6] L. Lukšan, J. Vlček, *Sparse and partially separable test problems for unconstrained and equality constrained optimization*, Report V-767, Prague, ICS AS CR, 1998.

[7] L. Lukšan, J. Vlček, *Numerical experience with iterative methods for equality constrained nonlinear programming problems*, Report V-779, Prague, ICS AS CR, 2000.

[8] L. Lukšan, J. Vlček, *Introduction to nonsmooth analysis. Theory and algorithms*, Technical Report DMSIA 00/1 (serie didattica), Universita degli Studi di Bergamo, 2000.

[9] L. Lukšan, J. Vlček, *Interior-point method for nonlinear nonconvex optimization*, Numer. Linear Algebra Appl. (to appear).

[10] L. Qi, J. Sun, *A nonsmooth version of Newton's method*, Math. Programming **58** (1993), 353–367.

[11] L.N. Vicente, S.J. Wright, *Local convergence of a primal-dual method for degenerate nonlinear programming*, SIAM J. Optim. (to appear).

# Part III

# Finite Element Meshes

# Nonstandard Nonobtuse Refinements of Planar Triangulations

Sergey Korotov,[1] Jacek Stańdo[2]

[1]Department of Mathematical Information Technology, University of Jyväskylä, P.O. Box 35, FIN–40014 Jyväskylä, Finland korotov@mit.jyu.fi

[2]Institute of Mathematics, Technical University of Łódź, Al. Politechniki 11, 90-924 Łódź, Poland standoj@ck-sg.p.lodz.pl

## 1 Introduction

Throughout the paper, we consider conforming planar triangulations of a given polygon into triangles, i.e., the union of all triangles of any particular triangulation is always equal to the polygon, and any two different triangles in any particular triangulation may only have a common edge, or a common vertex, or no common point (cf. [12]). In most of cases namely such conforming triangulations are used in the finite element modelling and analysis.

A planar triangulation is called *nonobtuse* if all interior angles of all triangles in the triangulation are less than or equal to $\frac{\pi}{2}$. This condition was first introduced in [4] and [16], where it was particularly used to get the monotone stiffness matrix and prove the discrete maximum principle for the piecewise linear finite element approximations to the 2D Poisson equation with the Dirichlet boundary condition. Later this condition was generalized and imposed on the interior angles between faces of tetrahedra in tetrahedral triangulations, and it was again used to prove the discrete maximum principle for the finite element solutions to one class of 3D nonlinear elliptic problems for which the continuous maximum principle holds, see [13], and also [14]. Nonobtuse triangulations are also employed to provide the validity of the discrete maximum principle for the parabolic problems [6] and to derive $L^\infty$-convergence of the finite element method for a linear problem with trailing boundary conditions [5].

Note that in nonobtuse triangulations, all triangles satisfy the maximum angle condition with the same constant, $\frac{\pi}{2}$. This condition is very important for the finite element analysis, since the optimal interpolation properties of the finite elements are preserved [12, Chapter 4].

The construction of a conforming triangulation of a polygon into arbitrary triangles is, obviously, simple; moreover, there are proofs of the existence of nonobtuse conforming triangulations of arbitrary polygons [1], [7]. Thus,

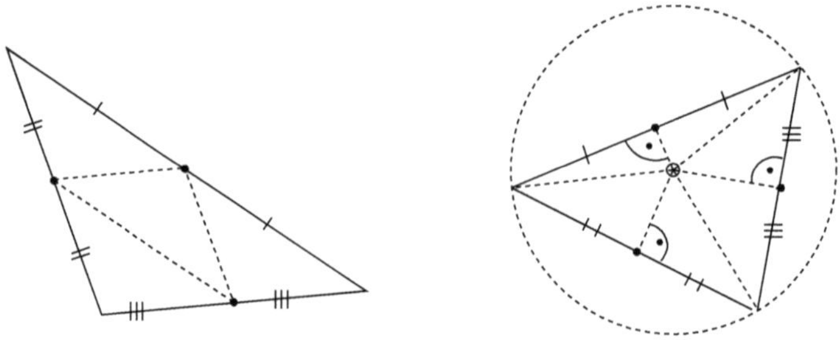

a) red refinement b) yellow refinement

Figure 1

having a nonobtuse initial triangulation, and applying the standard red refinement [2] and/or the standard yellow refinement [9] (see Fig. 1) we can easily build a family of globally refining triangulations where no obtuse angles occur.

Nevertheless, in practice, the initial triangulation of the polygonal solution domain is usually performed by some "black-box" grid generator, and it is most probably that such an "unpredictably made" (initial) triangulation is not a nonobtuse one. One way to change such an "undesirable" triangulation into a nonobtuse one is to move locally several vertices in it in order to get locally nonobtuse triangles. Another approach, presented in [11], is to use special nonobtuse refinement techniques for small polygonal zones in the initial triangulations around the obtuse triangles, where each such a zone is made of the obtuse triangle and several nonobtuse triangles adjacent to it. The nonobtuse refinement of such a zone shall only use the midpoints of the edges bordering the zone and several points inside of it together with the original vertices of the zone, therefore we can use the standard red and/or yellow refinements for the other (nonobtuse) triangles outside of such zones. It means that after one global (overall) refinement step, we obviously get a nonobtuse conforming triangulation of the solution domain.

In this paper, we consider two types of such special zones with corresponding refinement techniques. In Section 2, we describe the yellow-red refinement of a quadrilateral made by an obtuse triangle and a nonobtuse triangle adjacent to the obtuse one along its longest edge. In Section 3, we analyze the case of a zone which is a union of an obtuse triangle and three adjacent to it nonobtuse triangles, and introduce the quasi-red and quasi-yellow refinements. For both the cases, the conditions on when the corresponding refinements can be used are given in terms of coordinates of the vertices involved.

**Remark 1.1.** We note that the nonobtusity condition is only sufficient to prove the discrete maximum principle: it is possible to observe its validity also in situations when certain interior angles in triangulations are slightly obtuse. This issue, leading to so-called weakened acute type conditions, is addressed in [15] and [10], for the two and three-dimensional cases, respectively.

**Remark 1.2.** Another known refinements of triangles are green [2], and blue [8], they are used for local refinement procedures. For a comprehensive survey of various topics in grid generation, we refer to monograph [3].

## 2 Yellow-red refinement

First, we give a simple criterion for a triangle to be nonobtuse, and introduce the idea of the yellow-red refinement.

Let $ABC$ denote a triangle with the vertices $A = (x_A, y_A)$, $B = (x_B, y_B)$, and $C = (x_C, y_C)$ (see Fig. 2).

**Lemma 2.1.** *The triangle ABC is nonobtuse iff the following inequalities hold*

$$(x_A - 2x_B + x_C)^2 + (y_A - 2y_B + y_C)^2 \geq (x_C - x_A)^2 + (y_C - y_A)^2,$$
$$(x_C - 2x_A + x_B)^2 + (y_C - 2y_A + y_B)^2 \geq (x_B - x_C)^2 + (y_B - y_C)^2,$$
$$(x_B - 2x_C + x_A)^2 + (y_B - 2y_C + y_A)^2 \geq (x_A - x_B)^2 + (y_A - y_B)^2.$$

Proof. For the proof see [11] (the geometrical interpretation of these relations is given in Fig. 2a). □

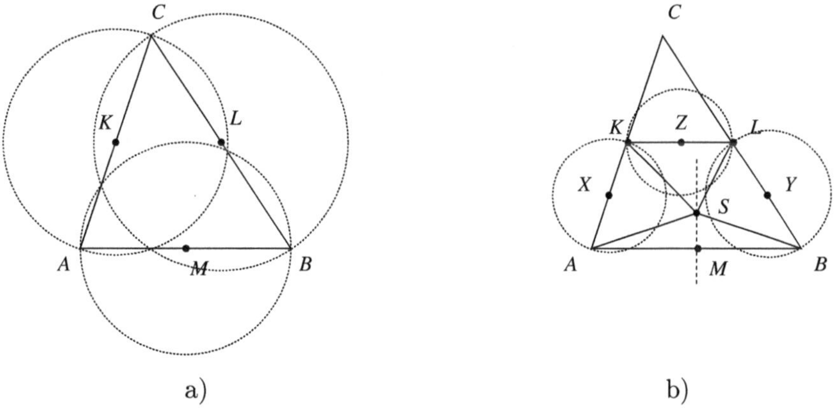


Figure 2

### 2.1 Yellow-red refinement of a nonobtuse triangle

The red refinement (cf. Fig. 1a) involves the midpoints of the sides of a triangle: connecting them, we get four subtriangles similar to the original one; the red refinement can be applied to any triangle.

The idea of the yellow refinement essentially uses the assumption that the center of the circumscribed circle of the triangle lies in the closure of it. Using this center and the midpoints of sides of the given triangle, we get a partition of it into four or six right subtriangles, depending on the value of the maximum angle in the given triangle (cf. Fig. 1b). Thus, the yellow refinement can only be applied to a nonobtuse triangle.

We shall now show that it is possible to "combine" both refinement techniques inside of a (single) nonobtuse triangle. The idea of such a construction is illustrated in Fig. 2b. Let $K$, $L$, $M$ be the midpoints of the sides $AC$, $BC$, $AB$, respectively, let the dashed line via point $M$ be the mid-perpendicular to side $AB$, let points $X$, $Y$, $Z$ be the midpoints of $AK$, $BL$, $KL$, respectively (these three points are also centers of three dotted circles having $AK$, $BL$, and $KL$ as their diameters). If we can take point $S$ from trapezoid $AKLB$ ($\subset ABC$), lying on the mid-perpendicular via $M$, so that it lies outside of all three (dotted) circles, then triangle $ABC$ can, obviously, be partitioned into six nonobtuse subtriangles – $KLC$, $KSA$, $KLS$, $LSB$, $SMA$, and $SMB$ (the last two are, moreover, right ones at the vertex $M$). We shall call such a partition of the nonobtuse triangle $ABC$ as the *yellow-red refinement.*

**Remark 2.2.** In general, there exist many choices of the point $S$. Nevertheless, there can also occur a situation when the only possible choice of $S$ is to be at point $M$, it happens e.g. if $ABC$ is the right triangle. In this case, the yellow-red refinement coincides with the standard red refinement of triangle $ABC$, and produces only four subtriangles.

## 2.2 Nonobtuse refinement of a zone formed by two adjacent triangles

Let us consider two adjacent triangles, where (in denotations of Fig. 3a) the triangle $ADB$ is obtuse at $D = (x_D, y_D)$ and the triangle $ABC$ is nonobtuse. Let $X = (x_X, y_X)$, $Y = (x_Y, y_Y)$, and $Z = (x_Z, y_Z)$ be the midpoints of $AK$, $BL$, and $KL$, respectively.

**Lemma 2.3.** *The coordinates of $X$, $Y$, and $Z$ are given by the formulae*

$$x_X = \frac{3}{4}x_A + \frac{1}{4}x_C, \qquad y_X = \frac{3}{4}y_A + \frac{1}{4}y_C; \tag{2.1}$$

$$x_Y = \frac{3}{4}x_B + \frac{1}{4}x_C, \qquad y_Y = \frac{3}{4}y_B + \frac{1}{4}y_C; \tag{2.2}$$

$$x_Z = \frac{1}{2}x_C + \frac{1}{4}x_A + \frac{1}{4}x_B, \quad y_Z = \frac{1}{2}y_C + \frac{1}{4}y_A + \frac{1}{4}y_B. \tag{2.3}$$

Further, let $S = (x_S, y_S)$ be a center of the circumscribed circle of $ADB$, which lies outside of $ADB$.

**Lemma 2.4.** *The coordinates of $S$ are given by the formulae*

$$x_S = \frac{(y_B - y_D)(x_A^2 + y_A^2 - x_D^2 - y_D^2)}{2[(x_A - x_D)(y_B - y_D) - (x_B - x_D)(y_A - y_D)]} + \frac{(y_D - y_A)(x_B^2 + y_B^2 - x_D^2 - y_D^2)}{2[(x_A - x_D)(y_B - y_D) - (x_B - x_D)(y_A - y_D)]}, \tag{2.4}$$

$$y_S = \frac{(x_D - x_B)(x_A^2 + y_A^2 - x_D^2 - y_D^2)}{2[(x_A - x_D)(y_B - y_D) - (x_B - x_D)(y_A - y_D)]} + \frac{(x_A - x_D)(x_B^2 + y_B^2 - x_D^2 - y_D^2)}{2[(x_A - x_D)(y_B - y_D) - (x_B - x_D)(y_A - y_D)]}. \tag{2.5}$$

The calculation in both above lemmas is straightforward.

Further, the following theorem holds, where $|\cdot|$ denotes the standard distance between two points in $\mathbb{R}^2$.

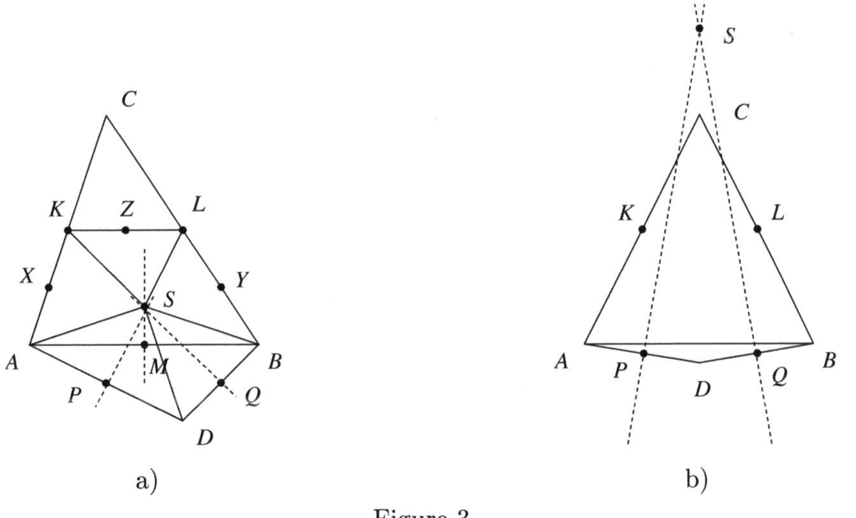


Figure 3

**Theorem 2.5.** *Let the triangle $ABC$, adjacent to the obtuse triangle $ADB$, be nonobtuse. Let $P$ and $Q$ be the midpoints of the sides $AD$ and $DB$, respectively. If*

$$[(x_C - x_A)(y_A - y_B) + (y_C - y_A)(x_B - x_A)] \times [(x_S - \frac{1}{2}(x_C + x_A))(y_A - y_B) + (y_S - \frac{1}{2}(y_C + y_A))(x_B - x_A)] \leq 0, \tag{2.6}$$

*and*

$$|SX| \geq \frac{1}{4}|AC|, \quad |SZ| \geq \frac{1}{4}|AB|, \quad |SY| \geq \frac{1}{4}|BC|, \tag{2.7}$$

*where coordinates of points* $S$, $X$, $Y$, *and* $Z$ *are given by* (2.1)–(2.5), *then the triangles* $KLC$, $KSA$, $KLS$, $LSB$, $SAP$, $SPD$, $SDQ$, *and* $SQB$ *are nonobtuse (last four are, moreover, right triangles).*

Proof. The proof is presented in [11]. □

**Remark 2.6.** If the conditions of Theorem 2.5 are fulfilled, we can refine the quadrilateral $ADBC$ into nonobtuse triangles employing its vertices, the midpoints $K$, $P$, $Q$, and $L$ of its sides, and the point $S$ inside of the quadrilateral.

**Remark 2.7.** Obviously, the conditions of Theorem 2.5 cannot be fulfilled in a general case. For example, in Fig. 3b, two adjacent (obtuse and nonobtuse) triangles are such that the point $S$ is completely outside of the quadrilateral formed by the triangles. Therefore, a partition of the quadrilateral in the manner similar to Fig. 3a is not possible.

**Remark 2.8.** One of another possibilities to get nonobtuse triangles, particularly, for the case illustrated in Fig. 3b, would be to use the swapping of diagonals in the quadrilateral. Thus, in $ADBC$ we may consider two triangles $ACD$ and $BCD$ (which could be both nonobtuse) instead of $ABC$ and $ABD$. However, the analysis of such an approach, is beyond the scope of the present paper.

**Remark 2.9.** The nonobtuse partition of a quadrilateral in a manner of Fig. 3a is also called *yellow-red refinement.*

# 3 Quasi-red and quasi-yellow refinements

In this section, we introduce the ideas of quasi-red and quasi-yellow refinements of triangles (see Fig. 4), which present natural generalizations of the standard red and yellow refinement techniques.

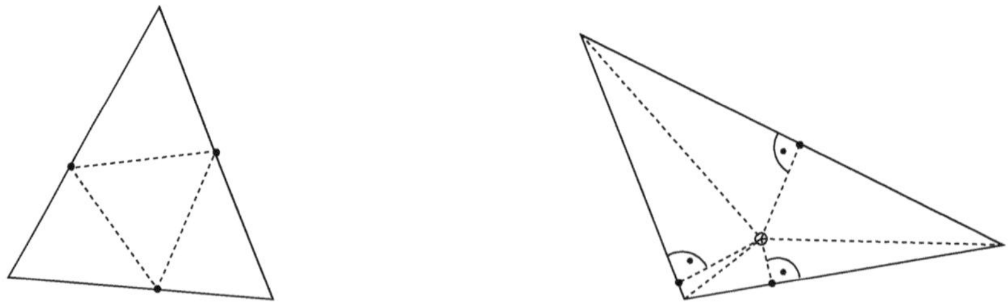

a) quasi-red refinement b) quasi-yellow refinement

Figure 4

### 3.1 Quasi-red nonobtuse refinement of a triangle

As mentioned in Section 2.1, the standard red refinement emploies the midpoints of the sides of a triangle (see Fig. 1a). But, under certain circumstances, we could use, for example, one, or two, or even three points, on sides of the given triangle, which are different from the midpoints. A refinement of "red-type" (called *quasi-red refinement*) could be applied to such a case as well, producing four nonobtuse subtriangles (cf. Fig. 4a).

In this paper, we employ the idea of quasi-red refinement for the case when only one auxiliary point is allowed to be taken arbitrarily (and the other two are still the midpoints of two sides of the triangle). Our purpose is to find the set of points from which this third auxiliary point can be taken, so that the quasi-red refinement, applied to a nonobtuse triangle, produces four nonobtuse subtriangles.

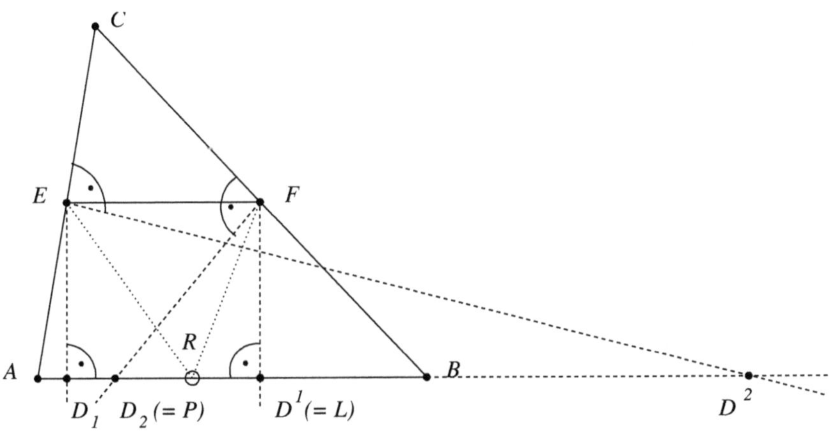


Figure 5

**Lemma 3.1.** *Let $ABC$ be a nonobtuse triangle with vertices $A = (x_A, y_A)$, $B = (x_B, y_B)$, and $C = (x_C, y_C)$ (see Fig. 5), and let $E$ and $F$ be midpoints of $AC$ and $CB$, respectively. We define four real numbers $k_1$, $k_2$, $k^1$, and $k^2$ as follows*

$$k_1 = \frac{1}{2}\frac{(x_C - x_A)(x_B - x_A) + (y_C - y_A)(y_B - y_A)}{(x_B - x_A)^2 + (y_B - y_A)^2}, \tag{3.1}$$

$$k_2 = \frac{1}{2}\frac{x_B + x_C - 2x_A)(x_B - x_C) + (y_B + y_C - 2y_A)(y_B - y_C)}{(x_B - x_A)(x_B - x_C) + (y_B - y_A)(y_B - y_C)}, \tag{3.2}$$

$$k^1 = \frac{1}{2}\frac{(x_B + x_C - 2x_A)(x_B - x_A) + (y_B + y_C - 2y_A)(y_B - y_A)}{(x_B - x_A)^2 + (y_B - y_A)^2}, \tag{3.3}$$

$$k^2 = \frac{1}{2}\frac{(x_C - x_A)^2 + (y_C - y_A)^2}{(x_B - x_A)(x_C - x_A) + (y_B - y_A)(y_C - y_A)}. \tag{3.4}$$

*If* $(x_B - x_A)(x_B - x_C) + (y_B - y_A)(y_B - y_C) = 0$, *we set* $k_2 = -\infty$, *if* $(x_B - x_A)(x_C - x_A) + (y_B - y_A)(y_C - y_A) = 0$, *we set* $k^2 = +\infty$.

*Let*

$$k = \frac{1}{2}(k_1 + k^1), \quad k_* = \max\{k_1, k_2\} \quad \textit{and} \quad k^* = \min\{k^1, k^2\},$$

*we define*

$$P = (k_* x_B + (1 - k_*)x_A, k_* y_B + (1 - k_*)y_A), \tag{3.5}$$

$$L = (k^* x_B + (1 - k^*)x_A, k^* y_B + (1 - k^*)y_A), \tag{3.6}$$

*and require, in addition to the nonobtuseness of* $ABC$, *that scalar product* $(\overline{DE}, \overline{DF}) \geq 0$, *where the point* $D := (kx_B + (1 - k)x_A, ky_B + (1 - k)y_A)$. *Then, for any point* $R \in \overline{PL}$ $(\subset \overline{AB})$, *the triangles* $AER$, $ERF$, $FRB$, *and* $CEF$ *are nonobtuse.*

Proof. The proof is straightforward (cf. Fig. 5). $\square$

### 3.2 Quasi-yellow nonobtuse refinement of a triangle

As far as it concerns the "yellow-type" refinement (see Fig. 4b), it is obvious, that if we do not necessarily have to use only midpoints for the refinement into right subtriangles, we can, in fact, employ some point in the given triangle, for example, a centre of the inscribed sphere, and three projections of this point onto the sides, and get again a partition of the triangle into six right subtriangles (cf. Fig. 4b). This way of partitioning is referred to as *quasi-yellow refinement.* Actually, we are able to define the set, called *admissible area of triangle*, of points in the triangle so that the above-described partition is possible. In case of a nonobtuse triangle, this admissible area coincides with the whole triangle; if there is an obtuse angle in the triangle $ABC$ (see Fig. 6a), then the admissible area is the subtriangle $AEF$, where the points $E$ and $F$ are taken on the segment $CB$ in such a way that $AE$ is perpendicular to $AB$, and $AF$ is perpendicular to $AC$ (cf. Fig. 6b).

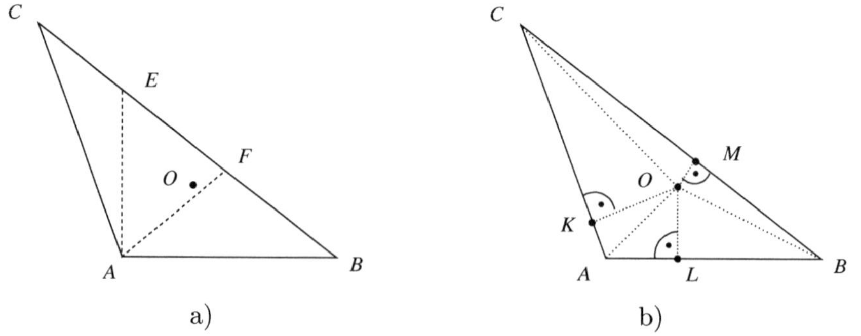


Figure 6

Obviously, in the case when $ABC$ is nonobtuse and $O$ is the centre of circumscribed sphere, the quasi-yellow refinement coincides with the standard yellow refinement.

### 3.3 Nonobtuse refinement of a zone formed by four triangles

Let $BAG$ be an obtuse triangle at vertex $A$, surrounded by three adjacent nonobtuse triangles $ABC$, $AGE$, and $BGD$ (see Fig. 7a).

Using Lemma 3.1, we are able to find coordinates of points $L = (x_L, y_L)$ and $P = (x_P, y_P)$ (using the coordinates of the vertices $A$, $B$, $C$), so that for any point from the segment $\overline{LP}$ and the midpoints $F$ and $E$, the quasi-red refinement of $ABC$ produces nonobtuse subtriangles.

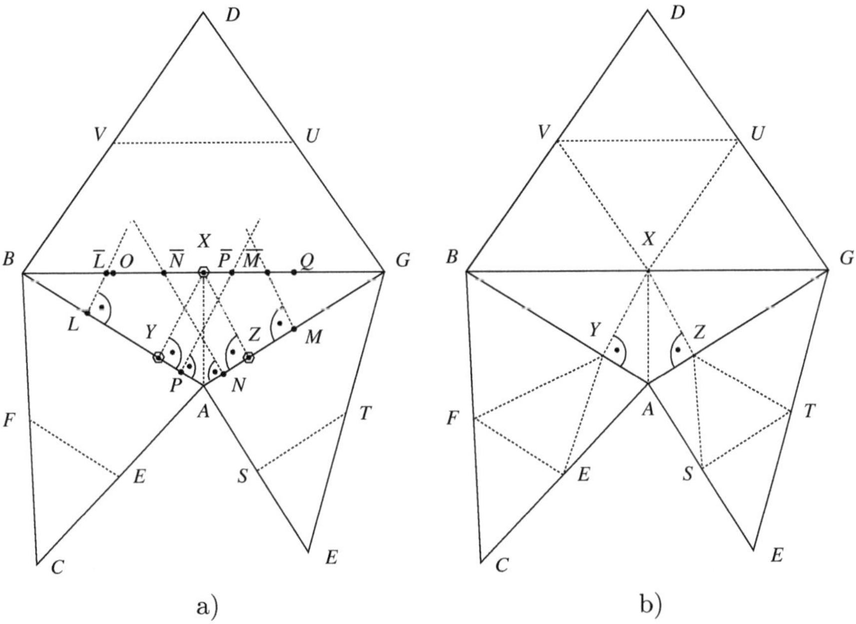


Figure 7

We can similarly find the points $N = (x_N, y_N)$ and $M = (x_M, y_M)$ for the triangle $AGE$; and the points $O = (x_O, y_O)$ and $Q = (x_Q, y_Q)$ – for the triangle $BGD$. Our goal is to find conditions on all the vertices $A$, $B$, $C$, $G$, $E$, and $D$, under which there exists some point $X \in \overline{OQ}$, so that its orthogonal projection $Y$ on the side $AB$ belongs to $\overline{LP}$, and its orthogonal projection $Z$ on the side $AG$ belongs to $\overline{NM}$. It shall mean then that we can conformly divide the zone formed by the triangles $ABC$, $AGE$, $BGD$, and $BAG$ using the quasi-red and the quasi-yellow refinements, into nonobtuse subtriangles (see Fig. 7b).

To find conditions on the existence of such point $X$, we define point $\overline{L}$ as the intersection of perpendicular to $BA$ at point $L$ with side $BG$, and point $\overline{P}$ as the intersection of perpendicular to $BA$ at point $P$ with side $BG$. Similarly, let $\overline{N}$ denote the intersection of perpendicular to $AG$ at point $N$ with side $BG$, and $\overline{M}$ – the intersection of perpendicular to $AG$ at point $M$

with side $BG$. In this way, we get six points on side $BG$: $\overline{L}$, $\overline{P}$, $\overline{N}$, $\overline{M}$, $O$, and $Q$.

Using Lemma 3.1, we can easily find representations

$$O = (q_1 x_G + (1-q_1)x_B \ , \ q_1 y_G + (1-q_1)y_B), \tag{3.7}$$
$$Q = (q^1 x_G + (1-q^1)x_B \ , \ q^1 y_G + (1-q^1)y_B), \tag{3.8}$$

where $q_1$ and $q^1$ are some real numbers. Analogously, we have

$$\overline{L} = (q_2 x_G + (1-q_2)x_B, q_2 y_G + (1-q_2)y_B), \tag{3.9}$$
$$\overline{N} = (q_3 x_G + (1-q_3)x_B, q_3 y_G + (1-q_3)y_B), \tag{3.10}$$
$$\overline{P} = (q^2 x_G + (1-q^2)x_B, q^2 y_G + (1-q^2)y_B), \tag{3.11}$$
$$\overline{M} = (q^3 x_G + (1-q^3)x_B, q^3 y_G + (1-q^3)y_B), \tag{3.12}$$

where the precise values of $q_2$, $q_3$, $q^2$ and $q^3$ are presented below in Lemma 3.3.

**Lemma 3.2.** *Let* $\underline{q} = \max\{q_1, q_2, q_3\}$ *and* $\overline{q} = \min\{q^1, q^2, q^3\}$. *If* $\underline{q} \leq \overline{q}$, *then point* $X$ (*cf. Fig.* 7b) *always exists; for example, one proper choice is*

$$X = (dx_G + (1-d)x_B, dy_G + (1-d)y_B), \tag{3.13}$$

*where* $d = \frac{1}{2}(\underline{q} + \overline{q})$.

**Lemma 3.3.** *The values of numbers in* (3.9)–(3.12) *are*

$$q_2 = \frac{(x_L - x_B)(x_B - x_A) + (y_L - y_B)(y_B - y_A)}{(x_G - x_B)(x_B - x_A) + (y_G - y_B)(y_B - y_A)}, \tag{3.14}$$
$$q^2 = \frac{(x_P - x_B)(x_B - x_A) + (y_P - y_B)(y_B - y_A)}{(x_G - x_B)(x_B - x_A) + (y_G - y_B)(y_B - y_A)}, \tag{3.15}$$
$$q_3 = \frac{(x_N - x_B)(x_G - x_A) + (y_N - y_B)(y_G - y_A)}{(x_G - x_B)(x_G - x_A) + (y_G - y_B)(y_G - y_A)}, \tag{3.16}$$
$$q^3 = \frac{(x_M - x_B)(x_G - x_A) + (y_M - y_B)(y_G - y_A)}{(x_G - x_B)(x_G - x_A) + (y_G - y_B)(y_G - y_A)}. \tag{3.17}$$

Proof. The proofs in both above lemmas are straightforward. □

## 4 Nonobtuse refinement of the coarse triangulations

In this section, we shortly describe how the ideas of the previous sections can be used to make a nonobtuse triangulation out of the initial triangulation, containing obtuse triangles.

1) First, using the criterion from Section 2, we check all the triangles for being nonobtuse and mark all obtuse ones up.

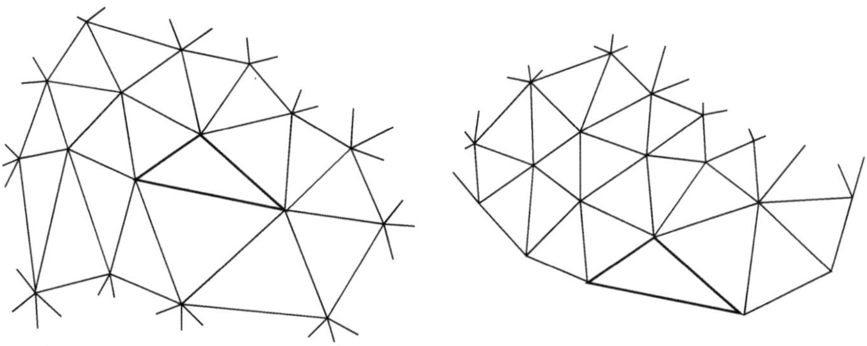

a) obtuse triangle is inside of the solution domain

b) obtuse triangle is adjacent to the boundary

Figure 8

2) If the marked obtuse triangle is adjacent to the boundary along its longest edge (see Fig. 8b), we let the perpendicular from the interior vertex of this triangle onto the boundary, and add a new node and two righ-angled triangles to the triangulation.

3) Further, if we have an obtuse triangle inside of the solution domain (see Fig. 8a), then we try the ideas of Section 2 or Section 3 in order to extract and nonobtusely refine a certain polygonal zone around such a triangle.

4) For all the remaining (nonobtuse) triangles outside of the polygonal zones, made in the previous step, we use the standard red and/or the standard yellow refinement techniques.

If Step 3 is performable for all the obtuse triangles, then in view of the way of partitioning, the resulting triangulation will be nonobtuse and conforming.

**Remark 4.1.** The above-procedure seems to work the most effectively in the case when the obtuse triangles are "isolated" in the initial triangulations, and do not have very big obtuse angles.

**Acknowledgements.** The first author was partly supported by the Agora Center under the Grant InBCT of TEKES. The authors are grateful to Prof. Michal Křížek and Prof. István Faragó for fruitful discussions of the paper.

## References

[1] B. S. Baker, E. Grosse, and C. S. Rafferty, *Nonobtuse triangulation of polygons*, Discrete Comput. Geom. **3** (1988), 147–168.

[2] R. E. Bank, *PLTMG: A Software Package for Solving Partial Differential Equations: Users' Guide 7.0*, SIAM, Philadelphia, 1994.

[3] G. F. Carey, *Computational Grids. Generation, Adaptation, and Solution Strategies*, Series in Computational and Physical Processes in

Mechanics and Thermal Sciences, Taylor & Francis, Washington, DC, 1997.

[4] P. G. Ciarlet, P. A. Raviart, *Maximum principle and uniform convergence for the finite element method*, Comput. Methods Appl. Mech. Engrg. **2** (1973), 17–31.

[5] M. Feistauer, J. Felcman, M. Rokyta, and Z. Vlášek, *Finite-element solution of flow problems with trailing conditions*, J. Comput. Appl. Math. **44** (1992), 131–165.

[6] H. Fujii, *Some remarks on finite element analysis of time-dependent field problems*, Theory and Practice in Finite element Structural Analysis, Univ. Tokyo Press, Tokyo, 1973, pp. 91–106.

[7] J. L. Gerver, *The dissection of a polygon into nearly equilateral triangles*, Geom. Dedicata **16** (1984), 93–106.

[8] R. Kornhuber, R. Roitzsch, *On adaptive grid refinement in the presence of internal or boundary layers*, Impact Comput. Sci. Engrg. **2** (1990), 40–72.

[9] S. Korotov, M. Křížek, *Acute type refinements of tetrahedral partitions of polyhedral domains*, SIAM J. Numer. Anal. **39** (2001), 724–733.

[10] S. Korotov, M. Křížek, and P. Neittaanmäki, *Weakened acute type condition and the discrete maximum principle*, Math. Comp. **70** (2001), 107–119.

[11] S. Korotov, J. Stańdo, *Yellow-red and nonobtuse refinements of planar triangulations*, Math. Notes (Miskolc) **3** (2002), 39–46.

[12] M. Křížek, P. Neittaanmäki, *Mathematical and Numerical Modelling in Electrical Engineering: Theory and Applications*, Kluwer Academic Publishers, Dordrecht, 1996.

[13] M. Křížek, Qun Lin, *On diagonal dominancy of stiffness matrices in 3D*, East-West J. Numer. Math. **3** (1995), 59–69.

[14] M. Křížek, J. Šolc, *Acute versus nonobtuse tetrahedralizations*, in: Conjugate Gradients Algorithms and Finite Element Methods, Springer-Verlag, Berlin, 2004, pp. 161–170.

[15] V. Ruas Santos, *On the strong maximum principle for some piecewise linear finite element approximate problems of non-positive type*, J. Fac. Sci. Univ. Tokyo Sect. IA Math. **29** (1982), 473–491.

[16] G. Strang, G. J. Fix, *An Analysis of the Finite Element Method*, Prentice Hall, Englewood Cliffs, N. J., 1973.

# Acute Versus Nonobtuse Tetrahedralizations

Michal Křížek,[1] Jakub Šolc[2]

[1]Mathematical Institute, Academy of Sciences of the Czech Republic, Žitná 25, CZ-115 67 Praha 1, Czech Republic krizek@math.cas.cz

[2]Department of Mathematics, Faculty of Civil Engineering, Czech Technical University, Thákurova 7, CZ-166 29 Praha 6, Czech Republic solc@math.cas.cz

*Don't be afraid to look and think before you understand.*

Michael Starbird

*Dedicated to Prof. Karel Segeth on his 60th birthday.*

## 1 Introduction

Let $\Omega \subset \mathbb{R}^d$, $d \in \{1, 2, \dots\}$, be a bounded domain with a Lipschitz boundary $\partial\Omega$. Consider the problem

$$-\Delta u = f \quad \text{in } \Omega, \tag{1.1}$$

$$u = 0 \quad \text{on } \partial\Omega, \tag{1.2}$$

where $f \in L^2(\Omega)$. It is well known that the weak solution $u$ of (1.1)–(1.2) satisfies the maximum principle, i.e., if $f \le 0$ then the maximum (essential supremum) of $u$ over $\overline{\Omega}$ is attained on the boundary $\partial\Omega$. A similar maximum principle holds for a wide class of nonlinear second order elliptic problems [9].

From now on assume that the domain $\Omega$ is polytopic (in particular, polygonal for $d = 2$ or polyhedral for $d = 3$). By [4, p. 150], linear nonobtuse triangular elements guarantee the validity of the discrete maximum principle. This result can be generalized into three-dimensional space. Namely, in [15] linear finite tetrahedral elements are used to solve a relatively large class of nonlinear second order elliptic problems. It is shown that if all dihedral angles between triangular faces are nonobtuse, then the discrete maximum principle holds. In particular, for problem (1.1)–(1.2) we have: if $f \le 0$ then the standard Galerkin approximation $u_h$ of $u$ attains its maximum on the boundary $\partial\Omega$. According to [15], the associated stiffness matrix is irreducibly diagonally dominant (and hence, monotone) even if a numerical integration is employed for a class of nonlinear elliptic boundary value problems. This property is a consequence of the following lemma (see [15, p. 63]).

**Lemma 1.1.** *Let $T \subset \mathbb{R}^3$ be an arbitrary tetrahedron with vertices $A$, $B$, $C$, and $D$. Let $p$ and $q$ be two linear functions such that*

$$p(A) = 1, \quad p(B) = p(C) = p(D) = 0,$$
$$q(B) = 1, \quad q(A) = q(C) = q(D) = 0.$$

*Then*

$$\nabla p \cdot \nabla q = -C_T \cos \varphi,$$

*where $C_T > 0$ depends on $T$, and $\varphi$ is the inner angle between the faces $ACD$ and $BCD$.*

In view of this lemma, we observe that the contribution to off-diagonal entries of the element stiffness matrix associated to $T$ will be nonpositive if $\varphi$ is nonobtuse.

Note that the discrete maximum principle can be preserved even if some dihedral angles between faces are slightly bigger than 90° (see [14]). Anyway, the condition number of the stiffness matrix essentially depends on the maximum angle in a given partition. Generally speaking, the smaller the maximum angle is, the smaller the condition number is for a fixed number of interior nodes (degrees of freedom). Therefore, it is important to investigate partitions, where no large angles appear.

If we replace equation (1.1) by $-\Delta u + bu = f$, where $b \in L^\infty(\Omega)$ and $b > 0$, then acute triangles are required to guarantee the validity of the discrete maximum principle (see [4, p. 150]). The term $bu$ yields positive contributions to off-diagonal entries of the associated stiffness matrix [10].

## 2 Preliminaries

Consider an arbitrary tetrahedron $T \subset \mathbb{R}^3$. The six inner angles between its four faces are called *dihedral*. Inner angles between edges of any face of $T$ are called *solid*.

A tetrahedron is said to be *acute* if all its six dihedral angles are acute (i.e., less than 90°). Similarly, we say that a tetrahedron is *nonobtuse* if all its dihedral angles are nonobtuse (i.e., not greater than 90°).

**Lemma 2.1.** *Let $T$ be an acute tetrahedron. Then all its solid angles are also acute.*

Proof. Let $A$, $B$, $C$, and $D$ be vertices of an acute tetrahedron $T$. Assume that $\omega$ is an arbitrary solid angle in one of the four triangular faces of $T$, say, the angle between the edges $DB$ and $DC$. Let $\alpha$, $\beta$, and $\gamma$ be the dihedral angles between the pairs of faces of $T$ that have common edge $DA$, $DB$, and $DC$, respectively. Considering a sufficiently small sphere with the centre $D$ and radius less than $\min(|DA|, |DB|, |DC|)$, the Cosine Theorem for the angles has the form (see [19, p. 85])

$$\cos \alpha = -\cos \beta \cos \gamma + \sin \beta \sin \gamma \cos \omega. \tag{2.1}$$

Since $\alpha, \beta, \gamma$ are positive and acute angles, we conclude that $\cos\omega > 0$, and thus, $\omega$ is positive and acute as well. □

**Remark 2.2.** The converse implication in Lemma 2.1 is not true. To see this consider the tetrahedron with vertices $A = (-1,0,0)$, $B = (1,0,0)$, $C = (0,-1,\varepsilon)$, and $D = (0,1,\varepsilon)$ for an $\varepsilon \in (0,1)$. Then all solid angles of all triangular faces are acute, whereas the dihedral angle between the two faces at the edge $AB$ and also at the edge $CD$ is obtuse.

**Lemma 2.3.** *Let $T$ be a nonobtuse tetrahedron. Then all its solid angles are also nonobtuse.*

The proof can be done in a similar way as for Lemma 2.1 (see also [2], [15]). The converse implication in Lemma 2.3 again does not hold. To see this it is enough to consider the same tetrahedron as in Remark 2.2.

**Remark 2.4.** According to (2.1), all solid angles of any tetrahedron are uniquely defined by its six dihedral angles.

Throughout the paper we shall consider only face-to-face partitions (of bounded polytopic domains or of the whole space $\mathbb{R}^d$) into $d$-simplexes, which will be called for simplicity just *partitions*. For $d = 3$ such a partition is called a *tetrahedralization.*

**Definition 2.5.** A tetrahedralization $\mathcal{T}$ is said to be *acute* (*nonobtuse*) if all tetrahedra from $\mathcal{T}$ are acute (nonobtuse).

**Remark 2.6.** It is evident that in any acute (nonobtuse) tetrahedralization each inner edge is surrounded by at least five (four) tetrahedra.

**Lemma 2.7.** *Let $\alpha, \beta, \gamma, \delta, \varepsilon, \varphi$ be all the dihedral angles of an acute tetrahedron in any order. Then we have $\alpha+\beta+\gamma > 90°$, $\alpha+\beta+\gamma+\delta > 180°$, and $\alpha+\beta+\gamma+\delta+\varepsilon > 270°$.*

Proof. Assume to the contrary that $\alpha + \beta + \gamma \leq 90°$. It is well known that the sum of all the dihedral angles in an arbitrary tetrahedron lies in the interval $(360°, 540°)$. Consequently, $\delta + \varepsilon + \varphi \geq 270°$, and thus at least one of the dihedral angles has to be greater than or equal to $90°$, which is a contradiction.

The other cases can be investigated analogously. □

**Remark 2.8.** For a nonobtuse tetrahedron the strict inequalities in Lemma 2.7 change to nonstrict.

## 3 Why do acute and nonobtuse tetrahedralizations differ so much?

In Fig. 3.1 we observe dissections of an obtuse triangle (see [17]) and the unit square (see [7]) into acute triangles. We can easily generate uniform partitions consisting of congruent acute triangles, since each triangle is a plane filler. Note also that each quadrangle is a plane filler and it can be divided into two triangles. The famous Penrose rhombic tiles (see [18]) can be divided into two acute isosceles triangles that can be used to get a nonperiodic tiling of the plane by triangles with the maximum angle 72°. Bern et al. introduced in [3] an efficient algorithm which generates two-dimensional partitions with maximum angle at most 80°. In [8] another algorithm is given that enables us to divide special polygonal domains into almost equilateral triangles with angles not exceeding 72°. We easily find that each inner vertex of such partitions (i.e., the vertex which does not lie on the boundary $\partial\Omega$) has to be surrounded by at least 5 triangles (cf. Fig. 3.1). On the other hand, each inner vertex of a partition consisting of nonobtuse triangles is surrounded by at least 4 triangles.

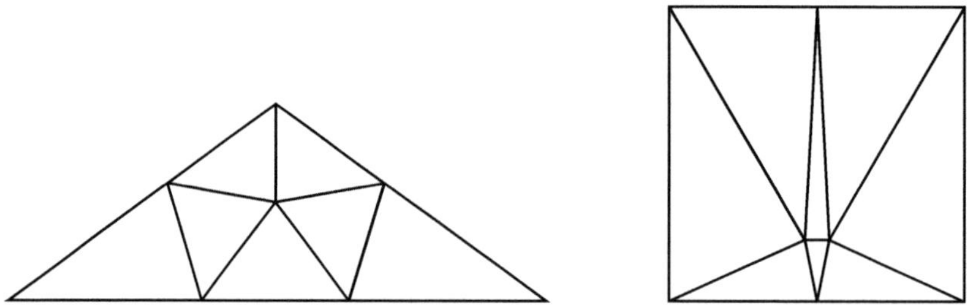

Figure 3.1

To generate acute or nonobtuse tetrahedralizations is considerably more difficult than in the case $d = 2$. Typical difficulties are described, e.g., in [10], [12]. Theorems 3.1 and 3.2 below show why an inner vertex of an acute tetrahedralization cannot be surrounded, e.g., by 18 or 31 tetrahedra.

**Theorem 3.1.**

a) *In any acute tetrahedralization each inner vertex is surrounded by at least* 20 *tetrahedra.*

b) *In any nonobtuse tetrahedralization each inner vertex is surrounded by at least* 8 *tetrahedra.*

c) *These numbers are attainable.*

Proof. a) Let $A$ be an arbitrary inner vertex of a given acute tetrahedralization and let $A$ be surrounded by $t$ tetrahedra $T_1, T_2, \ldots, T_t$ from this tetrahedralization. Define a polyhedron $P$ by the formula

$$P = \bigcup_{i=1}^{t} T_i. \tag{3.1}$$

Assume that $P$ has $v$ vertices. Since $T_i$ are acute tetrahedra, faces of $P$ are evidently triangles. Their number is obviously $t$ and let $e$ be the number of all edges on the surface $\partial P$. Then we have

$$3\,t = 2\,e, \tag{3.2}$$

since each triangle has 3 edges and each edge is shared by exactly two triangles.

The polyhedron $P$ is evidently convex, and thus the well-known Euler formula holds:

$$v + t = e + 2. \tag{3.3}$$

Since each vertex has to be surrounded by at least 5 triangles (compare with Remark 2.6), we have

$$5\,v \leq 3\,t. \tag{3.4}$$

From this, (3.2), and (3.3) we get

$$t = 6\,t - 5\,t \geq 10\,v + 10\,t - 15\,t = 10\,v + 10\,t - 10\,e = 20. \tag{3.5}$$

b) Consider now a nonobtuse tetrahedralization with an inner vertex $A$. Let $P$ be again defined by (3.1). We divide $\partial P$ into triangles such that each triangle is a face of only one of the tetrahedra $T_1, \ldots, T_t$. Since each vertex on the surface $\partial P$ is surrounded by at least 4 triangles (cf. Remark 2.6), we find as in (3.4) and (3.2) that

$$4\,v \leq 3\,t = 2\,e$$

holds. From this and (3.3), we obtain

$$t = 3\,t - 2\,t \geq 4\,v + 4\,t - 6\,t = 4\,v + 4\,t - 4\,e = 8. \tag{3.6}$$

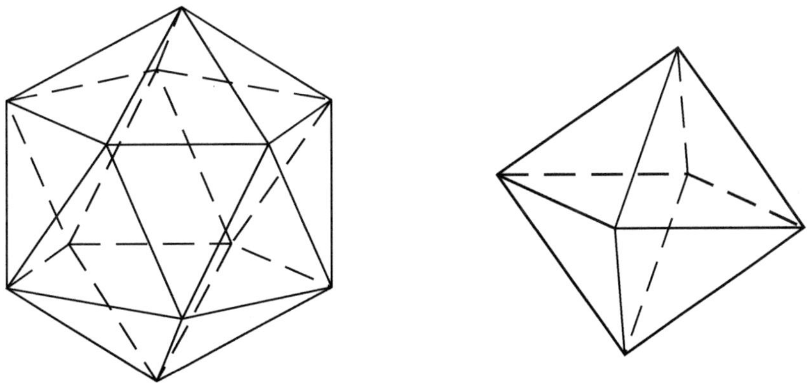

Figure 3.2

c) Consider the regular icosahedron with faces $F_1, F_2, \ldots, F_{20}$ (cf. Fig. 3.2) and let $A$ be its centre of gravity. Setting

$$T_i = \operatorname{conv}\{A, F_i\}, \tag{3.7}$$

we deduce that $\mathcal{T} = \{T_i\}_{i=1}^{20}$ forms an acute tetrahedralization of the icosahedron into congruent tetrahedra. Note that three dihedral angles of each $T_i$ are exactly $72°$ and the remaining three dihedral angles are acute (approximately $69°$).

The regular octahedron can be similarly subdivided into 8 congruent nonobtuse tetrahedra. Note that three dihedral angles of each tetrahedron are exactly $90°$ and the remaining three dihedral angles are acute (approximately $55°$). □

**Theorem 3.2.** *In any tetrahedralization each inner vertex is surrounded by an even number of tetrahedra.*

Proof. Apply equality (3.2). □

**Example 3.3.** Let $T_\varepsilon$ be a tetrahedron with vertices $(0,0,0)$, $(1,\varepsilon,\varepsilon)$, $(\varepsilon,1,\varepsilon)$, and $(\varepsilon,\varepsilon,1)$ for $\varepsilon \in [0,1)$. Consider a nonobtuse tetrahedralization containing $T_\varepsilon$ such that the point $(0,0,0)$ is its inner vertex. According to Theorem 3.1, this point can be surrounded by 8 nonobtuse tetrahedra when $\varepsilon = 0$. On the other hand, if $\varepsilon \to 0^+$ then $T_\varepsilon$ is arbitrarily close to $T_0$, but at least 20 tetrahedra have to surround the point $(0,0,0)$ to get an acute tetrahedralization.

**Theorem 3.4.** *In any acute and nonobtuse tetrahedralization each inner vertex is shared by at least* 12 *and* 6 *edges, respectively.*

Proof. According to (3.3), (3.2), Theorem 3.2, and (3.5), for an inner vertex of an acute tetrahedralization we find that

$$v = 2 + \frac{t}{2} \geq 2 + \frac{20}{2} = 12.$$

Analogously, for a nonobtuse tetrahedralization we have that $v = 2 + t/2 \geq 6$ thanks to (3.6). □

**Remark 3.5.** It is known that the three-dimensional Euclidean space $\mathbb{R}^3$ cannot be filled by regular tetrahedra, even though Aristotle in his book *On the heaven* (350 BC, Vol. 3, Chapt. 8) asserted that this is possible and that each edge in such a tetrahedralization is surrounded by 5 tetrahedra. However, this would require that the dihedral angle between any two faces of the regular tetrahedron is $72°$. This mistake was repeated until the 16th century, when it was found that this angle is $\arccos \frac{1}{3}$, which is approximately $71°$.

**Remark 3.6.** To the authors' knowledge there is no algorithm which would generate an acute tetrahedralization of the whole space $\mathbb{R}^3$. On the other hand, there are several efficient techniques concerning how to construct nonobtuse tetrahedralizations (see, e.g., [1], [2], [11], [12], [13], [15]).

## 4 Extensions to $\mathbb{R}^d$

A $d$-simplex from $\mathbb{R}^d$ is said to be *acute* (*nonobtuse*) if all dihedral angles between its $(d-1)$-dimensional faces are acute (nonobtuse). A partition $\mathcal{T}$ is said to be *acute* (*nonobtuse*) if all $d$-simplexes from $\mathcal{T}$ are acute (nonobtuse). By [2, Sec. 1.2] no large angle at a $(d-i)$-face implies no large angle at a $(d-i-1)$-face for $i = 2, \ldots, d-1$ (cf. Lemma 2.3).

**Remark 4.1.** The regular $d$-simplex is a space filler only for $d \leq 2$. This follows from the fact that the dihedral angle $\alpha = \alpha(d)$ between its two $(d-1)$-dimensional faces satisfies for $d > 1$ the relation

$$\alpha(d) = \arccos \frac{1}{d} < 90°.$$

Therefore, $\alpha(2) = 60° < \alpha(3) < 72° < \alpha(4) \approx 76°$. Since $\{\alpha(d)\}_{d=2}^{\infty}$ is an increasing sequence, it is true that $k\alpha(d) \neq 360°$ for any $d > 2$ and any integer $k$.

**Remark 4.2.** The unit $d$-cube $[0,1]^d$ can be partitioned into $d!$ nonobtuse $d$-simplexes $S_\sigma$ for $\sigma \in \Sigma^d$, where $\Sigma^d$ is the symmetric group of all $d!$ permutations of the numbers $1, 2, \ldots, d$ and

$$S_\sigma = \{x = (x_1, \ldots, x_d) \in \mathbb{R}^d \mid 0 \leq x_{\sigma(1)} \leq \cdots \leq x_{\sigma(d)} \leq 1\}.$$

The space $\mathbb{R}^d$ can be thus partitioned into nonobtuse $d$-simplexes. Note that any $d$-rectangle can also be partitioned into $d!$ nonobtuse $d$-simplexes as the unit $d$-cube. Other examples can be found in [6].

**Remark 4.3.** According to [20], there does not exist any tiling of the whole space $\mathbb{R}^d$, $d \notin \{2,4\}$, by regular polytopes except for $d$-cubes. On the other hand, the space $\mathbb{R}^2$ can be, moreover, partitioned into equilateral triangles and also into regular hexagons. The space $\mathbb{R}^4$ can be partitioned into congruent regular 4-orthoplexes (the four-dimensional analogue of the regular octahedron), and also into congruent regular 24-cells whose three-dimensional faces are regular octahedra. Since these two kinds of regular polytopes can be partitioned into nonobtuse 4-simplexes, there exist other partitions of $\mathbb{R}^4$ into nonobtuse 4-simplexes than the one described in Remark 4.2 (see also the proof of Theorem 4.5 below).

**Remark 4.4.** In planar partitions each inner vertex is surrounded by at least 5 acute triangles or by at least 4 nonobtuse triangles. From Theorem 3.1 we know that for $d = 3$ this ratio is bigger, namely $20:8$. However, for $d = 4$ this ratio seems to be even much bigger, namely $600:16$. Below we prove that these numbers are attainable (see Theorem 4.5 and Remark 4.7).

To this end first consider the Euler-Poincaré formula (see [16, p. 147]) which is a generalization of (3.3). This formula states that

$$f_0 - f_1 + f_2 - \cdots + (-1)^{d-1} f_{d-1} = 1 + (-1)^{d-1}, \tag{4.1}$$

where $f_m$ is the number of $m$-dimensional faces of a convex $d$-dimensional polytope. For instance, we have $f_0 = f_3 = 5$ and $f_1 = f_2 = 10$ for any 4-simplex.

**Theorem 4.5.** *In any nonobtuse partition in* $\mathbb{R}^4$ *each inner vertex is surrounded by at least sixteen 4-simplexes. This number is attainable.*

Proof. Setting $v = f_0$, $e = f_1$, $t = f_2$, and $s = f_3$, the Euler-Poincaré formula (4.1) for $d = 4$ has the form

$$v + t = e + s. \tag{4.2}$$

Let $A$ be an arbitrary inner vertex of a given nonobtuse partition. Consider a bounded convex polytopic four-dimensional domain $P$ defined as in (3.1), where tetrahedra are replaced by 4-simplexes. Its three-dimensional surface $\partial P$ is thus covered by $s$ tetrahedra (each tetrahedron being the exterior three-dimensional face of the corresponding 4-simplex). It is easy to see that

$$t = 2\,s, \tag{4.3}$$

since each triangular face is shared by exactly two tetrahedra and each tetrahedron has 4 triangular faces. Hence, the Euler-Poincaré formula (4.2) reduces to the form

$$v + s = e. \tag{4.4}$$

By a contradiction argumentation, we find that for a nonobtuse partition at least 6 edges meet in each vertex of $\partial P$. This yields that $v \geq 7$ and $6v \leq 2e$, i.e.,

$$3\,v \leq e. \tag{4.5}$$

We shall distinguish two cases (excluding the first case):

1) Let $v = 7$. Then the associated graph is the complete graph $K_7$, which has

$$t = \binom{7}{3} = 35$$

triangular faces. Since this is an odd number, this case is impossible due to (4.3).

2) Let $v \geq 8$. Then by (4.5) we conclude that

$$e - v \geq 2\,v \geq 16,$$

and by (4.4) we get $s \geq 16$.

For the regular 4-orthoplex we have $v = 8$, $e = 24$, $t = 32$, and $s = 16$. Its three-dimensional surface is thus covered by 16 regular tetrahedra (see [20]), and hence, it can be partitioned into 16 nonobtuse 4-simplexes defined as in (3.7). □

**Remark 4.6.** By (4.3) the total number of triangular faces of the three-dimensional surface $\partial P$ from the above proof is always even (cf. (3.2)).

**Remark 4.7.** There exists the regular four-dimensional 600-cell (see, e.g., [5], [20]) for which $v = 120$, $e = 720$, $t = 1200$, $s = 600$, and its three-dimensional faces are the regular tetrahedra. It can be, therefore, partitioned into 600 congruent 4-simplexes defined as in (3.7). An easy calculation shows that these 4-simplexes are acute.

**Remark 4.8.** For each $d > 1$ there exists exactly one integer $n(d)$ such that each inner vertex in any nonobtuse partition is surrounded by at least $n(d)$ simplexes and this number is attainable. Similarly we can define a number $a(d)$ (possibly infinite) for acute partitions in $\mathbb{R}^d$. Clearly,

$$a(d) \geq n(d).$$

Due to Theorems 3.1 and 4.5 and Remark 4.4, we have

$$n(2) = 4, \quad n(3) = 8, \quad n(4) = 16.$$

Note that an analogue of the regular octahedron exists in any dimension $d \geq 1$. It is defined via the formula

$$P = \left\{ x = (x_1, \ldots, x_d) \mid \sum_{i=1}^{d} |x_i| \leq 1 \right\}.$$

Since it can be partitioned into $2^d$ nonobtuse congruent $d$-simplexes, we have $n(d) \leq 2^d$ for $d > 1$, and the converse inequality can be obtained by a contradiction argumentation.

From Theorem 3.1 and Remarks 4.4 and 4.7, we conclude that

$$a(2) = 5, \quad a(3) = 20, \quad a(4) \leq 600.$$

**Open problem**: *Which are the values of $a(d)$ for $d \geq 4$?*

**Acknowledgement.** This paper was supported by grant A 1019201 of the Grant Agency of the Academy of Sciences of the Czech Republic and the grant FRVŠ obtained from the Czech Technical University. The authors thank Prof. Lawrence Somer (Washington, D.C.) and Karel Kolman (Prague) for valuable suggestions.

## References

[1] L. Beilina, S. Korotov, and M. Křížek, *Application of the local nonobtuse tetrahedral refinement techniques near Fichera-like corner*, Preprint No. 2003-02, Chalmers Finite Element Center, Göteborg (2003), 1–17.

[2] M. Bern, P. Chew, D. Eppstein, and J. Ruppert, *Dihedral bounds for mesh generation in high dimensions*, Proc. of the Sixth Annual ACM-SIAM Sympos. on Discrete Algorithms (San Francisco, 1995), ACM, New York, 1995, pp. 189–196.

[3] M. Bern, D. Eppstein, and J.R. Gilbert, *Provably good mesh generation*, J. Comput. System Sci. **48** (1994), 384–409.

[4] P. G. Ciarlet, *Basic error estimates for elliptic problems*, Handbook on Numer. Anal., vol. II (P. G. Ciarlet, J. L. Lions, eds.), North Holland, Amsterdam, 1991.

[5] H. S. M. Coxeter, *Regular Polytopes*, Methuen, London, 1948.

[6] H. E. Debrunner, *Tiling Euclidean d-space with congruent simplexes*, Proc. Conf. Discrete Geometry and Convexity, New York, 1982, Ann. N. Y. Acad. Sci. Vol. 440, 1985, pp. 230–261.

[7] M. Gardner, *Mathematical games*, Scient. Amer. **202** (1960), 172–186.

[8] J. L. Gerver, *The dissection of a polygon into nearly equilateral triangles*, Geom. Dedicata **16** (1984), 93–106.

[9] D. Gilbarg, T. D. Trudinger, *Elliptic Partial Differential Equations of Second Order*, Springer Velag, Berlin-New York, 1977.

[10] J. Karátson, S. Korotov, *Discrete maximum principles for finite element solutions of nonlinear elliptic problem with mixed boundary conditions*, Preprint of Eötvös Loránd University (2003), 1–24.

[11] S. Korotov, M. Křížek, *Local nonobtuse tetrahedral refinements of a cube*, Appl. Math. Lett. **16** (2003), 1101–1104.

[12] S. Korotov, M. Křížek, *Techniques for global and local nonobtuse refinements of tetrahedral partitions*, to appear in Comput. Math. Appl. in 2004.

[13] S. Korotov, M. Křížek, *Dissection of an arbitrary polyhedron into nonobtuse tetrahedra*, Preprint no. B 3/2002, Dept. of Math. Inform. Technology, Univ. of Jyväskylä, 1–6.

[14] S. Korotov, M. Křížek, and P. Neittaanmäki, *Weakened acute type condition for tetrahedral triangulations and the discrete maximum principle*, Math. Comp. **70** (2001), 107–119.

[15] M. Křížek, Qun Lin, *On diagonal dominance of stiffness matrices in 3D*, East-West J. Numer. Math. **3** (1995), 59–69.

[16] L. A. Lyusternik, *Convex figures and polyhedra*, Dover Publications, Inc., New York, 1963; Moscow, 1956.

[17] W. Manheimer, J. P. Federico et al, *Dissecting an obtuse triangle into acute triangles*, Amer. Math. Monthly **67** (1960), 923.

[18] R. Penrose, *Pentaplexity: a class of nonperiodic tilings of the plane*, Math. Intelligencer **2** (1979/80), 32–37.

[19] K. Rektorys et al., *Survey of Applicable Mathematics, Vol. 1*, Kluwer Academic, Dordrecht, 1994.

[20] J. Stillwell, *The story of the 120-cell*, Notices Amer. Math. Soc. **48** (2001), 17–24.

# A Posteriori Error Estimation of "Quantities of Interest" on "Quantity-Adapted" Meshes

Pekka Neittaanmäki,[1] Sergey Korotov,[2] Janne Martikainen[3]

[1]Department of Mathematical Information Technology, University of Jyväskylä, P.O. Box 35, FIN-40014 Jyväskylä, Finland pn@mit.jyu.fi
[2]Department of Mathematical Information Technology, University of Jyväskylä, P.O. Box 35, FIN-40014 Jyväskylä, Finland korotov@mit.jyu.fi
[3]Department of Mathematical Information Technology, University of Jyväskylä, P.O. Box 35. FIN-40014 Jyväskylä, Finland jamartik@mit.jyu.fi

## 1 Introduction

A posteriori error estimation plays an important role in the modern numerical analysis. The direction from which this topic initially started was the estimation of the computational error in global (energy) norm. Concerning various approaches for this type of estimation we refer to works [1], [2], [3], [4], [5], [6], [13], [15], [16], [17], [18]. For approximations of linear elliptic problems, they examine error estimates in the global (energy) norm and suggest error indicators/estimators that are further used in various mesh adaptive procedures (see, e.g., [10]). Global error estimates give a general presentation on the quality of an approximate solution and a stopping criteria.

However, for engineering purposes, such an information is often not sufficient. In many cases, analysts are mainly interested not in the value of the total error, but in errors (so-called "quantities of interest") over various "zones of interests": certain subdomains, lines, or at special points. A possible way of estimating such errors is to introduce a linear functional $\ell$ associated with a "quantity of interest" and to obtain an estimate for the value of $\langle \ell, u - v \rangle$, where $u$ is the exact solution and $v$ is an approximate one. The methods developed so far (see, e.g., [1], [7], [8], [14]) find estimates of $\langle \ell, u - u_h \rangle$ for a Galerkin approximation $u_h$ by employing an additional (adjoint) problem, whose right-hand side is formed by the functional $\ell$. If the Galerkin approximation of the adjoint problem is computed on the same mesh as $u_h$, then the quantity $\langle \ell, u - u_h \rangle$ is expressed via a certain integral functional, which can be estimated by using, e.g., "equilibrated residual method" (see, e.g., [1], [14]).

A new way of estimation of "quantities of interest" was recently proposed in [11]. It is based on two principles: (a) the original and adjoint problems are solved on non-coinciding meshes, and (b) the term presenting the product of errors arising in the primal and adjoint problems is estimated by the gradient recovery technique widely used in various applied problems (see [9], [12], [17], [18]). This differs the approach in [11] from the others, where it is usually assumed that the Galerkin approximations of the primal and adjoint problems are computed on the same finite-dimensional subspaces. However, the question of choice of optimal (computationally cheap and reliable) mesh for solving the adjoint problem (so-called *adjoint mesh*) appears. In [11], it was suggested to build the adjoint meshes by local refining the *primal meshes* (meshes used for solving the primal problems) in the neigbourhoods of the corresponding "zones of interest". This idea is justified numerically in the present paper for one and two-dimensional model problems of elliptic type.

## 2 The test problem and the error estimator

### 2.1 Formulation of the test problem

The test problem in a classical setting reads as follows: Find a function $u$ such that

$$-\operatorname{div}(A\nabla u) = f \quad \text{in } \Omega, \tag{2.1}$$

$$u = 0 \quad \text{on } \partial\Omega, \tag{2.2}$$

where $\Omega$ is a bounded polygonal domain in $\mathbb{R}^2$ with a Lipschitz boundary $\partial\Omega$, the matrix of coefficients $A = \{a_{ij}(x)\}_{i,j=1}^2$ is symmetric and meets the conditions

$$a_{ij}(x) \in L_\infty(\Omega), \quad A(x)\xi \cdot \xi \geq C\|\xi\|^2 \quad \forall \xi \in \mathbb{R}^2 \qquad \text{for a. e. } x \in \overline{\Omega}. \tag{2.3}$$

It is also assumed that

$$f \in L_2(\Omega). \tag{2.4}$$

We formulate first the primal and adjoint problems as follows.

**Primal Problem** ($\mathcal{P}$): Find $u \in V_0 := H_0^1(\Omega)$ such that

$$\int_\Omega A\nabla u \cdot \nabla w \,\mathrm{d}x = \int_\Omega f w \,\mathrm{d}x \quad \forall w \in V_0. \tag{2.5}$$

Let $\tilde{u} \in V_0$ be an approximation of $u$, and assume that we are interested in the value of the following "quantity of interest":

$$\langle \ell, u - \tilde{u} \rangle = \int_\Omega \varphi(u - \tilde{u}) \,\mathrm{d}x, \tag{2.6}$$

where $\varphi \in L_2(\Omega)$ and $\operatorname{supp}\varphi \subset \omega \subseteq \Omega$. Estimation of such a quantity for one (or several) a priori given functions $\varphi$ gives information about the behaviour of $u - \tilde{u}$ in a subdomain $\omega$, which is called later on as *zone of interest*.

**Adjoint Problem** $(\mathcal{P}_a)$: Find $v \in V_0$ such that

$$\int_\Omega A\nabla v \cdot \nabla w \, dx = \langle \ell, w \rangle \quad \forall w \in V_0. \tag{2.7}$$

Let now $V_h$ and $V_\tau$ be two finite-dimensional subspaces in $V_0$, constructed by the Courant type finite element discretization on triangular meshes $\mathcal{T}_h$ and $\mathcal{T}_\tau$ of the solution domain $\Omega$, respectively. Consider the corresponding finite-dimensional problems:

**Problem** $(\mathcal{P}^h)$: Find $u_h \in V_h$ such that

$$\int_\Omega A\nabla u_h \cdot \nabla w_h \, dx = \int_\Omega f w_h \, dx \quad \forall w_h \in V_h, \tag{2.8}$$

and

**Problem** $(\mathcal{P}_a^\tau)$: Find $v_\tau \in V_\tau$ such that

$$\int_\Omega A\nabla v_\tau \cdot \nabla w_\tau \, dx = \langle \ell, w_\tau \rangle \quad \forall w_\tau \in V_\tau. \tag{2.9}$$

Then, as proved in [11],

$$\langle \ell, u - u_h \rangle =: E(u_h, v_\tau) = E_0(u_h, v_\tau) + E_1(u_h, v_\tau), \tag{2.10}$$

where

$$E_0(u_h, v_\tau) = \int_\Omega f v_\tau \, dx - \int_\Omega A\nabla v_\tau \cdot \nabla u_h \, dx, \tag{2.11}$$

$$E_1(u_h, v_\tau) = \int_\Omega A\nabla(v - v_\tau) \cdot \nabla(u - u_h) \, dx. \tag{2.12}$$

In a particular case of $V_h \equiv V_\tau$, the term $E_0(u_h, v_\tau) = 0$ due to the orthogonality condition in (2.8). Thus, if the meshes used in Problems $(\mathcal{P}^h)$ and $(\mathcal{P}_a^\tau)$ coincide, then equality (2.10) has only one term containing the product of the (unknown) energy errors. On the contrary, using non-coinciding meshes leads to another equality that has two terms. Obviously, the directly calculable term $E_0(u_h, v_\tau)$ dominates if $v_\tau$ is "sufficiently close" to the exact solution $v$ of Problem $(\mathcal{P}_a)$, and, thus, in this case, $E_0(u_h, v_\tau)$ contains the major part of the "quantity of interest".

## 2.2 Construction of the estimator

Finding a sharp approximation for the solution of the adjoint problem may require high computational costs. A more economical way is to use an approximate solution of the adjoint problem having approximately the same size as the approximate solution of the primal one and recover unknown functions $\nabla u$ and $\nabla v$ by some post-processing techniques.

On $\mathcal{T}_h$, we define the gradient averaging operator

$$G_h\colon [L_\infty(\Omega)]^2 \to [W_2^1(\Omega)]^2$$

as the operator that defines a vector-valued piecewise linear function by setting each its nodal value as the mean value of $\nabla u_h$ on all elements incident with the corresponding nodal point. The averaging operator $G_\tau\colon [L_\infty(\Omega)]^2 \to [W_2^1(\Omega)]^2$ is defined in the same manner. Such averaging procedures are widely used in the mathematical modelling (see, e.g., [1], [4], [9], [12], [17]), and found to be very efficient if the superconvergence phenomena take place.

Further, we replace $E(u_h, v_\tau)$ by the functional

$$\widetilde{E}(u_h, v_\tau) := E_0(u_h, v_\tau) + \widetilde{E}_1(u_h, v_\tau), \tag{2.13}$$

where

$$\widetilde{E}_1(u_h, v_\tau) = \int_\Omega A(\nabla u_h - G_h(\nabla u_h)) \cdot (\nabla v_\tau - G_\tau(\nabla v_\tau))\,\mathrm{d}x. \tag{2.14}$$

The functional $\widetilde{E}(u_h, v_\tau)$ is directly computable once the approximations $u_h$ and $v_\tau$ are defined.

We further observe that $E_0 = \mathcal{O}(h)$, $E_1 = \mathcal{O}(h\tau)$, and $\widetilde{E}_1 = \mathcal{O}(h\tau)$. The following result is proved in [11].

**Theorem 1.** *Under usual assumptions leading to superconvergence of the averaged gradients (see, e.g., [9]), we have for sufficiently small $h$ and $\tau$ ($\tau < h$)*

$$\begin{aligned} &|E_1(u_h, v_\tau) - \widetilde{E}_1(u_h, v_\tau)| \\ &\qquad \leq C(h\tau^{\frac{3}{2}} + \tau h^{\frac{3}{2}} + h\tau(h+\tau)^{\frac{1}{2}-\frac{1}{m}} + \tau^2) + \mu(h,\tau), \end{aligned} \tag{2.15}$$

*where $m$ is any positive integer greater than 2, $\mu(h,\tau)$ contains higher order terms, and the constant $C$ does not depend on $h$ and $\tau$.*

In view of Theorem 1, it follows that the procedure of changing $E_1(u_h, v_\tau)$ to $\widetilde{E}_1(u_h, v_\tau)$ is assymptotically correct, and the following computational strategy can be proposed:

a) Define $V_\tau$ taking into an account the nature of the functional $\ell$ and calculate $v_\tau$;
b) Define $V_h$ and calculate $u_h$;
c) Calculate $E_0(u_h, v_\tau)$ directly and use post-processed values of $\nabla u_h$ and $\nabla v_\tau$ to estimate $E_1(u_h, v_\tau)$ replacing gradients by the averaged gradients.

This strategy has been first checked in [11]. However, not much efforts was devoted to numerical justification of item a), therefore, in the present work, we test numerically the idea of construction of the adjoint mesh by local refinements of the corresponding primal mesh in the neighbourhood of the zone of interest, which is equivalent to putting extra trial functions in a subdomain associated with it.

# 3 Numerical experiments

The effectivity index is defined as follows

$$I_{\text{eff}} = \frac{|\tilde{E}(u_h, v_\tau)|}{|\langle \ell, u - u_h \rangle|}. \tag{3.1}$$

This index shows the effectivity of the estimation, the closer it is to 1, the better the estimate is.

In what follows, we always denote the number of nodes in the primal mesh by $N$, and the number of nodes in the corresponding adjoint mesh by $M$.

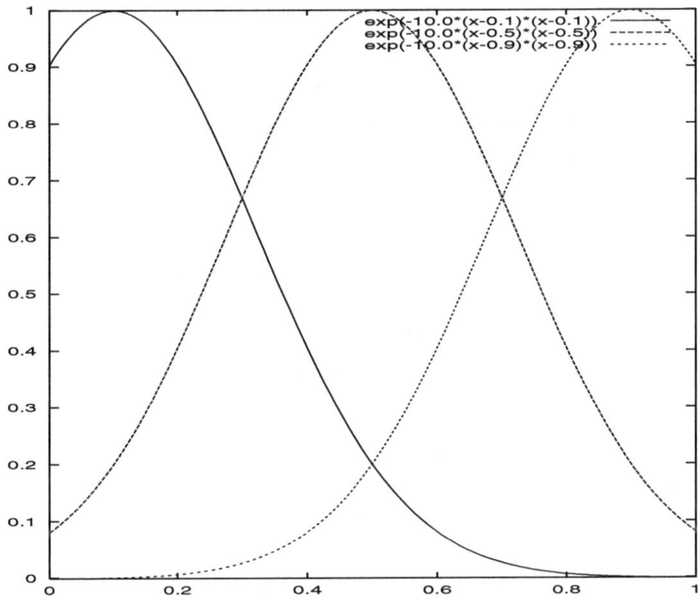


Figure 1. The shapes of the righ-hand side functions for problem (3.2).

## 3.1 One-dimensional case

**Test Problem 1**: Find $u$ such that

$$-u'' = f \quad \text{in } (0,1), \tag{3.2a}$$

$$u(0) = u(1) = 0. \tag{3.2b}$$

The function $\varphi$ in (2.6) is defined to be 1 in "zone of interest" $\omega_1 = (0, 0.125)$ and 0 outside of it in all one-dimensional computations. In the corresponding series of tests, we employ 4 types of adjoint meshes (see Fig. 2). The first adjoint mesh coincides with the primal one, thus giving $E_0 = 0$. In the second case, we perform double refinement in the zone of interest and make mesh coarsening outside of it. The third adjoint mesh is obtained from the primal one by single refining only in the zone of interest. The fourth adjoint mesh is obtained by uniform global refining the primal one. We also vary the right-hand side function $f$ in (3.2) (see Fig. 1). Note that we need to solve the adjoint problem only once for each type of the adjoint meshes and use it for various choices of the right-hand side function $f$.

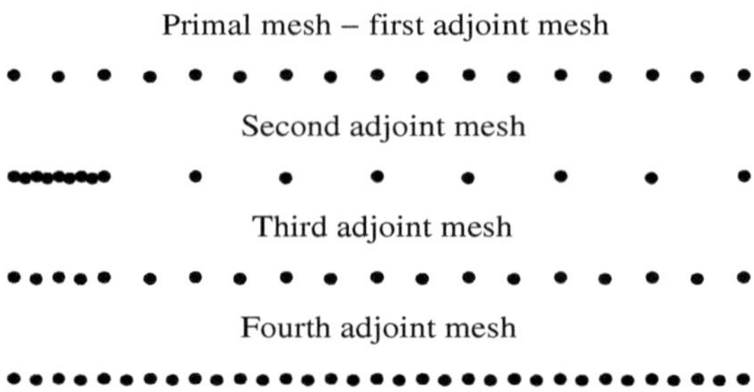


Figure 2. Primal and adjoint meshes.

Table 1. The results of Test 1a: $f = \mathrm{e}^{-10(x-0.1)^2}$.

| $N$ | Adj. mesh | $M$ | $E_0$ | $\tilde{E}_1$ | $\tilde{E}$ | $\langle \ell, u - u_h \rangle$ | $I_{\text{eff}}$ |
|---|---|---|---|---|---|---|---|
| 17 | 1st | 17 | 0.00e-0 | 4.02e-5 | 4.02e-5 | 3.97e-5 | 1.01 |
| 17 | 2nd | 16 | 3.72e-5 | 2.71e-6 | 3.99e-5 | 3.97e-5 | 1.01 |
| 17 | 3rd | 19 | 2.98e-5 | 8.90e-6 | 3.87e-5 | 3.97e-5 | 0.97 |
| 17 | 4th | 33 | 2.98e-5 | 8.78e-6 | 3.85e-5 | 3.97e-5 | 0.97 |

Table 2. The results of Test 1b: $f = \mathrm{e}^{-10(x-0.5)^2}$.

| $N$ | Adj. mesh | $M$ | $E_0$ | $\tilde{E}_1$ | $\tilde{E}$ | $\langle \ell, u - u_h \rangle$ | $I_{\text{eff}}$ |
|---|---|---|---|---|---|---|---|
| 17 | 1st | 17 | 0.00e-0 | 5.94e-6 | 5.94e-6 | 6.20e-6 | 0.96 |
| 17 | 2nd | 16 | 5.81e-6 | 1.33e-7 | 5.94e-6 | 6.20e-6 | 0.96 |
| 17 | 3rd | 19 | 4.65e-6 | 1.19e-6 | 5.84e-6 | 6.20e-6 | 0.94 |
| 17 | 4th | 33 | 4.65e-6 | 1.44e-6 | 6.09e-6 | 6.20e-6 | 0.98 |

Table 3. The results of Test 1c: $f = \mathrm{e}^{-10(x-0.9)^2}$.

| $N$ | Adj. mesh | $M$ | $E_0$ | $\tilde{E}_1$ | $\tilde{E}$ | $\langle \ell, u - u_h \rangle$ | $I_{\text{eff}}$ |
|---|---|---|---|---|---|---|---|
| 17 | 1st | 17 | 0.00e-0 | 2.80e-8 | 2.80e-8 | 4.38e-8 | 0.64 |
| 17 | 2nd | 16 | 4.00e-8 | 8.92e-9 | 3.11e-8 | 4.38e-8 | 0.71 |
| 17 | 3rd | 19 | 3.20e-8 | 1.77e-9 | 3.37e-8 | 4.38e-8 | 0.77 |
| 17 | 4th | 33 | 3.30e-8 | 9.35e-9 | 4.23e-8 | 4.38e-8 | 0.97 |

The results of the tests are exposed in Tables 1–3. It is immediately seen that the calculations performed on coinciding meshes give vanishing term $E_0$ as expected. The best results in the one-dimensional case are obtained for the fourth adjoint mesh, i.e., when $\frac{M}{N} \approx 2$. However, sufficiently reliable results are obtained on adjoint meshes with considerably smaller ratio $\frac{M}{N} \approx 0.95 - 1.1$, which clearly demonstrate the high effectivity of our estimator. It is obvious that the ratio can be even more close to 1 if the length of the zone of interest is smaller with respect to the length of the solution interval than in the tests presented here. Moreover, the tests with the second mesh show that a clever choice of the adjoint mesh can provide with a reasonably cheap estimation (number of nodes in the adjoint mesh is less than the number of points in the primal one).

### 3.2 Two-dimensional case

In the results below, we expose the performance of our estimator for the standard cases of the problems posed in square and $L$-shaped domains.

**Test Problem 2**: Find $u$ such that

$$-\Delta u = f \quad \text{in } \Omega, \tag{3.3a}$$

$$u = 0 \quad \text{on } \partial\Omega, \tag{3.3b}$$

where $\Omega = (0,1) \times (0,1)$ and $f \equiv 1$.

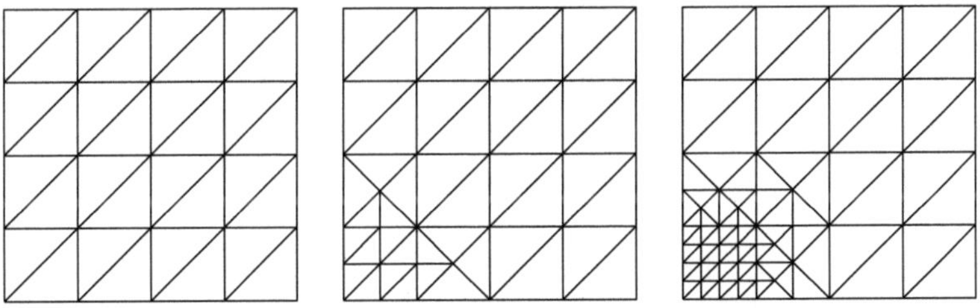

Figure 3. No local refinement (*left*), local refinement done 1 time (*center*), local refinement done 2 times (*right*) for Test Problem 2.

The function $\varphi$, defining the "quantity of interest" is equal to 1 in the "zone of interest"

$$\omega_2 = (0.0, 0.25) \times (0.0, 0.25),$$

and is equal to 0 ouside of it. In Fig. 3, we present the construction of the adjoint meshes from the primal ones by local refining one and two times in the neigbourhood of the zone of interest.

From the tests presented in Table 4, we observe that the meshes chosen as adapted to the zone of interest give very good results for reasonably cheap price. Thus we get the effectivety index between 1.04 and 1.44 with ratio

$\frac{M}{N} \approx 1.25$, which is obtained by a single local refinement. As could be expected, more local refinements give better results.

Table 4. The results of calculations for Test Problem 2.

| $N$ | Loc. ref. | $M$ | $E_0$ | $\tilde{E}_1$ | $\tilde{E}$ | $\langle \ell, u - u_h \rangle$ | $I_{\text{eff}}$ |
|---|---|---|---|---|---|---|---|
| 25 | - | 25 | 0.00e-0 | 1.84e-4 | 1.84e-4 | 8.50e-5 | 2.16 |
| 25 | 1 time | 32 | 6.12e-5 | 6.03e-5 | 1.22e-4 | 8.50e-5 | 1.44 |
| 25 | 2 times | 57 | 4.84e-5 | 4.73e-5 | 9.56e-5 | 8.50e-5 | 1.12 |
| 81 | - | 81 | 0.00e-0 | 4.95e-5 | 4.95e-5 | 3.51e-5 | 1.41 |
| 81 | 1 time | 101 | 1.91e-5 | 1.92e-5 | 3.83e-5 | 3.51e-5 | 1.09 |
| 81 | 2 times | 172 | 2.07e-5 | 1.49e-5 | 3.57e-5 | 3.51e-5 | 1.02 |
| 289 | - | 289 | 0.00e-0 | 1.27e-5 | 1.27e-5 | 1.03e-5 | 1.23 |
| 289 | 1 time | 353 | 5.28e-6 | 5.38e-6 | 1.07e-5 | 1.03e-5 | 1.04 |
| 289 | 2 times | 588 | 6.23e-6 | 4.06e-6 | 1.03e-5 | 1.03e-5 | 1.00 |

**Test Problem 3**: We solve the problem (3.3) in $L$-shaped domain

$$\Omega = (0,1) \times (0,1) \setminus (0.75,1) \times (0.75,1),$$

with $f \equiv 1$.

The function $\varphi$, defining the quantity of interest, is equal to 1 in the subdomain

$$\omega_3 = (0, 0.25) \times (0.75, 1),$$

and is 0 ouside of it. The zone of interest is, thus, far from the re-entrant corner, where the solution are know to behave "badly". In Fig. 4, we present the construction of the adjoint meshes.

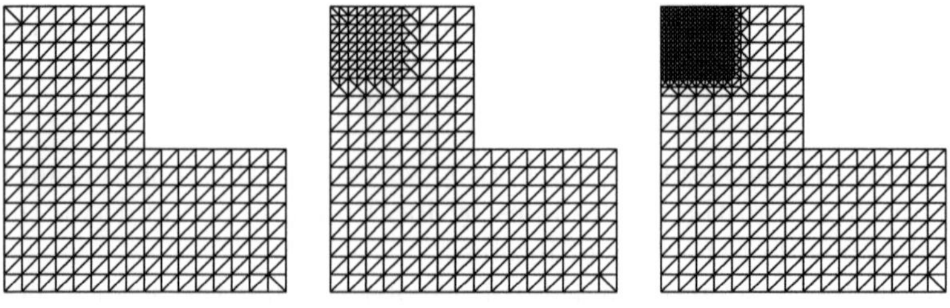

Figure 4. No local refinement (*left*), local refinement done 1 time (*center*), local refinement done 2 times (*right*) for Test Problem 3.

Table 5. The results of calculations for Test Problem 3.

| $N$ | Loc. ref. | $M$ | $E_0$ | $\tilde{E}_1$ | $\tilde{E}$ | $\langle \ell, u - u_h \rangle$ | $I_{\text{eff}}$ |
|---|---|---|---|---|---|---|---|
| 21 | - | 21 | 0.00e-0 | 2.67e-4 | 2.67e-4 | 1.34e-4 | 1.99 |
| 21 | 1 time | 28 | 4.77e-5 | 1.16e-4 | 1.64e-4 | 1.34e-4 | 1.22 |
| 21 | 2 times | 53 | 8.88e-5 | 5.21e-5 | 1.41e-4 | 1.34e-4 | 1.05 |
| 65 | - | 65 | 0.00e-0 | 1.05e-4 | 1.05e-4 | 7.92e-5 | 1.33 |
| 65 | 1 time | 85 | 4.38e-5 | 3.54e-5 | 7.91e-5 | 7.92e-5 | 1.00 |
| 65 | 2 times | 156 | 5.74e-5 | 2.28e-5 | 8.02e-5 | 7.92e-5 | 1.01 |
| 225 | - | 225 | 0.00e-0 | 3.31e-5 | 3.31e-5 | 3.02e-5 | 1.10 |
| 225 | 1 time | 289 | 1.81e-5 | 1.15e-5 | 2.96e-5 | 3.02e-5 | 0.98 |
| 225 | 2 times | 524 | 2.29e-5 | 7.49e-6 | 3.04e-5 | 3.02e-5 | 1.01 |

**Test Problem 4**: We solve the problem (3.3) in $L$-shaped domain

$$\Omega - (0,1) \times (0,1) \setminus (0.75,1) \times (0.75,1),$$

with $f \equiv 1$.

The function $\varphi$ is equal to 1 in the zone

$$\omega_4 = (0.375, 0.625) \times (0.375, 0.625) \setminus (0.5, 0.625) \times (0.5, 0.625),$$

and 0 ouside of it. Now, the zone of interest covers the re-entrant corner. In Fig. 5, we illustrate the construction of the adjoint meshes from the primal ones.

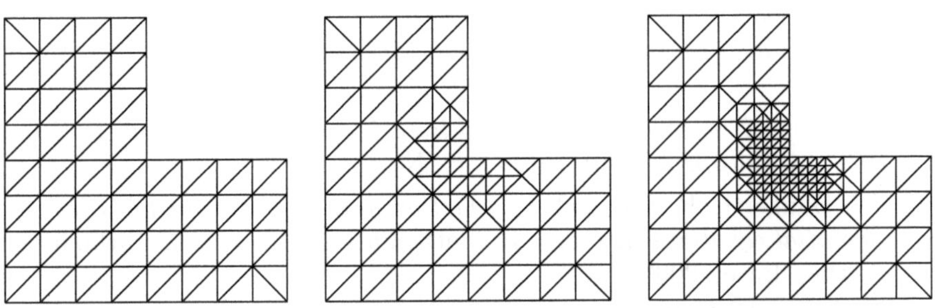

Figure 5. No local refinement (*left*), local refinement done 1 time (*center*), local refinement done 2 times (*right*) for Test Problem 4.

The observations on the performance of the estimator for the above two problems are actually the same as for Test Problem 2, and clearly demonstrate the effectivity of the optimal choice of adjoint meshes as adapted to the defined "quantity of interest".

Table 6. The results of calculations for Test Problem 4.

| $N$ | Loc. ref. | $M$ | $E_0$ | $\tilde{E}_1$ | $\tilde{E}$ | $\langle \ell, u - u_h \rangle$ | $I_{\rm eff}$ |
|---|---|---|---|---|---|---|---|
| 65 | - | 65 | 0.00e-0 | 1.87e-4 | 1.87e-4 | 1.96e-4 | 0.95 |
| 65 | 1 time | 84 | 1.27e-4 | 5.19e-5 | 1.78e-4 | 1.96e-4 | 0.91 |
| 65 | 2 times | 153 | 1.63e-4 | 2.76e-5 | 1.91e-4 | 1.96e-4 | 0.97 |
| 255 | - | 225 | 0.00e-0 | 7.43e-5 | 7.43e-5 | 6.90e-5 | 1.08 |
| 255 | 1 time | 281 | 4.28e-5 | 2.08e-5 | 6.36e-5 | 6.90e-5 | 0.92 |
| 255 | 2 times | 484 | 5.68e-5 | 9.84e-6 | 6.66e-5 | 6.90e-5 | 0.97 |
| 833 | - | 833 | 0.00e-0 | 2.65e-5 | 2.65e-5 | 2.44e-5 | 1.09 |
| 833 | 1 time | 1017 | 1.51e-5 | 7.61e-6 | 2.27e-5 | 2.44e-5 | 0.93 |
| 833 | 2 times | 1704 | 2.04e-5 | 3.24e-6 | 2.36e-5 | 2.44e-5 | 0.97 |

**Acknowledgements.** The authors would like to thank Prof. S. Repin for the valuable discussions and comments during preparation of the paper. The authors were supported by the Agora Center under Grant InBCT of TEKES, Finland.

## References

[1] M. Ainsworth, J.T. Oden, *A Posteriori Error Estimation in Finite Element Analysis*, John Wiley & Sons, Inc., New York, 2000.

[2] I. Babuška, W.C. Rheinbold, *Error estimates for adaptive finite element computations*, SIAM J. Numer. Anal. **15** (1978), 736–754.

[3] I. Babuška, W.C. Rheinbold, *A posteriori error estimates for the finite element method*, Internat. J. Numer. Methods Engrg. **12** (1978), 1597–1615.

[4] I. Babuška, T. Strouboulis, *The Finite Element Method and its Reliability*, Oxford University Press Inc., New York, 2001.

[5] W. Bangerth, R. Rannacher, *Adaptive finite element methods for differential equations*, Lectures in Mathematics ETH Zürich, Birkhäuser-Verlag, Basel, 2003.

[6] R.E. Bank, A. Weiser, *Some a posteriori error estimators for elliptic partial differential equations*, Math. Comp. **44** (1985), 283–301.

[7] R. Becker, R. Rannacher, *Weighted a posteriori error control in finite element methods*, Preprint 96–1, SFB 359, Heidelberg, 1996.

[8] R. Becker, R. Rannacher, *A feed-back approach to error control in finite element methods: Basic approach and examples*, East-West J. Numer. Math. **4** (1996), 237–264.

[9] I. Hlaváček, M. Křížek, *On a superconvergent finite element scheme for elliptic systems. I. Dirichlet boundary conditions*, Apl. Mat. **32** (1987), 131–154.

[10] S. Korotov, M. Křížek, *Global and local refinement techniques yielding nonobtuse tetrahedral partitions*, Comput. Math. Appl. (to appear).

[11] S. Korotov, P. Neittaanmäki, and S. Repin, *A posteriori error estimation of goal-oriented quantities by the superconvergence patch recovery*, J. Numer. Math. **11** (2003), 33–59.

[12] M. Křížek, P. Neittaanmäki, *Superconvergence phenomenon in the finite element method arising from averaging gradients*, Numer. Math. **45** (1984), 105–116.

[13] P. Ladevéze, D. Leguillon, *Error estimate procedure in the finite element method and applications*, SIAM J. Numer. Anal. **20** (1983), 485–509.

[14] J. T. Oden, S. Prudhomme, *Goal-oriented error estimation and adaptivity for the finite element method*, Comput. Math. Appl. **41** (2001), 735–756.

[15] S. Repin, *A unified approach to a posteriori error estimation based on duality error majorants*, Math. Comput. Simulation **50** (1999), 305–321.

[16] S. Repin, *A posteriori error estimation for variational problems with uniformly convex functionals*, Math. Comp. **69** (2000), 481–500.

[17] R. Verfürth, *A Review of A Posteriori Error Estimation and Adaptive Mesh-Refinement Techniques*, Wiley-Teubner, 1996.

[18] O. C. Zienkeiewicz, J. Z. Zhu, *A simple error estimator and adaptive procedure for practical engineering analysis*, Internat. J. Numer. Methods Engrg. **24** (1987), 337–357.

Part IV

# Applications to the Solution of Linear and Nonlinear Partial Differential Equations

# Inversion of Block-Tridiagonal Matrices and Nonnegativity Preservation in the Numerical Solution of Linear Parabolic PDEs

István Faragó,[1] Sergey Korotov[2]

[1]Department of Applied Analysis, Eötvös Loránd University, P.O. Box 120, H–1518, Budapest, Hungary faragois@cs.elte.hu

[2]Department of Mathematical Information Technology, University of Jyväskylä, P.O. Box 35, FIN–40014 Jyväskylä, Finland korotov@mit.jyu.fi

## 1 Introduction

Besides the covergence, another natural requirement in the process of numerical solution of partial differential equations, is a presevation of basic qualitative properties of the original (physical) solution, assuming that they are inherent to the continuous mathematical model. As an example, consider the advection-diffusion-reaction equation arising, e.g., in the large air-pollution modeling [17]:

$$\frac{\partial c}{\partial t} = -\operatorname{div}(vc) + \operatorname{div}(D \operatorname{grad} c) + R(c), \quad t \in [0, L], \ c(0, x) = c_0(x), \tag{1}$$

where the vector-valued function $c(t, x)$ denotes the concentration of compounds, $v = v(t, x)$ presents the current velocity of the medium, $D$ is the so-called diffusion coefficient matrix, and the function $R$ describes the chemical reactions between the compounds and includes the parametrized deposition and emission. (As a rule, we use equation (1) in the componentwise sense.) We solve the above problem by suitably chosen numerical method. Since $c$ denotes the concentration, which is always nonnegative, it is natural to require the nonnegativity from the numerical approximations of $c$ as well.

The technique, often used to solve problems like (1) (based on several physical processes) is the so-called *operator splitting method*, see e.g. [4], [7]: we choose a suitable time-step $\Delta t > 0$, and for $j = 1, 2, \ldots$, solve the sequence of subproblems

$$\frac{\partial u_1^{(j)}}{\partial t} = -\operatorname{div}(v u_1^{(j)}), \ u_1^{(j)}((j-1)\Delta t, x) = u_3^{(j-1)}((j-1)\Delta t, x), \tag{2}$$

$$\frac{\partial u_2^{(j)}}{\partial t} = \operatorname{div}(D \operatorname{grad} u_2^{(j)}), \ u_2^{(j)}((j-1)\Delta t, x) = u_1^{(j)}(j\Delta t, x), \tag{3}$$

$$\frac{\partial u_3^{(j)}}{\partial t} = R(u_3^{(j)}), \ u_3^{(j)}((j-1)\Delta t, x) = u_2^{(j)}(j\Delta t, x), \tag{4}$$

on the intervals $[(j-1)\Delta t, j\Delta t]$, where $u_3^{(0)}(0,x) = c_0(x)$. In this procedure, $u_3^{(j)}$ yields an approximation to $c(j\Delta t, x)$.

For numerical solution of each of the above subproblems, we can choose a certain suitable numerical method (naturally, all three may be completely different each from other). Obviously, if all numerical techniques used for solving (2)–(4) are nonnegativity preserving, then the whole computational scheme, used for solving (1), is nonnegativity preserving.

In the above sequence of subproblems (2)–(4), the central role belongs to subproblem (3), which, in the simplest setting, has the following form

$$\frac{\partial u}{\partial t} = \triangle u \quad \text{in } Q_L = (0,L) \times \Omega, \tag{5}$$

$$u = 0 \quad \text{on } \Gamma_L = (0,L) \times \partial\Omega, \tag{6}$$

$$u|_{t=0} = u_0 \quad \text{on } \Omega, \tag{7}$$

where $\Omega \subset \mathbb{R}^d$ is a segment ($d = 1$, and the problem is reffered to as the one-dimensional), or a rectangle ($d = 2$, and the problem is called two-dimensional), the symbol $\triangle$ denotes the Laplace operator, and $u_0$ is the initial function defined from the splitting procedure.

This equation has an importance of its own. For instance, it describes the heat conduction process, therefore hereafter we refer to it as the heat conduction equation. For this equation, the nonnegativity preservation principle reads as follows: for any nonnegative initial function $u_0$, the solution $u$ has to be nonnegative in the whole $Q_L$ as well, see, e.g. [15].

A typical numerical technique for solving (5)–(7) presents a combination of separate discretizations in space and time. For the first one, we can employ the finite element method (based e.g. on the linear elements in the one-dimensional case, and on bilinear elements – for the two-dimensional case), or the standard finite difference method with the mesh-size $h$. As a result, we get the following Cauchy problem for the *semidiscrete solution* $u_h$

$$\frac{du_h}{dt}(t) = \triangle_h u_h(t), \quad t \in (0,L), \tag{8}$$

where the initial value $u_h(0)$ is given and $\triangle_h$ denotes the corresponding discrete Laplace operator (represented by a matrix). Applying some suitable time discretization method with the time-step $\tau$ and parameter $\theta$ to the problem (8), we finally arrive at the *fully discretized problem*, presenting the following algebraic iterative procedure

$$X_1 y^{j+1} = X_2 y^j, \tag{9}$$

where $X_1$ and $X_2$ are given matrices, and the vector $y^j$ represents the approximation to $u_h(j\tau)$.

Our aim is to formulate, for a fixed $\theta \in [0,1]$, such conditions on $h$ and $\tau$, under which the corresponding semidiscrete and fully discretized solutions preserve the nonnegativity property. This problem for both kinds of fully discretized problems (FEM and FDM) is studied in the one-dimensional case and the exact (necessary and sufficient) conditions with respect to $\tau$ and $h$ for any fixed $\theta$ are found, see e.g. [5], [8]. The obtained results are based on finding the *exact represenation* of the matrix $X = X_1^{-1}X_2$, where the crucial point consists of finding the inverse $X_1^{-1}$.

In the two-dimensional case, a similar problem is obviously more difficult, due to more complicated structures of the corresponding matrices. A certain sufficient condition is given in [8], and it is based on the following requirements: $X_2 \geq 0$ and $X_1$ is a monotone matrix. However, finding the necessary and sufficient condition is still an open problem.

In our paper, we first analyse the nonnegativity preservation for the semidiscrete solutions of (8), and additionally establish a direct connection of the nonnegativity preservation for the semidiscrete solutions (8) between the one and two-dimensional cases. Further, we shall derive explicit formulae for the calculation of entries of the matrix $X_1^{-1}$, appearing in the numerical scheme for solving the two-dimensional problems. Our results make possible the exact calculation of the entries of the iteration matrix $X = X_1^{-1}X_2$ in (9), which is a basic point in obtaining the exact (necessary and sufficient) conditions for the nonnegativity property.

Finally, we note that the nonegativity preservation also follows from the discrete maximum principle, which is addressed, e.g., in [11], [16] – for the parabolic problems, and in [12], [13] – for the elliptic problems.

## 2 Nonnegativity preservation for the semidiscrete solutions

In this section, we consider the nonnegativity preservation for the Cauchy problem (8), where $\triangle_h$ arises from the space uniform discretization of the Laplace operator on rectangular mesh. The solution has a form

$$u_h(t) = \exp(\triangle_h t)u_0, \quad t \in (0, L), \tag{10}$$

where $\exp(\triangle_h t)$ denotes the exponent of the matrix $\triangle_h t$. Therefore, the problem of the nonnegativity preservation is equivalent to finding conditions on dicretization methods for which the matrix exponential $\exp(\triangle_h t)$ is nonnegative.

The following lemma holds (cf. [1, p. 172]).

**Lemma 2.1.** *Let $A$ be an arbitrary square matrix with the entries $a_{ij}$. Then $\exp(At)$ is nonnegative for any $t \geq 0$ iff*

$$a_{ij} > 0 \quad \textit{for all } i \neq j. \tag{11}$$

Proof. By the definition of the exponential, we have

$$\exp(At) = I + At + \ldots,$$

where $I$ denotes the identity matrix. This series immediately shows the necessity of condition (11). Let now $s$ be a scalar such that $A + sI$ is a nonnegative matrix. Then, obviously, $\exp((A + sI)t)$ is nonnegative if $t \geq 0$. Moreover, $\exp(-sIt)$ is also nonnegative, and the matrices $(A + sI)t$ and $-sIt$ commute. Therefore, due to the identity

$$\exp(At) = \exp((A + sI)t - sIt) = \exp((A + sI)t) \cdot \exp(-sIt),$$

the sufficiency of condition (11) is also proven. □

In the following we consider the application of this lemma to the one-dimensional case.

### 2.1 One-dimensional case

If problem (5)–(7) is considered in the one-dimensional case, the structure of the matrix $\triangle_h$ is well-known for the both finite difference and finite element (space) discretizations. Namely, using the standard denotations

$$Q = \text{tridiag}(1, -2, 1), \quad M = \frac{1}{6}\,\text{tridiag}(1, 4, 1), \tag{12}$$

we have:

for the finite difference method

$$\triangle_h = \frac{1}{h^2} Q, \tag{13}$$

for the linear finite element method

$$\triangle_h = \frac{1}{h^2} M^{-1} Q. \tag{14}$$

The matrix $\triangle_h$ from (13) obviously satisfies the condition (11), at the same time, the matrix $\triangle_h$ from (14) (which can be computed explicitly) is known to have its entries changing the sign chessboard-likely [9], i.e., it does not satisfy the condition (11).

Thus, using Lemma 2.1, we obtain the following result.

**Theorem 2.1.** *For the one-dimensional problem* (8), *the semidiscrete numerical solutions, obtained by the finite difference discretization, preserve the nonnegativity property. However, this property is not preserved, in general, for the numerical solutions resulting from the linear finite element discretization.*

**Remark 2.1.** The lumped mass method for the linear finite element method results in the semidiscrete problem which coincides with the finite difference semidiscrete problem. Therefore, this approach makes possible to improve the qualitative property of the finite element discretization.

**Remark 2.2.** In [3] there is given the condition of the positivity of the time derivative of the semidiscrete solutions. Clearly, this condition can be regarded as a sufficient condition of the nonnegativity preservation of the semidiscrete solutions.

**Remark 2.3.** Let us introduce the following denotation

$$T_1(p) = \text{tridiag}(1, p, 1), \tag{15}$$

where $p \in \mathbb{R}$, i.e., $Q = T_1(-2)$. Then, the semidiscretization in the form

$$\frac{du_h}{dt}(t) = T_1(p)u_h(t), \quad t \in (0, L), \tag{16}$$

is nonnegativity preserving for any value of the parameter $p$. Therefore, instead of (5), we can consider a more general equation

$$\frac{\partial u}{\partial t} = \triangle u + ku \quad \text{in } Q_L = (0, L) \times \Omega, \tag{17}$$

where $k$ is an *arbitrary* constant, and prove that the finite difference semidiscretization for such an equation is nonnegativity preserving, because the approximation of the new term effects only the diagonal elements of the matrix in (16).

**Remark 2.4.** However, the nonnegativity property is known to hold only for some $k \geq k_0$ for the original continuous problem (see, e.g., [15]), and, of course, the preservation of this property in the numerical realization is somewhat meaningless for certain values of $k$.

### 2.2 Two-dimensional case

We now consider problem (5)–(7) in the two-dimensional case.

We introduce the following denotation

$$T_2(p) = \text{tridiag}(I, T_1(p), I) \tag{18}$$

for a block tridiagonal matrix (from $\mathbb{R}^{n^2 \times n^2}$), where $p \in \mathbb{R}$.

We consider the both semidiscretization methods. Clearly, if we apply the usual finite difference method, then in the Cauchy-problem (8) the matrix $\triangle_h$ has the form $\triangle_h = T_2(-4)$, that is, all its off-diagonal elements are nonnegative. If we use the linear finite element method, then we get $\triangle_h = M^{-1}T_2(-4)$. Here $M$ denotes the corresponding block tridiagonal nonnegative mass matrix. Due to its structure, its inverse consists of a negative off-diagonal block. Therefore, in the above approximation the disccrete Laplacian has some negative off-diagonal entries. Summarizing the above results, on base of Lemma 2.1, we obtain

**Theorem 2.2.** *For the two-dimensional problem on rectangular mesh, the semidiscrete numerical solution, obtained by the regular finite difference discretization, preserves the nonnegativity property. However, this property is not preserved, in general, for the linear finite element discretization.*

In what follows, we establish the relation between the exponentials of the matrices $T_1(p)$ from $\mathbb{R}^{n\times n}$ and $T_2(p)$ from $\mathbb{R}^{n^2\times n^2}$, i.e., the matrix exponential between the 1D and 2D discrete Laplacians. Using the denotation

$$A(p) = \text{tridiag}(1, p+2, 1) = T_1(p) + 2I, \tag{19}$$

we have

$$\begin{aligned} T_2(p) &= \text{tridiag}(I, A(p) - 2I, I) = \text{diag}(A(p)) + \text{tridiag}(I, -2I, I) \\ &= I \otimes A(p) + Q \otimes I, \end{aligned} \tag{20}$$

where $\otimes$ denotes the Kronecker product of matrices, see e.g. [10], [14]. In order to attribute the matrix exponential of the matrix $T_2(p) \in \mathbb{R}^{n^2\times n^2}$ to the matrix exponentials of the matrices $A(p)$ and $Q$ from $\mathbb{R}^{n\times n}$, i.e., the two-dimensional problem to the one-dimensional one, we prove the folowing lemma.

**Lemma 2.2.** *For the matrices $T_2(p)$, $A(p)$, and $Q$, the relation*

$$\exp(T_2(p)) = \exp(Q) \otimes \exp(A(p)) \tag{21}$$

*holds.*

Proof. For any matrices $A$, $B$, $C$, $D$ of the same size, we have [10, p. 228],

$$(A \otimes B)(C \otimes D) = AC \otimes BD. \tag{22}$$

Therefore

$$\begin{aligned} (I \otimes A(p))(Q \otimes I) &= Q \otimes A(p), \\ (Q \otimes I)(I \otimes A(p)) &= Q \otimes A(p). \end{aligned} \tag{23}$$

Consequently, the corresponding exponential can be written by use of the binomial rule as follows

$$\begin{aligned} \exp(T_2(p)) &= \exp((I \otimes A(p)) + (Q \otimes I)) \\ &= \sum_{n=0}^{\infty} \frac{1}{n!}(I \otimes A(p) + Q \otimes I)^n \\ &= \sum_{n=0}^{\infty}\sum_{k=0}^{n} \frac{1}{k!(n-k)!}(I \otimes A(p))^k (Q \otimes I)^{n-k} \\ &= \sum_{n=0}^{\infty}\sum_{k=0}^{n} \frac{1}{k!(n-k)!} Q^k \otimes A(p)^{n-k}. \end{aligned} \tag{24}$$

On the other hand, by the definition of the tensor product, we have

$$\exp(Q) \otimes \exp(A(p)) = \sum_{i=0}^{\infty} \frac{1}{i!} Q^i \otimes \sum_{j=0}^{\infty} \frac{1}{j!} A(p)^j$$
$$= \sum_{i=0}^{\infty} \sum_{j=0}^{\infty} \frac{1}{i!j!} Q^k \otimes A(p)^j.$$

Since the right-hand side is equal to the right-hand side of (24), we get relation (21). □

Obviously, the tensor product of two matrices is nonnegative iff both involved matrices are nonnegative. Moreover, $\exp(Q)$ is nonnegative and $\exp(A(p))$ is nonnegative iff $\exp(T_1(p))$ is nonnegative. This repeatly shows the validity of Theorem 2.2.

**Remark 2.5.** Clearly, the statement of Theorem 2.2 is also valid for the more general equation (17).

# 3 Inverting the block-tridiagonal matrices

Applying some simple but efficient method to the numerical solution of the semidiscrete problem (8) (like the well-known $\theta$-method), we get a system of linear algebraic equation in the form

$$X_1 y^{j+1} = X_2 y^j, \tag{25}$$

where $X_1$ and $X_2$ are given matrices with special structures, and $y^j$ is the approximation to $u_h(j\tau)$. In order to analyse the nonnegativity preservation for the fully discretized problem (25), the main task is the inversion of the matrix $X_1$. For the one-dimensional case this matrix is a uniformly tridiagonal, and in the two-dimension case, it is a uniformly block-tridiagonal matrix. The inversion for the first case is well-known (see, e.g., [6], [8], [14]), and it is natural to ask whether the second (block-tridiagonal case) is related to the first (scalar) case? In this section, we establish the relation between these two cases.

## 3.1 Tridiagonal and block-tridiagonal matrices

We introduce a new denotation

$$T(p) = \text{tridiag}(-1, p, -1), \tag{26}$$

which differs from the similar denotation in (15) in the sign. In the sequel, we always assume that the matrix $T(p)$ is invertible, i.e., its eigenvalues are (see, e.g., [14])

$$\lambda_k = p - 2\cos\frac{k\pi}{n+1} \neq 0, \quad k = 1, 2, \ldots, n. \tag{27}$$

Following the standard technique of matrix inversion, each entry of the inverse for $T(p)$ is calculated as a fraction of two polynomials

$$(T(p)^{-1})_{ij} = g^n_{ij}(p) = \frac{p^n_{ij}(p)}{q^n(p)}, \tag{28}$$

where $p^n_{ij}(p)$ is the determinant of the adjoint matrix with respect to the element $(T(p))_{ij}$, and $q^n(p)$ denotes the determinant of $T(p)$.

In fact, using the $l$-th Chebyshev polynomials of the second kind

$$U_l(x) = \sum_{k=0}^{[\frac{l}{2}]} (-1)^k \binom{l-k}{k} x^{l-2k},$$

the entries in (28) can be expressed explicitly as folows from [14, p. 49],

$$(T(p)^{-1})_{ij} = \begin{cases} \dfrac{U_{i-1}(p) \cdot U_{n-j}(p)}{U_n(p)}, & \text{if } i \leq j, \\ \dfrac{U_{j-1}(p) \cdot U_{n-i}(p)}{U_n(p)}, & \text{if } i > j. \end{cases}$$

The block-tridiagonal matrices present, in some sense, the generalization of the tridiagonal matrices. Thus, we consider the matrix

$$B(T) = \text{tridiag}(-I, T, -I) \in \mathbb{R}^{n^2 \times n^2}, \tag{29}$$

where $T$ is defined in (26). (We remark that, in general, the blocks in the block tridiagonal matrices need not be square matrices, but we will consider only the latter case). In our case, the hypermatrix $B(T)$ consists of the *commutative blocks*, i.e., $B(T)_{ij} \cdot B(T)_{kl} = B(T)_{kl} \cdot B(T)_{ij}$ for all $i, j, k, l = 1, 2, \ldots, n$. Similarly to the scalar matrices, the representation

$$\text{Hdet}\,(B(T)) = \sum_{(n!)} (-1)^I B(T)_{1v_1} B(T)_{2v_2} \ldots B(T)_{nv_n},$$

where $v_k \in \{1, 2, \ldots n\}$, can be introduced and it is called the hyperdeterminant of the hypermatrix $B(T)$.

Obviously, $\text{Hdet}\,(B(T)) \in \mathbb{R}^{n \times n}$, i.e., it is a matrix itself. Further, the determinant of the block tridiagonal matrix $B(T)$ is equal to the determinant of the hyperdeterminant $\text{Hdet}\,(B(T))$:

$$\det B(T) = \det(\text{Hdet}\,(B(T)). \tag{30}$$

The proof can be found in [14, p. 324].

## 3.2 Inverting the block-tridiagonal matrices

We consider the inversion of the block-tridiagonal matrix $B(T)$. The idea is to show that the operator calculus works in the following sense: substituting the argument $p$ by the matrix $T$ in scalar formula (28), we obtain the $(i,j)$-th block of the inverse matrix.

First of all, we have to show that $q^n(T)$ is invertible under the assumption (27). Obviously, due to the definitions, the equality

$$\text{Hdet}\,(B(T)) = q^n(T)$$

holds. Using (30), we get

$$\det q^n(T) = \det B(T),$$

which means that, if $B(T)^{-1}$ exists, i.e., $\det B(T) \neq 0$, then $\det q^n(T) \neq 0$.

**Theorem 3.1.** *Assume that $B(T)$ is an invertible matrix, then $B(T)^{-1}$ consists of the blocks of size $n \times n$, having the form*

$$(B(T)^{-1})_{ij} = g_{ij}^n(T) = p_{ij}^n(T)(q^n(T))^{-1}, \quad i,j = 1,2,\ldots,n. \tag{31}$$

Proof. For simplicity, we shall omit in the proof the superscripts for the functions $g, p$ and $q$.

Let $G$ denote the block-matrix $\{g_{ij}(T)\}_{i,j=1}^n$, and, for the later use, for $i=0$, $j=0$, $i=n+1$, $j=n+1$ we define the matrices $g_{ij}(T)$ as the zero matrix of the size $n \times n$.

We will show the following: multiplying the $i$-th block-row of $B(T)$ with the $j$-th block-column of $G$, we get the identity matrix $I$ for $i=j$, and the zero matrix for $i \neq j$. Obviously,

$$\begin{aligned} -I \cdot g_{i-1,j}(T) &+ T \cdot g_{ij}(T) - I \cdot g_{i+1,j}(T) \\ &= -g_{i-1,j}(T) + T \cdot g_{ij}(T) - g_{i+1,j}(T) \\ &= (-p_{i-1,j}(T) + T \cdot p_{ij}(T) - p_{i+1,j}(T)) \cdot q(T)^{-1}. \end{aligned}$$

According to (28), we have

$$\begin{aligned} -g_{i-1,j}(s) + s \cdot g_{ij}(s) - g_{i+1,j}(s) &= \frac{-p_{i-1,j}(s) + s \cdot p_{ij}(s) - p_{i+1,j}(s)}{q(s)} \\ &= \begin{cases} 0, & \text{if } i \neq j, \\ 1, & \text{if } i = j, \end{cases} \end{aligned}$$

for all $s$, if $\det(T) = q(s) \neq 0$. So, for $i \neq j$, we have

$$P_1(s) = -p_{i-1,j}(s) + s \cdot p_{ij}(s) - p_{i+1,j}(s) = 0,$$

and, for $i = j$ we get

$$P_2(s) = -p_{i-1,j}(s) + s \cdot p_{ij}(s) - p_{i+1,j}(s) - q(s) = 0.$$

Since the previous formulae are valid for all $s$ except for a finite number of values (we have to exclude those values of $s$ for which $q(T) = 0$) and each polynomial has only a finite number of roots, we can see that both, $P_1(s)$ and $P_2(s)$ are zero polynomials. If we use the matrix $T$ instead of a scalar variable, this property remains valid, since the scalar zeros are obviously replaced by its matrix equivalent.

Thus, we can prove that the matrix $G$ is the right-hand inverse of $B(T)$. Using the fact that $B(T)$ is made of commutative blocks and the commutativity of real numbers, we easily get a similar result by calculating the product $GB(T)$. $\square$

## 4 Conclusions

In this paper, we have considered the nonnegativity preservation property for the linear parabolic PDEs. We established the direct connection of the nonnegativity preservation for the semidiscrete solutions between the one and two-dimensional cases. Namely, we proved that the two-dimensional problem has this property if the corresponding one-dimensional problem is nonnegativity preserving. The conditions posed are satisfied for all arbitrary (linear FEM and FDM) methods. Moreover, we gave the explicit formula for the inversion of the block-tridiagonal matrices with scalar tridiagonal matrices placed along their diagonals. This result may be useful in obtaining some sharper sufficient condition on the choice of the time discretization parameter in order to preserve the nonnegativity property of the fully discretized problems.

**Acknowledgement.** The first author was supported by Hungarian National Research Fund OTKA No. T031807 and No. T043765. The second author was supported by Grant no. 49051 of the Academy of Finland and by the Agora Center under Grant InBCT of TEKES.

## References

[1] R. Bellman, *Introduction to Matrix Analysis*, Mcgraw-Hill, 1960.

[2] A. Berman, R. J. Plemmons, *Nonnegative Matrices in the Mathematical Sciences*, Academic Press, 1979.

[3] M. Berzins, *Modified mass matrices and positivity preservation for hyperbolic and parabolic PDE's*, Commun. Numer. Meth. Engng. **17** (2001), 659–666.

[4] I. Dimov, I. Faragó, A. Havasi, and Z. Zlatev, *L-commutativity of the operators in splitting methods for air pollution models*, Ann. Univ. Sci. Budapest Eötvös Sect. Math. **44** (2001), 129–150.

[5] I. Faragó, H. Farkas, and P. Simon, *Qualitative properties of conductive heat transfer*, Thermodynamics of energy conversion and transport (S. Sienuitycz, A. De Vos, eds.), Springer Verlag, 2000, pp. 199–239.

[6] I. Faragó, H. Hariton, N. Komáromi, and T. Pfeil, *Heat transfer and qualitative properties of its numerical solutions, II. (Nonnegativity preservation of the second order approximation, the maximum principle and the nonoscillation)*, (in Hungarian), Alkalmaz. Mat. Lapok **17** (1994), 123–141.

[7] I. Faragó, A. Havasi, and S. Korotov, *Mathematical analysis of the weighted splitting*, Preprint B12, Department of Mathematical Information Technology, University of Jyväskylä, Finland, 2002.

[8] I. Faragó, R. Horváth, *On the nonnegativity conservation of finite element solutions of parabolic problems*, Finite Element Methods: Three-Dimentional Problems (P. Neittaanmäki, M. Křížek, eds.), GAKUTO Sci. Appl., Tokyo, 2001, pp. 78–86.

[9] I. Faragó, N. Komáromi, *Nonnegativity of the numerical solution of the parabolic problems*, Numerical Methods. Colloquia Math. Soc. J. Bolyai, 59 (D. Greenspan, P. Rózsa, eds.), North-Holland, 1991, pp. 173–179.

[10] M. Fiedler, *Special Matrices and Their Application in the Numerical Mathematics*, Martinus Nijhoff Publishers, 1986.

[11] H. Fujii, *Some remarks on finite element analysis of time-dependent field problems*, Theory and Practice in Finite element Structural Analysis, Univ. Tokyo Press, Tokyo, 1973, pp. 91–106.

[12] S. Korotov, M. Křížek, and P. Neittaanmäki, *Weakened acute type condition for tetrahedral triangulations and the discrete maximum principle*, Math. Comp. **70** (2001), 107–119.

[13] M. Křížek, Qun Lin, *On diagonal dominance of stiffness matrices in 3D*, East-West J. Numer. Math. **3** (1995), 59–69.

[14] P. Rózsa, *Linear Algebra and its Applications (in Hungarian)*, Műszaki Kőnyvkiadó, 1976.

[15] J. Smoller, *Shock Waves and Reaction-Diffusion Equations*, Springer Verlag, 1981.

[16] T. Vejchodský, *On the nonnegativity conservation in semidiscrete parabolic problems*, in: Conjugate Gradient Algorithms and Finite Element Methods, Springer-Verlag, Berlin, 2004, pp. 197–210.

[17] Z. Zlatev, *Computer Treatment of Large Air Pollution Models*, Kluwer Academic Publishers, 1995.

# On the Nonnegativity Conservation in Semidiscrete Parabolic Problems

Tomáš Vejchodský

Mathematical Institute, Academy of Sciences of the Czech Republic, Žitná 25, 115 67 Prague 1, Czech Republic vejchod@math.cas.cz

## 1 Introduction

Certain physical quantities, as concentration, density, absolute temperature etc., naturally attains only nonnegative values. This fact is mirrored in the mathematical formulation of a particular physical problem. We take advantage of maximum and comparison principles in these situations. Maximum and comparison principles are fundamental properties of partial differential equations of second order. There are different formulations of these principles. They hold for a variety of linear and nonlinear problems, see e.g., [1], [3], [4], [10], [13], [16], [17], [20].

The term *nonnegativity conservation (or preservation)*, has a tight connection to the comparison principle and will be defined below. Papers [5], [6], [7], [12], [14] deal with fully discrete schemes and give conditions under which the nonnegativity of numerical solution is conserved. Our work is devoted to the question whether the comparison principle is satisfied also for semidiscrete approximations.

Let us choose as a model problem the following prototype of a linear parabolic problem with the homogeneous right-hand side and Dirichlet boundary condition:

$$\begin{aligned} \partial_t u(x,t) - \Delta u(x,t) &= 0 && \text{in } \Omega \times (0,T), \\ u(x,t) &= 0 && \text{on } \partial\Omega \times [0,T], \\ u(x,0) &= u_0(x) && \text{in } \Omega, \end{aligned} \tag{1.1}$$

where $\partial_t$ denotes the partial derivative with respect to $t$, $T > 0$, and the unknown function $u$ and the initial condition $u_0$ are supposed to be sufficiently smooth. The initial and boundary conditions are considered to be consistent. The domain $\Omega \subset \mathbb{R}^d$, $d = 1, 2, 3$, is assumed to be polyhedral, for the sake of simplicity.

It is well known that problem (1.1) satisfies the following comparison principle.

**Theorem 1.1** (Comparison principle). *Let $u_1$, $u_2$ be two classical solutions of* (1.1) *corresponding to the initial conditions $u_{01}$, $u_{02}$, respectively. If $u_{01} \le u_{02}$ in $\Omega$, then $u_1 \le u_2$ in $\Omega \times (0,T)$.*

Proof. It follows from the classical maximum principle for parabolic problems, see, e.g., [16, p. 161]. □

**Definition 1.1.** We say that problem (1.1) satisfies the *nonnegativity conservation* (or *conserves nonnegativity*) if any nonnegative initial condition yields a nonnegative solution of (1.1), cf. [5], [7], [12].

It is obvious that the comparison principle and nonnegativity conservation are equivalent for linear problem (1.1). Note that this equivalence does not hold, in general, for nonlinear problems.

We obtain the semidiscrete solution by the method of lines. The spatial variable is discretized by the standard finite element method of the first order and the time variable is kept continuous. We will call this procedure the *finite element method of lines.*

The definition of the semidiscrete solution is based on the weak formulation: find $u \in L^2([0,T], H_0^1(\Omega))$ with $\partial_t u \in L^2(\Omega \times (0,T))$ such that

$$\int_\Omega \partial_t u\, v \,\mathrm{d}x + \int_\Omega \nabla u \cdot \nabla v \,\mathrm{d}x = 0 \quad \forall v \in H_0^1(\Omega), \text{ for a.a. } t \in [0,T],$$
$$u(x,0) = u_0(x) \text{ a.e. in } \Omega,$$

with the initial condition $u_0 \in H_0^1(\Omega)$.

Let the space $X_{h0} \subset H_0^1(\Omega)$ be a finite element space consisting of piecewise linear continuous functions based on a face-to-face simplicial partition of the domain $\Omega$. Denote by $N$ the dimension of $X_{h0}$ and by $\varphi_j$, $j = 1, 2, \dots, N$, the standard finite element basis functions. Let us emphasize that $\varphi_j$ are nonnegative.

The semidiscrete solution $\bar{U} \in C^1([0,T], X_{h0})$ is defined by the following Galerkin problem

$$\int_\Omega \partial_t \bar{U}\, V \,\mathrm{d}x + \int_\Omega \nabla \bar{U} \cdot \nabla V \,\mathrm{d}x = 0 \quad \forall V \in X_{h0}, \tag{1.2}$$
$$\bar{U}(x,0) = \bar{U}_0(x) \quad \text{in } \Omega,$$

where $\bar{U}_0(x)$ is the elliptic projection of $u_0$ into $X_{h0}$, given by the problem: find $\bar{U}_0 \in X_{h0}$ such that

$$\int_\Omega \nabla \bar{U}_0 \cdot \nabla V \,\mathrm{d}x = \int_\Omega \nabla u_0 \cdot \nabla V \,\mathrm{d}x \quad \forall V \in X_{h0}.$$

The semidiscrete solution $\bar{U}(x,t)$ can be written as a linear combination of the basis functions $\varphi_i$ with time dependent coefficients $y_j(t)$, i.e.,

$$\bar{U}(x,t) = \sum_{j=1}^{N} y_j(t) \varphi_j(x). \tag{1.3}$$

Substituting (1.3) into (1.2) and taking $V = \varphi_i$, we obtain

$$\sum_{j=1}^{N} \dot{y}_j(t) \int_\Omega \varphi_i \varphi_j \, \mathrm{d}x + \sum_{j=1}^{N} y_j(t) \int_\Omega \nabla \varphi_i \cdot \nabla \varphi_j \, \mathrm{d}x = 0, \tag{1.4}$$
$$y_i(0) = y_{0i}, \quad i = 1, 2, \dots, N,$$

where the dot stand for the time derivative. This is a system of ordinary differential equations with initial condition given by $\bar{U}_0(x) = \sum_{j=1}^{N} y_{0j}\varphi_j(x)$. System (1.4) can be rewritten in a vector notation as follows,

$$M\dot{y}(t) + Ay(t) = 0, \tag{1.5}$$
$$y(0) = y_0,$$

where $y(t) = (y_1(t), y_2(t), \dots, y_N(t))^\top$ is a vector of unknown coefficients, $y_0 = (y_{01}, y_{02}, \dots, y_{0N})^\top$ is a vector of the initial condition, $M$ and $A$ are the *mass* and *stiffness matrices*, respectively, with entries

$$M_{ij} = \int_\Omega \varphi_i(x)\varphi_j(x) \, \mathrm{d}x, \qquad i, j - 1, 2, \dots, N,$$
$$A_{ij} = \int_\Omega \nabla\varphi_i(x) \cdot \nabla\varphi_j(x) \, \mathrm{d}x, \quad i, j = 1, 2, \dots, N.$$

System (1.5) is linear and has the following solution

$$y(t) = \exp(-M^{-1}At)y_0, \quad t > 0, \tag{1.6}$$

where the matrix exponential is defined in a standard way (cf. Definition 2.2).

The nonnegativity conservation of semidiscrete Galerkin problem (1.2) is equivalent to the nonnegativity conservation of system (1.5) in case of linear finite elements. This is clear because the elliptic projection of a nonnegative initial condition $u_0(x)$ into $X_{h0}$ gives nonnegative coefficients $y_{0j}$, $j = 1, 2, \dots, N$. This can be easily shown in one dimension. The initial condition $y_0$ given by the elliptic projection of $u_0$ can be computed as

$$y_0 = A^{-1}\vec{f}_0, \text{ where } \vec{f}_0 = \left( \int_\Omega \nabla u_0 \cdot \nabla \varphi_1 \, \mathrm{d}x, \dots, \int_\Omega \nabla u_0 \cdot \nabla \varphi_N \, \mathrm{d}x \right)^\top.$$

The components $f_{0j}$ of the right-hand side vector $\vec{f}_0$ can be easily computed in one-dimensional case as follows:

$$\begin{aligned} f_{0j} &= \int_\Omega u_0' \varphi_j' \, \mathrm{d}x = \int_{x_{j-1}}^{x_j} u_0' \frac{1}{h_j} \, \mathrm{d}x - \int_{x_j}^{x_{j+1}} u_0' \frac{1}{h_{j+1}} \, \mathrm{d}x \\ &= -\frac{1}{h_j} u_0(x_{j-1}) + \left( \frac{1}{h_j} + \frac{1}{h_{j+1}} \right) u_0(x_j) - \frac{1}{h_{j+1}} u_0(x_{j+1}) \\ &= a_{j,j-1}u_0(x_{j-1}) + a_{jj}u_0(x_j) + a_{j,j+1}u_0(x_{j+1}), \end{aligned}$$

where $x_j$, $j = 0, 1, \dots, N+1$, stand for the nodes of the partition, $h_j = x_j - x_{j-1}$, and $a_{ij}$ are the entries of $A$, $i, j = 1, 2, \dots, N$. Denoting

$$\vec{u}_0 = (u_0(x_1), u_0(x_2), \dots, u_0(x_N))^\top,$$

we have $\vec{f}_0 = A\vec{u}_0$, which implies

$$y_0 = A^{-1}\vec{f}_0 = \vec{u}_0 \geq 0.$$

Thus, the question is whether any positive initial condition $y_0$ leads to a positive solution $y(t)$ of system (1.5).

This question is answered in Section 3. Section 2 consists of several preliminary results especially from the matrix theory. An illustrative counterexample which shows that semidiscrete problem (1.5) does not conserve nonnegativity is given in Section 4. Several suggestions how to satisfy the nonnegativity conservation are presented in Section 5.

## 2 Preliminaries

In this section we present several results from the matrix analysis, which we will use in the following sections. Note that all matrices are supposed to be real in this paper.

**Definition 2.1.** We say that a matrix or a vector $Q$ is *nonnegative* (*positive*) if all its entries are nonnegative (positive), and we write $Q \geq 0$ ($Q > 0$).

**Definition 2.2.** Let $Q$ be a square matrix. The *matrix exponential* is given by the series

$$\exp Q = \sum_{n=0}^{\infty} \frac{1}{n!} Q^n.$$

**Theorem 2.1.** *Let $Q$ be a square matrix. The matrix exponential* $\exp(tQ)$ *is positive for all $t > 0$ if and only if all off-diagonal entries of $Q$ are nonnegative and $Q$ is irreducible.*

Proof. See [19, p. 257, Th. 8.1]. □

**Theorem 2.2.** *Let $A$ and $B$ be two symmetric and positive definite $N \times N$ matrices. Then their product, matrix $BA$, has all eigenvalues real and positive. Moreover, for each eigenvalue there exists a real corresponding eigenvector.*

Proof. This theorem is proven, e.g., in [11] using the square root of a matrix. It is also a direct consequence of the known facts from the matrix theory, which can be found in [9, Sec. X.7, p. 286]. We give here a different proof.

Let $\lambda = \lambda_\mathrm{r} + \mathrm{i}\lambda_\mathrm{i}$ and $v = v_\mathrm{r} + \mathrm{i}v_\mathrm{i} \neq 0$ be the complex eigenvalue and the corresponding eigenvector of the matrix $BA$. Thus, we have $BAv = \lambda v$, which can be rewritten as $Av = \lambda Mv$, where $M = B^{-1}$. This equality can be split into the real and imaginary part:

$$Av_{\mathrm{r}} = \lambda_{\mathrm{r}} M v_{\mathrm{r}} - \lambda_{\mathrm{i}} M v_{\mathrm{i}}, \tag{2.1}$$

$$Av_{\mathrm{i}} = \lambda_{\mathrm{r}} M v_{\mathrm{i}} + \lambda_{\mathrm{i}} M v_{\mathrm{r}}. \tag{2.2}$$

Let us multiply (2.1) by $v_{\mathrm{i}}$ and (2.2) by $v_{\mathrm{r}}$:

$$(Av_{\mathrm{r}}, v_{\mathrm{i}}) = \lambda_{\mathrm{r}}(Mv_{\mathrm{r}}, v_{\mathrm{i}}) - \lambda_{\mathrm{i}}(Mv_{\mathrm{i}}, v_{\mathrm{i}}), \tag{2.3}$$

$$(Av_{\mathrm{i}}, v_{\mathrm{r}}) = \lambda_{\mathrm{r}}(Mv_{\mathrm{i}}, v_{\mathrm{r}}) + \lambda_{\mathrm{i}}(Mv_{\mathrm{r}}, v_{\mathrm{r}}), \tag{2.4}$$

where $(\cdot,\cdot)$ is the Euclidean scalar product. Because $A$ and $M$ are real and symmetric, we have $(Av_{\mathrm{i}}, v_{\mathrm{r}}) = (v_{\mathrm{i}}, Av_{\mathrm{r}}) = (Av_{\mathrm{r}}, v_{\mathrm{i}})$, and similarly $(Mv_{\mathrm{i}}, v_{\mathrm{r}}) = (v_{\mathrm{i}}, Mv_{\mathrm{r}}) = (Mv_{\mathrm{r}}, v_{\mathrm{i}})$. Thus, we obtain from (2.3) and (2.4) that

$$\lambda_{\mathrm{i}}[(Mv_{\mathrm{r}}, v_{\mathrm{r}}) + (Mv_{\mathrm{i}}, v_{\mathrm{i}})] = 0$$

holds. The positive definiteness of the matrix $M$ gives us $(Mx, x) > 0$ for all $x \in \mathbb{R}^N$, $x \neq 0$, and we, therefore, conclude that $\lambda_{\mathrm{i}} = 0$.

Thus, we know that $\lambda$ is real and we can rewrite (2.1) and (2.2) as follows:

$$Av_{\mathrm{r}} = \lambda_{\mathrm{r}} M v_{\mathrm{r}}, \tag{2.5}$$

$$Av_{\mathrm{i}} = \lambda_{\mathrm{r}} M v_{\mathrm{i}}. \tag{2.6}$$

The equality $(Av_{\mathrm{r}}, v_{\mathrm{r}}) = \lambda_{\mathrm{r}}(Mv_{\mathrm{r}}, v_{\mathrm{r}})$ together with the positive definiteness of $A$ and $M$ give us $\lambda_{\mathrm{r}} > 0$.

We know that $v_{\mathrm{r}} \neq 0$ or $v_{\mathrm{i}} \neq 0$, and thus, (2.5) or (2.6) imply the existence of a real eigenvector. □

**Theorem 2.3.** *Let $Q$ be nonsingular square matrix. Then $Q$ is irreducible if and only if $Q^{-1}$ is irreducible.*

Proof. We prove that $Q$ is reducible if and only if $Q^{-1}$ is reducible. If $Q$ is reducible $N \times N$ nonsingular matrix, then there exists a permutation matrix $P$ such that

$$PQP^{\top} = \begin{pmatrix} A_1 & B \\ 0 & A_2 \end{pmatrix},$$

where $A_i$ are $N_i \times N_i$, $i = 1, 2$, nonsingular matrices with $N_i > 0$ and $N_1 + N_2 = N$ and $B$ is a $N_1 \times N_2$ matrix. It is easy to calculate the inverse matrix:

$$(PQP^{\top})^{-1} = P^{-\top} Q^{-1} P^{-1} = \begin{pmatrix} A_1^{-1} & -A_1^{-1} B A_2^{-1} \\ 0 & A_2^{-1} \end{pmatrix}.$$

Consequently, $Q^{-1}$ is also reducible. □

**Theorem 2.4.** *Let $P$ and $Q$ be two nonnegative $N \times N$ matrices. If $P$ is irreducible and if all diagonal entries of $Q$ are positive, then both $PQ$ and $QP$ are irreducible.*

Proof. It is well known that a matrix is irreducible if and only if its associated directed graph is strongly connected, see e.g. [19]. Denote by $D$ the diagonal matrix $D = \operatorname{diag}\{Q_{11}, Q_{22}, \dots, Q_{NN}\}$. It is clear that the directed graphs of the matrices $P$, $DP$, and $PD$ are the same. Adding any positive off-diagonal entry to the matrix $D$ can results, due to the nonnegativity of $P$, only to another edge in the directed graph. Thus, the directed graph of the matrices $PQ$ and $QP$ contains, possibly, more edges than the directed graph of $B$. Consequently, if the directed graph of $P$ is strongly connected then the directed graphs of $PQ$ and $QP$ are strongly connected, too. $\square$

**Theorem 2.5.** *Let $M$ be a nonnegative and nonsingular $N \times N$ matrix. If there exists a positive off-diagonal entry in $M$, then there exists a negative entry in matrix $M^{-1}$.*

Proof. Take indices $i \neq j$, $i, j \in \{1, 2, \dots, N\}$, such that $M_{ij} > 0$. Assume that $M^{-1} = (\bar{M}_{k\ell})$, $k, \ell \in \{1, 2, \dots, N\}$, and let $\bar{M}_{k\ell} \geq 0$ for all $k, \ell \in \{1, 2, \dots, N\}$. Choose any $m \neq i$, $m \in \{1, 2, \dots, N\}$. The equality

$$0 = (MM^{-1})_{im} = \underbrace{M_{ij}}_{>0} \bar{M}_{jm} + \sum_{\substack{p=1 \\ p \neq j}}^{N} \underbrace{M_{ip}}_{\geq 0} \underbrace{\bar{M}_{pm}}_{\geq 0}$$

can be satisfied only if $\bar{M}_{jm} = 0$ for all $m \in \{1, 2, \dots, N\}$, $m \neq i$. If $\bar{M}_{ji} = 0$ then $M^{-1}$ is singular, which is a contradiction. If $\bar{M}_{ji} > 0$ then $0 = (M^{-1}M)_{ji} = \bar{M}_{ji} M_{ij} > 0$, which is again a contradiction. $\square$

**Definition 2.3.** Let us define the following set of square matrices:

$$\mathcal{Z} = \{A; A \text{ is an } N \times N \text{ matrix and } A_{ij} \leq 0,\ i \neq j,\ N \in \mathbb{N}\}.$$

**Definition 2.4.** We say that $N \times N$ matrix $A$ is M-*matrix* if $A \in \mathcal{Z}$, $A$ is nonsingular and $A^{-1} \geq 0$. The set of all M-matrices is denoted by $\mathcal{K}$.

**Theorem 2.6.** *Let $A \in \mathcal{Z}$ be $N \times N$ matrix. Then $A \in \mathcal{K}$ if and only if all real eigenvalues of $A$ are positive.*

Proof. See [8, Th. 5.1]. $\square$

**Theorem 2.7.** *Let $A \in \mathcal{Z}$ be irreducible $N \times N$ matrix. Then the following conditions are equivalent:*

(i) $A^{-1} > 0$,
(ii) *there exist vector $x > 0$ such that $Ax > 0$,*
(iii) $A \in \mathcal{K}$.

Proof. See [8, Th. 5.12]. $\square$

**Theorem 2.8** (Perron-Frobenius). *Let $A$ be a nonnegative and irreducible $N \times N$ matrix. Then its spectral radius*

$$\varrho(A) = \max\{|\lambda|, \ \lambda \text{ is an eigenvalue of } A\}$$

*is a single positive eigenvalue of $A$ with a positive corresponding eigenvector.*

Proof. See e.g. [8, Th. 4.8]. □

**Theorem 2.9.** *Let $M$ be a nonnegative and irreducible $N \times N$ matrix. If there exists a zero entry in $M$, then there exists positive off-diagonal entry in $M^{-1}$, i.e., $M^{-1} \notin \mathcal{Z}$.*

Proof. Denote $K = M^{-1}$. According to Theorem 2.8 there exist $\lambda > 0$ and $v > 0$ such that $Mv = \lambda v$, i.e., $M^{-1}v = \lambda^{-1}v$. Thus, we have $v > 0$ such that $Kv > 0$. Clearly $K^{-1} \geq 0$ but $K \not> 0$. Moreover, $K$ is irreducible. Therefore, Theorem 2.7 gives $K \notin \mathcal{Z}$. □

## 3 Does the method of lines conserve nonnegativity?

By (1.6), the nonnegativity of $y(t)$ for all $y_0 \geq 0$ and all $t > 0$ is equivalent to the nonnegativity of the matrix $\exp(-M^{-1}At)$ for all $t > 0$.

According to Theorem 2.1,

$$\exp(-M^{-1}At) > 0 \quad \forall t > 0$$

if and only if

$$M^{-1}A \in \mathcal{Z} \quad \text{and} \quad M^{-1}A \text{ is irreducible.} \tag{3.1}$$

Thus, we have to examine the matrix $M^{-1}A$.

First, let us assume that the used triangulation satisfies the so-called *acute type condition*, which in two dimensions requires that all angles in the triangulation are not greater then $\pi/2$ and in three dimensions that all dihedral angles between faces of all tetrahedra are not greater then $\pi/2$. The stiffness matrix $A$ is M-matrix under this condition. For details see [2] and [15]. Note that no assumption on partition is needed in one dimension and the stiffness matrix $A$ is always M-matrix in this case.

Let us survey the most important properties of $M$ and $A$.

- $M \geq 0$ due to the positivity of finite element basis functions.
- Both $M$ and $A$ are the Gramm matrices. Thus, they are nonsingular, symmetric, and positive definite, (i.e., their diagonal entries are positive).
- Matrices $M$ and $A$ are irreducible and sparse provided that the mesh is sufficiently fine.
- All off-diagonal entries of $A$ are nonpositive due to the acute type condition.
- The stiffness matrix $A$ is M-matrix under the acute type condition, see Theorem 2.6. Consequently, using Theorem 2.7, we obtain $A^{-1} > 0$.

- Set $K = M^{-1}$. Although the matrix $K$ has many properties like M-matrices, mainly nonnegative inverse matrix and positive definiteness, it is not M-matrix, because it is not in $\mathcal{Z}$, see Theorem 2.9. Using Theorems 2.5 and positive definiteness of $M^{-1}$, we obtain that there are both positive and negative off-diagonal entries in $M^{-1}$.

Now, we present the list of properties of $M^{-1}A$.

- The matrix $M^{-1}A$ is not symmetric, in general. Anyway, there are several exceptions. Matrices $A$ and $M$ commute, for example, when $A$ and $M$ correspond to a one-dimensional equidistant partition – see [21, p. 394], or to a triangulation with equilateral triangles in two dimensions. In both these cases $M = c_1 I - c_2 A$ with suitable constants $c_1$ and $c_2$.
- The matrix $M^{-1}A$ is not positive definite, in general, but all its eigenvalues are real and positive, see Theorem 2.2.
- The inverse matrix to $M^{-1}A$ is nonnegative: $A^{-1}M \geq 0$.
- The matrix $M^{-1}A$ is irreducible: $A$ irreducible $\overset{\text{Th. 2.3}}{\Longrightarrow}$ $A^{-1}$ irreducible $\overset{\text{Th. 2.4}}{\Longrightarrow}$ $A^{-1}M$ irreducible $\overset{\text{Th. 2.3}}{\Longrightarrow}$ $M^{-1}A$ irreducible.
- The matrix $M^{-1}A$ exhibit many important properties of M-matrices, but it is not, unfortunately, M-matrix, in general, since it is not in $\mathcal{Z}$. The situation is similar as for the matrix $M^{-1}$. A one-dimensional example is given below.

These properties can be utilized in condition (3.1). The matrix $M^{-1}A$ is irreducible, but $M^{-1}A \notin \mathcal{Z}$, in general. Therefore, there exists $t > 0$ such that $\exp(-M^{-1}At)$ has a negative entry and, consequently, there exists a nonnegative initial condition $y_0$ such that at least one component of $y(t)$ is negative.

Thus, we conclude that the finite element method of lines does not conserve nonnegativity.

## 4 Counterexample of nonnegativity conservation

First, let us find another expression for the solution of system (1.5). According to [11, p. 40, Rem. 2.10.7], the product of two positive definite matrices is real diagonalisable. Therefore, there exist $N$ linearly independent eigenvectors $v_1, v_2, \dots, v_N$ corresponding to the positive eigenvalues $\lambda_1, \lambda_2, \dots, \lambda_N$ of $M^{-1}A$. Let us introduce the diagonal matrix $\Lambda = \operatorname{diag}\{\lambda_1, \lambda_2, \dots, \lambda_N\}$ and the square nonsingular matrix $V$ whose columns are formed by $v_1, v_2, \dots, v_N$. Clearly, $M^{-1}AV = V\Lambda$, which is equivalent to $M^{-1}A = V\Lambda V^{-1}$. Substituting this into (1.6), we obtain

$$y(t) = \exp(-V\Lambda V^{-1}t)y_0 = V\exp(-\Lambda t)V^{-1}y_0 = VE(t)V^{-1}y_0, \qquad (4.1)$$

where $E(t) = \operatorname{diag}\{\mathrm{e}^{-\lambda_1 t}, \mathrm{e}^{-\lambda_2 t}, \dots, \mathrm{e}^{-\lambda_N t}\}$.

Note that rows of $V^{-1}$ are the eigenvectors of the matrix $MA^{-1}$ corresponding to its eigenvalues $\lambda_1^{-1}, \lambda_2^{-1}, \dots, \lambda_N^{-1}$, because $MA^{-1}V^{-\top} = V^{-\top}\Lambda^{-1}$

holds. It also worth noting that system (1.5) can be decoupled by the substitution $z(t) = V^{-1}y(t)$.

Let us consider one dimensional problem (1.1) with $\Omega = (0,4)$. The partition of $\Omega$ consists of four elements separated by three inner points: $\{1,2,3\}$. Using the standard continuous and piecewise linear basis functions, we get the following mass and stiffness matrices:

$$M = \frac{1}{6}\begin{pmatrix} 4 & 1 & 0 \\ 1 & 4 & 1 \\ 0 & 1 & 4 \end{pmatrix}, \quad A = \begin{pmatrix} 2 & -1 & 0 \\ -1 & 2 & -1 \\ 0 & -1 & 2 \end{pmatrix}.$$

We introduce the inverse matrices to $M$ and $A$ just to illustrate that $A^{-1} > 0$ and that $M^{-1}$ has really both positive and negative off-diagonal entries:

$$M^{-1} = \frac{1}{28}\begin{pmatrix} 45 & -12 & 3 \\ -12 & 48 & -12 \\ 3 & -12 & 45 \end{pmatrix}, \quad A^{-1} = \frac{1}{4}\begin{pmatrix} 3 & 2 & 1 \\ 2 & 4 & 2 \\ 1 & 2 & 3 \end{pmatrix}.$$

To find the exact semidiscrete solution by the finite element method of lines according to (4.1), we have to calculate the matrix

$$M^{-1}A = \frac{1}{14}\begin{pmatrix} 51 & -36 & 9 \\ -36 & 60 & -36 \\ 9 & -36 & 51 \end{pmatrix},$$

its eigenvalues

$$\lambda_1 = \frac{6}{7}(5 - 3\sqrt{2}), \quad \lambda_2 = 3, \quad \lambda_3 = \frac{6}{7}(5 + 3\sqrt{2}),$$

and the associated eigenvectors, which form columns of the matrix $V$:

$$V = \frac{1}{2}\begin{pmatrix} 1 & \sqrt{2} & 1 \\ \sqrt{2} & 0 & -\sqrt{2} \\ 1 & -\sqrt{2} & 1 \end{pmatrix} = V^{-1}.$$

Equality (4.1) can be now rewritten componentwise

$$y_i(t) = \sum_{k=1}^{3} \gamma_{ik} e^{-\lambda_k t}, \quad i = 1,2,3.$$

The coefficients $\gamma_{ik}$ are given by the entries of matrices $V = \{v_{ik}\}$ and $V^{-1} = \{\bar{v}_{kj}\}$ and by the entries of the initial condition vector $y_0 = (y_{01}, y_{02}, y_{03})^{\top}$:

$$\gamma_{ik} = v_{ik} \sum_{j=1}^{3} \bar{v}_{kj} y_{0j}, \quad i,k = 1,2,3.$$

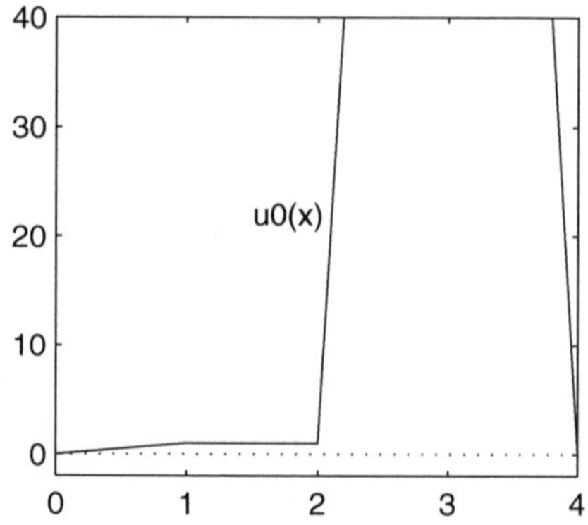


Figure 1. Continuous piecewise linear initial condition $u_0(x) \in H_0^1(0,4)$ with values 1, 1, 200 at the points 1, 2, 3, respectively.

The solution $y_j(t)$ would be definitely nonnegative, if the coefficients $\gamma_{ik}$ were nonnegative, but there is no reason for this. There really exists a positive initial condition $y_0$ such that some component of the semidiscrete solution is negative. For example, we can take

$$y_0 = (1, 1, 200).$$

This choice of the initial condition vector corresponds to the function

$$u_0(x) = \varphi_1(x) + \varphi_2(x) + 200\varphi_3(x) \tag{4.2}$$

sketched in Fig. 1.

The three components of the semidiscrete solution are shown in Fig 2. We can see that the first component

$$y_1(t) = \frac{201 + \sqrt{2}}{4} e^{-\lambda_1 t} - \frac{199}{2} e^{-\lambda_2 t} + \frac{201 - \sqrt{2}}{4} e^{-\lambda_3 t}$$

is negative in a small time interval near zero, which is in contradiction with the comparison principle.

Note that the finite element approximation of initial condition (4.2) is exact and even in spite of this fact the nonnegativity is not preserved.

## 5 How to satisfy nonnegativity conservation?

The situation with conservation of nonnegativity in semidiscrete case is not as hopeless as it seems. Although there exists a counterexample, the nonnegativity conservation is usually conserved. It seems that a strong gradient in the initial conditions and a quite coarse mesh are necessary to obtain

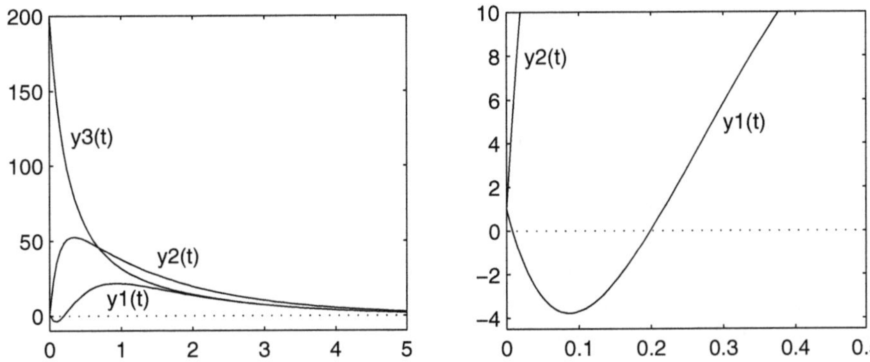


Figure 2. Three components $y_1(t)$, $y_2(t)$, and $y_3(t)$ of the semidiscrete solution. The interesting part is zoomed on the right.

a counterexample to nonnegativity conservation. It does not hold that the semidiscrete solution $y(t)$ is nonnegative for all $t \geq 0$ for any nonnegative initial condition, but we are able to prove, that there exists a time $t_0$ such that for all $t \geq t_0$ the inequality $y(t) \geq 0$ holds for all nonnegative initial conditions $y_0$. The exact formulation with a proof follows.

**Theorem 5.1.** *Consider problem* (1.1) *and the finite element method of lines based on linear elements. The semidiscrete solution* $y(t)$ *is given by system* (1.5). *Assume that* $A^{-1} \geq 0$. *Then there exists* $t_0 \in \mathbb{R}$ *such that* $y(t) > 0$ *holds for all* $t \geq t_0$ *and all initial conditions* $y_0 \geq 0$.

Proof. To prove the theorem we use formula (4.1). We know that $A^{-1}M \geq 0$ and that it is irreducible. Thus, Theorem 2.8 says that the largest eigenvalue of $A^{-1}M$ is single and positive and that the corresponding eigenvector is also positive. Consequently, the eigenvector of the matrix $M^{-1}A$ corresponding to the smallest eigenvalue of $M^{-1}A$ is positive, i.e., the first column of $V$ is positive, and hence $v_{i1} > 0$, $i = 1, 2, \ldots, N$. The same argument for the matrix $MA^{-1} \geq 0$ gives that the first row of the matrix $V^{-1}$ must be positive, i.e., $\bar{v}_{1j} > 0$, $j = 1, 2, \ldots, N$. The components of the matrix $VE(t)V^{-1}$ can be written in the following form:

$$[VE(t)V^{-1}]_{ij} = \underbrace{v_{i1}\bar{v}_{1j}}_{>0}\, \mathrm{e}^{-\lambda_1 t} + \sum_{k=2}^{N} v_{ik}\bar{v}_{kj}\mathrm{e}^{-\lambda_k t}, \quad i, j = 1, 2, \ldots, N.$$

It is clearly seen that for a large time $t$ the first term dominates, because $\lambda_1$ is the smallest of all $\lambda_k$, $k = 1, 2, \ldots, N$. Thus, the positivity of $v_{i1}\bar{v}_{1j}$ for all $i, j = 1, 2, \ldots, N$ ensures the positivity of all entries of the matrix $VE(t)V^{-1}$ for a sufficiently large $t$. Consequently, there exists $t_0 \in \mathbb{R}$ such that $y(t) = VE(t)V^{-1}y_0$ is nonnegative for all $y_0 \geq 0$ and all $t \geq t_0$. □

**Remark 5.1.** We calculated the value of $t_0$ in the example from Section 4. All solutions are nonnegative for all nonnegative initial conditions if $t > 0.213$.

The other and very simple way to save the nonnegativity conservation of a semidiscrete problem is to utilize *convergence* of the finite element method of lines. We know that continuous problem (1.1) satisfies the comparison principle. Therefore, it also conserves the nonnegativity. Taking a sufficiently fine partition of $\Omega$, we can get the semidiscrete solution $\bar{U}$ arbitrarily close to the exact solution $u$, which is positive for any positive initial condition. Thus, also the semidiscrete solution must be positive if it is sufficiently close to the exact one.

The numerical experiments indicate that this behaviour is usually achieved already for quite coarse partitions, e.g., in the example from the previous section it is enough to divide every element into four equal parts. The resulting semidiscrete solution will be positive for all $t > 0$ as it is indicated in Fig. 3, where we can also see unpleasant oscillations of some components $y_j(t)$ of the semidiscrete solution.

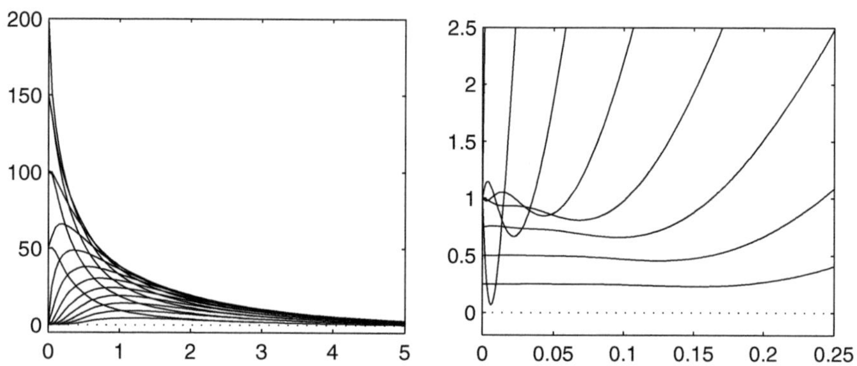


Figure 3. The graphs of coefficients $y_1(t), y_2(t), \ldots, y_{15}(t)$ of semidiscrete solution (see (1.3)). The detailed view of region around zero shows that all components $y_i(t)$, $i = 1, 2, \ldots, 15$ are positive. It also reveals unpleasant oscillations of some components of the semidiscrete solution.

The last and most often used way of ensuring the nonnegativity conservation is the *mass lumping* – for details about this method see, e.g., [18]. This method consists in replacing of the mass matrix $M$ in (1.5) by a diagonal matrix $D = \operatorname{diag}\{d_{11}, d_{22}, \ldots, d_{NN}\}$ with $d_{ii} = \sum_{j=1}^{N} m_{ij}$, i.e., we sum up all entries in a row of $M$ and put the result on the diagonal. The semidiscrete solution is then in the form

$$y(t) = \exp(-D^{-1}At)y_0.$$

The matrix $D^{-1}A$ has clearly all off-diagonal entries negative (if linear finite elements and acute type condition are used). Thus, according to Theorem 2.1, the matrix $\exp(-D^{-1}At)$ is positive for all $t > 0$, and consequently $y(t) \geq 0$ for all $y_0 \geq 0$ and all $t > 0$.

**Acknowledgement.** This paper was supported by grant No. A1019201 of the Academy of Sciences of the Czech Republic. This support is gratefully acknowledged.

## References

[1] M. Artola, *Sur une classe de problemes paraboliques quasi-linéaires (in French)*, Boll. Unione Mat. Ital., VI. Ser., B 5 (1986), 51–70.

[2] P. G. Ciarlet, P. A. Raviart, *Maximum principle and uniform convergence for the finite element method*, Comput. Methods Appl. Mech. Engrg. **2** (1973), 17–31.

[3] J. Douglas, T. Dupont, and J. Serrin, *Uniqueness and comparison theorems for nonlinear elliptic equations in divergence form*, Arch. Ration. Mech. Anal. **42** (1971), 157–168.

[4] L. C. Evans, *Weak Convergence Methods for Nonlinear Partial Differential Equations*, Expository lectures from the CBMS regional conference held at Loylola University of Chicago, June 27–July 1, 1988. Regional Conference Series in Mathematics, vol. 74, 1990.

[5] I. Faragó, R. Horváth, *On the nonnegativity conservation of finite element solutions of parabolic problems*, Proceedings of International Conference: Finite Elements Methods. Three-Dimensional Problems, Jyväskylä, Finland, June 28–July 1, 2000 (P. Neittaanmäki et al., ed.), Gakkotosho, Tokyo; GAKUTO Internat. Ser. Math. Sci. Appl. **15** (2001), 76–84.

[6] I. Faragó, R. Horváth, and S. Korotov, *Discrete maximum principle for linear parabolic problems solved on hybrid meshes*, preprint at the Dept. of Mathem. Inf. Technology, University of Jyväskylä, Finland, 2003; submitted to Appl. Numer. Math. (2003).

[7] I. Faragó, S. Korotov, *Inversion of block-tridiagonal matrices and nonnegativity preservation in the numerical solution of linear parabolic PDEs*, in: Conjugate Gradient Algorithms and Finite Element Methods, Springer-Verlag, Berlin, 2004, pp. 185–195.

[8] M. Fiedler, *Special Matrices and Their Applications in Numerical Mathematics*, SNTL – Publishers of Technical Literature, Prague, 1986.

[9] F. R. Gantmacher, *The Theory of Matrices, the second edition (in Russian)*, Nauka, Moscow, 1966.

[10] D. Gilbarg, N. S. Trudinger, *Elliptic Partial Differential Equations of Second Order*, Springer-Verlag, Berlin-Heidelberg-New York, 1977.

[11] W. Hackbusch, *Iterative Solution of Large Sparse Systems of Equations*, Applied Mathematical Sciences 95, Springer-Verlag, New York, 1994.

[12] Z. Horváth, *Consistency and stability for some nonnegativity-conserving methods*, Appl. Numer. Math. **13** (1993), 371–381.

[13] R. Jensen, *The maximum principle for viscosity solutions of fully nonlinear second order partial differential equations*, Arch. Ration. Mech. Anal. **101** (1988), 1–27.

[14] J. Karátson, S. Korotov, *Discrete maximum principles for finite element solutions of nonlinear elliptic problems with mixed boundary conditions*, preprint at Eotvos Lorán University, Budapest, Hungary, 2003; submitted to Numer. Math. (2003).

[15] M. Křížek, Qun Lin, *On diagonal dominance of stiffness matrices in 3D*, East-West J. Numer. Math. **3** (1995), 59–69.

[16] M. H. Protter, H. F. Weinberger, *Maximum Principles in Differential Equations*, Prentice-Hall, Englewood Cliffs, 1967.

[17] J. Serrin, *On the strong maximum principle for quasilinear second order differential inequalities*, J. Funct. Anal. **5** (1970), 184–193.

[18] V. Thomée, *Galerkin Finite Element Methods for Parabolic Problems*, Springer-Verlag, Berlin, 1997.

[19] R. S. Varga, *Matrix Iterative Analysis*, Prentice-Hall, Englewood Cliffs, 1962.

[20] T. Vejchodský, *Comparison principle for a nonlinear parabolic problem of a nonmonotone type*, Appl. Math. (Warsaw) **29** (2002), 65–73.

[21] E. Vitásek, *The Basics of the Theory of Numerical Methods for the Differential Equations (in Czech)*, Academia, Prague, 1994.

# Iterative Solution Methods of the Maxwell Equations Using Staggered Grid Spatial Discretization

Robert Horváth,[1] István Faragó,[2] Willy Schilders[3]

[1]Institute of Mathematics and Statistics, University of West-Hungary, Erzsébet u. 9, H–9400 Sopron, Hungary `rhorvath@ktk.nyme.hu`

[2]Department of Applied Analysis, Eötvös Loránd University, P.O. Box 120, H–1518 Budapest, Hungary `faragois@cs.elte.hu`

[3]Philips Research Laboratories Eindhoven, Prof. Holstlaan 4, 5656AA Eindhoven, Netherlands `wil.schilders@philips.com`

## 1 Introduction

The 3D Maxwell equations, which describe the behavior of time-dependent electromagnetic fields, in the absence of free charges and currents, can be written in the form

$$-\nabla \times \mathbf{H} + \varepsilon \partial_t \mathbf{E} = \mathbf{0}, \tag{1}$$

$$\nabla \times \mathbf{E} + \mu \partial_t \mathbf{H} = \mathbf{0}, \tag{2}$$

$$\nabla(\varepsilon \mathbf{E}) = 0, \tag{3}$$

$$\nabla(\mu \mathbf{H}) = 0, \tag{4}$$

where

$$\mathbf{E} = (E_x(t,x,y,z), E_y(t,x,y,z), E_z(t,x,y,z)) \tag{5}$$

is the electric field strength,

$$\mathbf{H} = (H_x(t,x,y,z), H_y(t,x,y,z), H_z(t,x,y,z)) \tag{6}$$

is the magnetic field strength, $\varepsilon$ is the electric permittivity and $\mu$ is the magnetic permeability (see [7] and [13] for more details). The two material parameters can depend on the spatial coordinates. It is well-known that the divergence equations (3) and (4) follow from the curl equations (1) and (2) if we suppose that the fields in question were divergence-free at the initial point of time. This means that we must solve only the curl equations applying divergence-free initial conditions for $\mathbf{E}$ and $\mathbf{H}$.

A lot of numerical Maxwell solution methods are known. Some of them are the finite difference method, the variational method, the method of moments, the finite element method, the transmission line matrix method, the Monte Carlo method and the method of lines ([7], [11], [13]). In this paper we investigate those finite difference methods that employ the staggered spatial discretization in definition of a system of linear first order ordinary differential equations.

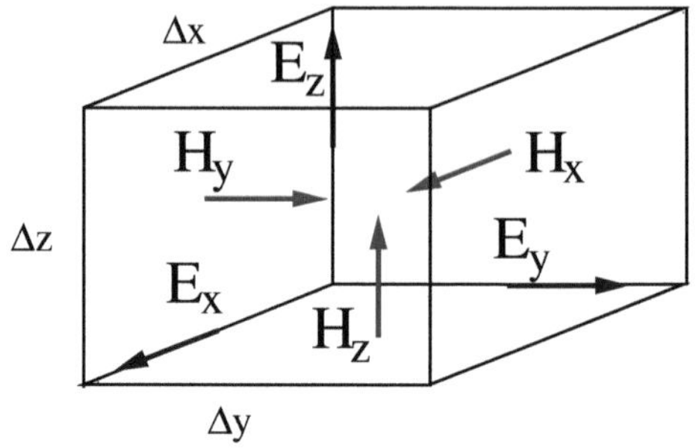


Figure 1. Standard Yee cell.

In order to obtain the above mentioned semi-discretized system we define a rectangular mesh with the step-sizes $\Delta x, \Delta y$ and $\Delta z$ for the electric field and another staggered (by $\Delta x/2, \Delta y/2$ and $\Delta z/2$) grid for the magnetic field in the computational domain. The building blocks of these meshes are the so-called Yee-cells (see Fig. 1). The staggered grid structure was firstly successfully applied by Yee ([16]) in 1966.

Let us consider the

$$\partial_t(\sqrt{\varepsilon}\mathbf{E}) = \frac{1}{\sqrt{\varepsilon}}\nabla \times \frac{(\sqrt{\mu}\mathbf{H})}{\sqrt{\mu}}, \tag{7}$$

$$\partial_t(\sqrt{\mu}\mathbf{H}) = -\frac{1}{\sqrt{\mu}}\nabla \times \frac{(\sqrt{\varepsilon}\mathbf{E})}{\sqrt{\varepsilon}}, \tag{8}$$

rearranged form of the curl equations. Discretizing these equations at the points shown in Fig. 1, we arrive at the system of ordinary differential equations

$$\frac{\mathrm{d}\Psi(t)}{\mathrm{d}t} = \mathbf{A}\Psi(t), \quad t > 0, \quad \Psi(0) \text{ is given} \tag{9}$$

(see [1] for details). $\Psi(t) \in \mathbb{R}^{6N}$, $N$ is the number of the Yee-cells in the computational domain, gives the approximations of the components $\sqrt{\varepsilon}E_x, \dots,$ $\sqrt{\mu}H_z$ at the discretization points at the time instant $t$. Every row of $\mathbf{A} \in \mathbb{R}^{6N\times 6N}$ consists at most four nonzero elements in the forms $\pm 1/(\sqrt{\varepsilon_{\cdot,\cdot,\cdot}\mu_{\cdot,\cdot,\cdot}}\,\Delta\cdot)$, that is $\mathbf{A}$ is a sparse matrix. Moreover, $\mathbf{A}$ is a skew-symmetric matrix ($\mathbf{A}^\top = -\mathbf{A}$). Because of the skew-symmetry of $\mathbf{A}$ the eigenvalues can be written in the form $\pm \mathrm{i}\lambda_k$, where $k = 1,\dots,3N$, $\lambda_k > 0$ and $\mathrm{i} = \sqrt{-1}$.

Applying the Gerschgorin theorem we obtain the upper bound $|\lambda_k| \leq 4c/h$ ($k = 1, \ldots, 3N$), where $h = \min\{\Delta x, \Delta y, \Delta z\}$ and $c$ is the maximal speed of light in the computational domain ($c = \max\{1/\sqrt{\varepsilon\mu}\}$).

We know from the theory of ordinary differential equations that the solution of (9) can be written in the form

$$\Psi(t) = \exp(t\mathbf{A})\Psi(0), \tag{10}$$

where $\exp(t\mathbf{A})$ denotes the matrix exponential and it is well-defined with the Taylor-series of the exponential function. This matrix exponential cannot be computed directly because $\mathbf{A}$ is, in real life problems, a very large matrix. According to the form (10), the numerical methods for the Maxwell equations are based on some approximation of the matrix exponential $\exp(t\mathbf{A})$. With the choice of a time-step $\Delta t > 0$

$$\Psi(t + \Delta t) = \exp(\Delta t\mathbf{A})\Psi(t) \tag{11}$$

follows from (10). Using this equality the one-step iteration

$$\Psi^{n+1} = U_n(\Delta t\mathbf{A})\Psi^n, \quad \Psi^0 \text{ is given} \tag{12}$$

can be defined, where $U_n(\Delta t\mathbf{A})$ is the approximation of the exponential $\exp(\Delta t\mathbf{A})$ (this approximation may depend on $n$) and $\Psi^n$ is the approximation of the function $\Psi$ at the time-level $n\Delta t$.

In this paper we investigate several time-integration schemes for (9). These methods will differ only in the definition of the exponential approximation $U_n(\Delta t\mathbf{A})$. At the end of this paper we will compare the schemes from the point of view of computational speed, stability and accuracy.

## 2 Time integration schemes for the semi-discretized Maxwell equations

In this section we list some possible time-integration methods for (9). In order to compare the methods, we calculate the number of operations per one iteration step, moreover, we discuss the question of stability of the methods. We call the numerical solution method stable, when the relation $\|\Psi^n\|_2 \leq K \cdot \|\Psi^0\|_2$ is valid for some fixed constant $K$ and for all natural number $n$. Let us introduce the notation $q = c\Delta t/h$.

### 2.1 Explicit Euler method

The most evident method, the explicit Euler method, is investigated first. The method approximates the exponential $\exp(\Delta t\mathbf{A})$ by the first two terms of the series of the exponential function. That is, $U_n(\Delta t\mathbf{A}) = \mathbf{I} + \Delta t\mathbf{A}$. The matrix $\mathbf{I}$ denotes the identity matrix. This iteration method is very fast (the number of operation per time-step is $36N$), but the method is not stable, so it is not usable in practice.

**Theorem 2.1.** *The numerical solution of the Maxwell equations using staggered spatial discretization and explicit Euler time-integration is unstable.*

Proof. It is sufficient to show that the modulus of the eigenvalues of $\mathbf{I} + \Delta t\mathbf{A}$ are greater than one. We have

$$|\Lambda_{\pm k}|^2 = |1 \pm \mathrm{i}\Delta t \lambda_k|^2 = 1 + \Delta t^2 \lambda_k^2 > 1. \qquad \square \tag{13}$$

## 2.2 Implicit Euler method

The second solution method is the implicit Euler method, where we employ the approximation $U_n(\Delta t\mathbf{A}) = (I - \Delta t\mathbf{A})^{-1}$.

**Theorem 2.2.** *The numerical solution of the Maxwell equations using staggered spatial discretization and implicit Euler time-integration is unconditionally stable.*

Proof. The unconditional stability can be shown with the inequality

$$\begin{aligned} \|\Psi^{n+1}\|_2^2 &= \|(\mathbf{I} - \Delta t\mathbf{A})^{-1}\Psi^n\|_2^2 \leq \|(\mathbf{I} - \Delta t\mathbf{A})^{-1}\|_2^2 \cdot \|\Psi^n\|_2^2 \\ &= \frac{1}{1 + \Delta t^2 \cdot \min_{k=1,\ldots,3N}\{\lambda_k^2\}} \|\Psi^n\|_2^2 \leq \|\Psi^n\|_2^2. \qquad \square \end{aligned} \tag{14}$$

Naturally, in practice, we do not compute the inverse of $\mathbf{I} - \Delta t\mathbf{A}$, but we solve a system of linear algebraic equations in the form

$$(\mathbf{I} - \Delta t\mathbf{A})\Psi^{n+1} = \Psi^n \tag{15}$$

in every time-step. Because the coefficient matrix of the system is a sparse one, we prefer the iterative solution method. Let us choose the simple gradient iteration

$$\Psi^{n+1}_{(l+1)} = [(1-\omega)\mathbf{I} + \omega\Delta t\mathbf{A}]\Psi^{n+1}_{(l)} + \omega\Psi^n, \quad l = 0, 1, \ldots, \quad \Psi^{n+1}_{(0)} = \Psi^n, \tag{16}$$

where $\omega$ is a suitably chosen positive constant. The method is convergent if and only if the spectral radius of the iteration matrix is less than one. So, we obtain the necessary and sufficient condition of the convergence

$$0 < \omega < \frac{2}{1 + \Delta t^2 \max_{k=1,\ldots,3N}\{\lambda_k^2\}}. \tag{17}$$

The smaller the spectral radius of the iteration matrix, the faster the convergence. Analyzing the second order (in $\omega$) form of the spectral radius we can find that to achieve the fastest convergence the parameter $\omega$ must be chosen according to the equality

$$\omega = \frac{1}{1 + \Delta t^2 \max_{k=1,\ldots,3N}\{\lambda_k^2\}}. \tag{18}$$

Inserting this parameter into the expression of the spectral radius we have

$$\begin{aligned}
&\varrho([(1-\omega)\mathbf{I}+\omega\Delta t\mathbf{A}]) \qquad (19)\\
&=\sqrt{\omega^2(1+\Delta t^2)\max_{k=1,\dots,3N}\{\lambda_k^2\}+1-2\omega}\\
&=1-\frac{1}{1+\Delta t^2\max_{k=1,\dots,3N}\{\lambda_k^2\}}\le 1-\frac{1}{1+16q^2}.
\end{aligned}$$

Although the implicit Euler method is unconditionally stable, which would make possible the choice of arbitrarily large time-steps, increasing $\Delta t$ the iteration method will be slower because of the relatively large spectral radius. When we would like to solve the system of linear equations decreasing the error of the initial approximation with a factor of $10^6$, then we have to perform about 75 iterations (choosing $q = 1/\sqrt{3}$, which will be the maximal value for $q$ in the Yee-method). So we would have $75 \cdot 7 \cdot 6N = 3150N$ operations per time-step. This number of operations increases dramatically increasing the time-step (for $q = 2$ we have $18816N$ operations per time-step).

## 2.3 The Yee-method

Yee derived the first efficient finite difference solution method for the Maxwell equations in 1966 (see [16]). This method uses a so-called leap-frog time integration scheme, for which the electric field at $t = 0$ and the magnetic field at $t = \Delta t/2$ must be given. Now we show that the Yee-method is also based on the matrix exponential approximation. Let us define two matrices, $\mathbf{A}_{1Y}$ and $\mathbf{A}_{2Y}$, as follows. The matrix $\mathbf{A}_{1Y}$ is composed from the matrix $\mathbf{A}$ changing the rows belonging to the electric field variables to zero rows. $\mathbf{A}_{2Y}$ can be derived, in similar manner, zeroing the rows belonging to the magnetic field variables. It can be easily shown that $\mathbf{A}_{1Y}$ and $\mathbf{A}_{2Y}$ do not commute and that the relations $\mathbf{A} = \mathbf{A}_{1Y} + \mathbf{A}_{2Y}$, $\exp(\Delta t\mathbf{A}_{1Y}) = \mathbf{I} + \Delta t\mathbf{A}_{1Y}$ and $\exp(\Delta t\mathbf{A}_{2Y}) = \mathbf{I}+\Delta t\mathbf{A}_{2Y}$ are fulfilled. In the case of the Yee-method we can apply the exponential approximation $U_n(\Delta t\mathbf{A}) = (\mathbf{I} + \Delta t\mathbf{A}_{1Y})(\mathbf{I} + \Delta t\mathbf{A}_{2Y})$, which comes from the relations

$$\begin{aligned}
\exp(\Delta t\mathbf{A}) &= \exp(\Delta t(\mathbf{A}_{1Y}+\mathbf{A}_{2Y})) \qquad (20)\\
&\approx \exp(\Delta t\mathbf{A}_{1Y})\exp(\Delta t\mathbf{A}_{2Y}) = (\mathbf{I}+\Delta t\mathbf{A}_{1Y})(\mathbf{I}+\Delta t\mathbf{A}_{2Y}).
\end{aligned}$$

It can be proven applying Von Neumann analysis, that the Yee-method can be kept to be stable choosing the time-step sufficiently small.

**Theorem 2.3** (e.g. [13]). *The numerical solution of the Maxwell equations using staggered spatial discretization and leap-frog time integration is stable if and only if the condition*

$$\Delta t < \frac{1}{c\sqrt{(1/\Delta x)^2+(1/\Delta y)^2+(1/\Delta z)^2}} \qquad (21)$$

*is fulfilled.*

From the theorem we obtain the upper bound $q < 1/\sqrt{3}$. The number of operations is $36N$ in one time-step, that is the same like in the explicit Euler method. Thus this method is very fast, but because of the strict stability condition it proceeds relatively slowly.

### 2.4 The Namiki-Zheng-Chen-Zhang method

A lot of effort has been invested during the last decade to bridge the stability problem of the Yee-method. The main goal was to construct methods, where $\Delta t$ can be chosen based on accuracy considerations instead of stability reason ([3], [4]). The first papers that described unconditionally stable methods were written by Namiki and by the triple Zheng, Chen and Zhang ([9], [15]).

The Namiki-Zheng-Chen-Zhang (NZCZ) method is based on the explicit and implicit Euler method. Let us define the matrices $\mathbf{A}_{1N}$ and $\mathbf{A}_{2N}$ such a way that $\mathbf{A}_{1N}$ comes from the discretization of the first items in the curl operator, and $\mathbf{A}_{2N}$ comes from the second ones. Then we can define the iteration process

$$\frac{\Psi^{n+1/2} - \Psi^n}{\Delta t/2} = \mathbf{A}_{1N}\Psi^{n+1/2} + \mathbf{A}_{2N}\Psi^n, \tag{22}$$

$$\frac{\Psi^{n+1} - \Psi^{n+1/2}}{\Delta t/2} = \mathbf{A}_{1N}\Psi^{n+1/2} + \mathbf{A}_{2N}\Psi^{n+1}, \tag{23}$$

which gives the exponential approximation

$$\begin{aligned} U_n(\Delta t\mathbf{A}) = (\mathbf{I} - (\Delta t/2)\mathbf{A}_{2N})^{-1} \cdot (\mathbf{I} + (\Delta t/2)\mathbf{A}_{1N}) \\ \cdot (\mathbf{I} - (\Delta t/2)\mathbf{A}_{1N})^{-1} \cdot (\mathbf{I} + (\Delta t/2)\mathbf{A}_{2N}). \end{aligned} \tag{24}$$

The unconditional stability of the method was previously demonstrated on test problems or its proof was given that used computer algebraic tools. Applying the fact $\mathbf{A}_{1N} + \mathbf{A}_{2N} = \mathbf{A}$ and the skew-symmetry of the matrices $\mathbf{A}_{1N}$ and $\mathbf{A}_{2N}$, a pure mathematical proof of the stability was given in [5].

**Theorem 2.4** (see [5]). *Let $h = \min\{\Delta x, \Delta y, \Delta z\}$ and let $q = c\Delta t/h$ be an arbitrary fixed number. The numerical solution of the Maxwell equations is unconditionally stable using staggered spatial discretization and using the Namiki-Zheng-Chen-Zhang time integration method.*

Let us notice that the constant $q$ must be chosen according to the inequality $q < 1/\sqrt{3}$ (here $h = \Delta x = \Delta y = \Delta z$) in 3D problems in the case of the classical Yee-method to guarantee the stability of the method. According to the previous theorem in the NZCZ-method the parameter $q$ can be set arbitrarily, which shows the unconditional stability of the method.

In every time-step we have to apply the explicit and implicit Euler method twice. The implicit method is used with a symmetric tridiagonal matrix, so the solution can be obtained by the so-called Thomas algorithm. The number of operation is $144N$ in one time-step. The NZCZ-method is four times slower than the Yee-method for a fixed time-step $\Delta t$. Because the NZCZ-method is unconditionally stable, we can choose time-steps beyond the stability bound of the Yee-method. Thus in the long run the NZCZ-method can be faster than the Yee-method.

## 2.5 The Kole-Figge-de Raedt method

Kole, Figge and de Raedt noticed (see [6]) that it is possible to split the matrix $\mathbf{A}$ into the sum of skew-symmetric matrices, for which the matrix exponential can be computed exactly (KFR-method). When we have such splitting in the form $\mathbf{A} = \mathbf{A}_1 + \cdots + \mathbf{A}_p$ ($p \in \mathbb{N}$), then we have $U_n(\Delta t\mathbf{A})$ as a product of exactly computed exponentials $\exp(\xi_{i_l}\Delta t\mathbf{A}_i)$, where $\xi_{i_l}$ is some suitable constant ($i \in \{1, \dots, p\}$). The computation of the matrix exponentials is based on the equality

$$\exp\left(\begin{bmatrix} 0 & \alpha \\ -\alpha & 0 \end{bmatrix}\right) = \begin{bmatrix} \cos\alpha & \sin\alpha \\ -\sin\alpha & \cos\alpha \end{bmatrix}, \tag{25}$$

where $\alpha$ is an arbitrary constant. Since the matrices $\mathbf{A}_1, \dots, \mathbf{A}_p$ are skew-symmetrical and only the products of the exponents of these matrices are used in the approximation, the iteration matrix will be orthogonal. That is its 2-norm is exactly one. Thus the KFR-method is also unconditionally stable.

**Theorem 2.5** (see [6]). *The numerical solution of the Maxwell equations using staggered spatial discretization and using products of exactly calculated matrix exponentials of skew-symmetric matrices in the time integration is unconditionally stable.*

With the splitting $\mathbf{A} = \mathbf{A}_1 + \cdots + \mathbf{A}_p$ we can define, for instance, the approximation

$$U_n^{(1)}(\Delta t\mathbf{A}) = \exp(\Delta t\mathbf{A}_1)\exp(\Delta t\mathbf{A}_2)\cdot\ldots\cdot\exp(\Delta t\mathbf{A}_p), \tag{26}$$

which is called KFR1-method, because this method has order one. A second order approximation can be achieved with

$$U_n^{(2)}(\Delta t\mathbf{A}) = \left(U_n^{(1)}\left(-\frac{\Delta t}{2}\mathbf{A}\right)\right)^\top \cdot \left(U_n^{(1)}\left(\frac{\Delta t}{2}\mathbf{A}\right)\right) \tag{27}$$

(KFR2-method), while we get a fourth order method with

$$\begin{aligned} U_n^{(4)}(\Delta t\mathbf{A}) = {} & U_n^{(2)}(\theta\Delta t\mathbf{A}) \cdot U_n^{(2)}(\theta\Delta t\mathbf{A}) \cdot U_n^{(2)}((1-4\theta)\Delta t\mathbf{A}) \\ & \times U_n^{(2)}(\theta\Delta t\mathbf{A}) \cdot U_n^{(2)}(\theta\Delta t\mathbf{A}), \end{aligned} \tag{28}$$

$\theta = 1/(4 - \sqrt[3]{4})$ (KFR4-method). For more details regarding the splitting methods consult [8] and [14].

The number of operations per time-step is $108N$ for the KFR1-, $216N$ for the KFR2-, and $1080N$ for the KFR4-method.

## 2.6 The Krylov-space method

In the previous methods we approximated the matrix exponential $\exp(\Delta t\mathbf{A})$ and used this approximation to generate a matrix iteration. Changing the philosophy of the matrix exponential approximation we can proceed as follows. We do not approximate the matrix exponential itself but the product of the matrix exponential and the previous state vector ([1], [2], [10], [12]).

Let us suppose that an initial vector $\Psi^0$ and a fixed natural number $m$ are given. We are interested in finding the best approximation to $\exp(\Delta t\mathbf{A})\Psi^0$ from the Krylov-space

$$\mathcal{K}(\Delta t\mathbf{A}, \Psi^0, m) = \text{span}\{\Psi^0, \Delta t\mathbf{A}\Psi^0, \ldots, (\Delta t\mathbf{A})^{m-1}\Psi^0\} \tag{29}$$

(span{} denotes the set of all possible linear combinations of the vectors). Because $\mathbf{A}$ is skew-symmetric, it is possible to find an orthonormal matrix $\mathbf{V}_m$ and a skew-symmetric tridiagonal matrix $\mathbf{T}_m$ such that the relation

$$\mathbf{V}_m^\top \mathbf{A}\mathbf{V}_m = \mathbf{T}_m \tag{30}$$

is satisfied. The matrices $\mathbf{V}_m$ and $\mathbf{T}_m$ can be calculated applying the modified Arnoldi-method. With the help of these matrices the best approximation for $\exp(\Delta t\mathbf{A})\Psi^0$ can be written in the form

$$\Psi^1 = \beta\mathbf{V}_m \exp(\Delta t\mathbf{T}_m)\mathbf{e}_1, \tag{31}$$

where $\beta = \|\Psi^0\|_2$ and $\mathbf{e}_1$ is the first unit vector. This method is also unconditionally stable.

**Theorem 2.6** ([5]). *The numerical solution of the Maxwell equations using staggered spatial discretization and using the Krylov-method with a modified Arnoldi orthogonalization in the time integration is unconditionally stable.*

The main advantage of the method is that choosing $m$ relatively small ($m \ll 6N$) we need to compute the matrix exponential only for the small matrix $\Delta t\mathbf{T}_m$ and we get the next approximation as a linear combination of the $m$ columns of $\mathbf{V}_m$. The number of operations per time-steps is $72mN$ (if we neglect the number of operations in the computation of $\exp(\Delta t\mathbf{T}_m)$).

**Remark 2.7.** Let us suppose that $\dim(\mathcal{K}(\Delta t\mathbf{A}, \Psi^0, m)) = m_0 < m$. Then in the above relations we must use $m_0$ instead of $m$. Moreover, in this case $\mathbf{A}^{m_0}\Psi^0 \in \text{span}\{\Psi^0, \Delta t\mathbf{A}\Psi^0, \ldots, (\Delta t\mathbf{A})^{m_0-1}\Psi^0\}$ and $\Psi^1$ gives the exact value of $\exp(\Delta t\mathbf{A})\Psi^0$, which means that $\exp(\Delta t\mathbf{A})\Psi^0$ can be computed exactly for arbitrary time-steps. This shows that the Krylov-method in special cases ($m_0 \ll 6N$) can be a very efficient one.

**Remark 2.8.** Considering Theorem 4 in [2] we can give an estimation for the error of this method in the form

$$\begin{aligned} &\| \exp(\Delta t \mathbf{A})\Psi^0 - \beta \mathbf{V}_m \exp(\Delta t \mathbf{T}_m)\mathbf{e}_1 \|_2 \\ &\qquad \le 12\mathbf{e}^{-(2q)^2/m}\left(\frac{2eq}{m}\right)^m, \quad m \ge 4q \;\; (q = c\Delta t/h). \end{aligned} \tag{32}$$

With this relation we are able to choose $m$ or $\Delta t$ to guarantee a certain accuracy level of the computations.

# 3 Comparison of the methods

In this section we compare the previously listed numerical schemes for the Maxwell equations. We are not going to present numerical tests here. Instead of this we summarize our experience in the use of the methods (see papers [1], [5], [6], [9], [10], [13]). We consider the Yee-method as a standard solution method, so we compare the methods with the Yee-method. Table 1 shows the number of operations per one time-step (100% = Yee-method) and indicates the stability properties for the methods.

Table 1. Number of computations and stability.

| | | |
|---|---:|---|
| Euler explicit | 100% | unstable |
| Euler implicit | min. 8750% | stable |
| YEE | 100% | stable iff $q < 1/\sqrt{3}$ |
| NZCZ | 400% | stable |
| KFR1 | 300% | stable |
| KFR2 | 600% | stable |
| KFR4 | 3000% | stable |
| Krylov | $200m$% | stable |

The explicit Euler method is unstable so it cannot be used in practice. The implicit Euler method is also unpractical because of the expenses of the computations.

The NZCZ-method is stable, so the time-step can be chosen arbitrarily. Of course, the increasing time-step decreases the accuracy of the method. We have to find the appropriate balance between the accuracy and the computational speed. We have found that with acceptable accuracy the NZCZ-method can be faster with a factor 10 than the Yee-method.

The KFR-method seems to be a very efficient one, because, like in the Yee-method, it computes the matrix exponentials exactly. Instead of this, the method appears to be very inaccurate. In order to lift the accuracy of the method we have to apply the fourth order version of it, which makes it slower than the Yee-method in the long run. However, because the 2-norm of the iteration matrix is exactly one, the method behaves nicely in spectrum computations.

The number of operations of the Krylov-method is about $2m$ times greater than in the Yee-method (the number of the iterations $m$ can be estimated from Remark 2.8). As we noticed in Remark 2.7 if $m_0$ is sufficiently small, then we can obtain the exact solution of the semi-discrtetized system for all $\Delta t$ step-sizes with an acceptable computational time.

## References


[1] I. Faragó, R. Horváth, and W. H. A. Schilders, *Investigation of Numerical Time Integrations of the Maxwell Equations Using the Staggered Grid Spatial Discretization*, RANA Report 02-15, University of Technology Eindhoven, Department of Mathematics and Computer Science, submitted to IEEE Transactions on CAD.

[2] M. Hochbruck, C. Lubich, *On Krylov-subspace approximations to the matrix exponential operator*, SIAM J. Numer. Anal. **34** (1997), 1911–1925.

[3] R. Holland, *Implicit three-dimensional finite differencing of Maxwell's equations*, IEEE Trans. Nuclear Science, Vol. NS-31 (1984), 1322–1326.

[4] R. Holland, K. S. Cho, *Alternating-direction implicit differencing of Maxwell's equations: 3D results*, Computer Science Corp., Albuquerque, NM, technical report to Harry Diamond Labs., Adelphi, MD, Contract DAAL02-85-C-0200, June 1, 1986.

[5] R. Horváth, *Uniform treatment of the numerical time-integration of the Maxwell equations*, Proceedings Scientific Computing in Electrical Engineering (to appear).

[6] J. S. Kole, M. T. Figge, H. De Raedt, *Unconditionally stable algorithms to solve the time-dependent Maxwell equations*, Phys. Rev. E 64, 066705 (2001).

[7] M. Křížek, P. Neittaanmäki, *Mathematical and Numerical Modelling in Electrical Engineering: Theory and Applications*, Kluwer, 1996.

[8] G. I. Marchuk, *Methods of Splitting*, Nauka, Moscow, 1988.

[9] T. Namiki, *A new FDTD algorithm based on alternating-direction implicit method*, IEEE Transactions on Microwave Theory and Techniques, Vol. 47, No. 10, 2003-2007, Oct. 1999.

[10] R. F. Remis, P. M. van den Berg, *A modified Lanczos algorithm for the computation of transient electromagnetic wavefields*, IEEE Transactions on Microwave Theory and Techniques, vol. 45, no. 12, 1997, pp. 2139–2149.

[11] M. N. O. Sadiku, *Numerical Techniques in Electromagnetics*, Second edition, CRC Press, 2000.

[12] R. B. Sidje, *EXPOKIT: Software Package for Computing Matrix Exponentials, ACM Transactions on Mathematical Software*, 1998.

[13] A. Taflove, *Computational Electrodynamics: The Finite-Difference Time-Domain Method*, Artech House, Boston, 2000.

[14] V. S. Varadarajan, *Lie Groups, Lie Algebras and Their Representations*, Prentice-Hall Inc., Englewood Cliffs, New Jersey, 1974.

[15] F. Zheng, Z. Chen, and J. Zhang, *Toward the development of a three-dimensional unconditionally stable finite-difference time-domain method*, IEEE Trans. Microwave Theory and Techniques, Vol. 48, No. 9, September 2000, pp. 1550–1558.

[16] K. S. Yee, *Numerical solution of initial boundary value problems involving maxwell equations in isotropic media*, IEEE Transactions on Antennas and Propagation, vol. 14, no. 3, March 1966, pp. 302–307.

# Iterative Solution of Linear Variational Problems in Hilbert Spaces: Some Conjugate Gradients Success Stories

Roland Glowinski,[1] Serguei Lapin[2]

[1]Department of Mathematics, University of Houston, Houston, TX 77204-3008, USA `roland@math.uh.edu`

[2]Department of Mathematics, University of Houston, Houston, TX 77204-3008, USA `slapin@math.uh.edu`

*To Joseph-Louis Lagrange and Jacques-Louis Lions for their contributions to Applied Mathematics and Variational Methods.*

## 1 Introduction

*Conjugate gradient* was introduced in the early 50s (see refs. [1], [2]) as an iterative method for the solution of finite dimensional linear systems associated to matrices with special properties such as symmetry and positive definiteness. If round-off errors are not taken into account, the conjugate gradient method enjoys the *finite termination property*, namely: for any initialization, conjugate gradient algorithms will converge in $d$ iterations at most, assuming that the matrix of the linear system, one is trying to solve, is a $d \times d$ one. Actually, for large problems (let say as soon as $d > 10^2$) the finite termination property has no practical interest, unlike an estimate of the speed of convergence involving the square-root of the condition number. In fact conjugate gradient algorithms can also be applied (theoretically, at least) to the solution of linear problems in *Hilbert spaces* and by the late 60s a solid foundation for the convergence properties of such algorithms was in place (see [3] and the references therein). Such development was most important, since it opened the door to the efficient iterative solution of *optimal control problems* and of large classes of *linear partial differential equations* via an appropriate *variational formulation* in a well-chosen Hilbert space. Examples of such applications can be found in, e.g., [4]–[7], and of course, in various parts of this volume.

Our goal in this article is to describe (in Section 2) the conjugate gradient solution of linear variational problems in Hilbert spaces and to apply the resulting methodology to the iterative solution of:

(i) A boundary control problem for the wave equation (Section 3).
(ii) A generalized Stokes problem which arises naturally when solving numerically the Navier-Stokes equations modeling incompressible viscous flow (Section 4).

Since this volume is dedicated to both *conjugate gradient* and *finite element methods*, here is the right place to mention that it is – almost always – by *combining* both methodologies that these variational problems in Hilbert spaces have been solved in practice.

# 2 On the conjugate gradient solution of variational problems in Hilbert spaces

## 2.1 A family of linear variational problems in Hilbert spaces

Our goal is to address the iterative solution, by conjugate gradient algorithms, of linear variational problems such as:

$$\left.\begin{aligned} &u \in V \\ &a(u,v) = L(v), \quad \forall v \in V, \end{aligned}\right\} \tag{LVP}$$

where, in (LVP), we have:

- $V$ is a real Hilbert space equipped with the scalar product $(\cdot,\cdot)$ and the corresponding norm $\|\cdot\|$.
- $a\colon V \times V \to \mathbb{R}$ is bilinear, continuous and $V$-elliptic, the last property meaning that there exists a positive constant $\alpha$ such that $a(v,v) \geq \alpha\|v\|^2$ for all $v \in V$.
- $L\colon V \to \mathbb{R}$ is linear continuous.

It follows from the *Lax-Milgram theorem* (see, e.g., refs. [7]–[10]) that (LVP) has a unique solution. The conjugate gradient solution of (LVP) will be discussed in Sections 2.2 ($a(\cdot,\cdot)$ symmetric) and 2.3 ($a(\cdot,\cdot)$ possibly non-symmetric).

## 2.2 Conjugate gradient solution of (LVP): (I) The symmetric case

Suppose that functional $a(\cdot,\cdot)$ is symmetric, i.e., $a(v,w) = a(w,v)$ for all $v, w \in V$. We have then equivalence between (LVP) and the following minimization problem

$$\left.\begin{aligned} &u \in V, \\ &J(u) \leq J(v), \quad \forall v \in V, \end{aligned}\right\} \tag{M}$$

with $J(v) = \frac{1}{2}a(v,v) - L(v)$ for all $v \in V$. Concerning the solution of problem (LVP)–(M), we advocate the following conjugate gradient algorithm:

**Step 0: Initialization**

$$u^0 \text{ is given in } V. \tag{2.1}$$

Solve

$$\left.\begin{array}{l} g^0 \in V, \\ (g^0, v) = a(u^0, v) - L(v), \quad \forall v \in V; \end{array}\right\} \tag{2.2}$$

if $\dfrac{\|g^0\|}{(\max \|u^0\|, \varepsilon_0)} \leq$ tol. take $u = u^0$; otherwise, set

$$w^0 = g^0. \tag{2.3}$$

Let $n \geq 0$, and assume that $u^n$, $g^n$ and $w^n$ are known, with $g^n$ and $w^n$ both different from 0. Then:

**Step 1: Steepest descent**

Solve

$$\left.\begin{array}{l} \bar{g}^n \in V, \\ (\bar{g}^n, v) = a(w^n, v), \quad \forall v \in V, \end{array}\right\} \tag{2.4}$$

and compute

$$\rho^n = \frac{\|g^n\|^2}{(\bar{g}^n, w^n)}, \tag{2.5}$$

$$u^{n+1} = u^n - \rho^n w^n, \tag{2.6}$$

$$g^{n+1} = g^n - \rho^n \bar{g}^n. \tag{2.7}$$

**Step 2: Convergence testing and updating of the descent direction**

If $\dfrac{\|g^{n+1}\|}{\|g^0\|} \leq$ tol. take $u = u^{n+1}$; otherwise compute

$$\gamma^n = \frac{\|g^{n+1}\|^2}{\|g^n\|^2} \tag{2.8}$$

and update $w^n$ via

$$w^{n+1} = g^{n+1} + \gamma^n w^n. \tag{2.9}$$

Set $n = n + 1$ and return to (2.4).

Despite its seeming simplicity (2.1)–(2.9) is one of the most powerful and versatile algorithms of Scientific Computing. It has been applied to the solution of a large variety of problems and new applications are found almost daily. Concerning the convergence of algorithm (2.1)–(2.9) the following results are proved in, e.g., [7, Chapter 3]:

**Theorem 2.1.** *Suppose that* tol. $= 0$ *in algorithm* (2.1)–(2.9). *Then for all* $u^0 \in V$ *we have*

$$\lim_{n \to +\infty} \|u^n - u\| = 0, \tag{2.10}$$

*where, in* (2.10), $u$ *is the solution of problem* (LVP). *Moreover, if* $V$ *is finite dimensional with* $\dim V = d$, *algorithm* (2.1)–(2.9) *converges in* $d$ *iterations at most.*

Actually, we have more since (cf., e.g., [3]) the following estimate holds:

$$a(u^n - u, u^n - u) \leq 4\left(\frac{\sqrt{\nu_a} - 1}{\sqrt{\nu_a} + 1}\right)^{2n} a(u^0 - u, u^0 - u), \quad \forall n \geq 1, \tag{2.11}$$

where, in (2.11), $\nu_a$ is the condition number of functional $a(\cdot, \cdot)$, namely,

$$\nu_a = \frac{\sup_{v \in S} a(v, v)}{\inf_{v \in S} a(v, v)}, \tag{2.12}$$

with $S = \{v \in V \mid \|v\| = 1\}$ in (2.12). Let $A$ be the symmetric *isomorphism* from $V$ onto $V$, uniquely defined (via the *Riesz representation theorem* for Hilbert spaces) by $(Av, w) = a(v, w)$ for all $v, w \in V$; we have then $\nu_a = \|A\| \, \|A^{-1}\|$, where

$$\|B\| = \sup_{v \in S} \|Bv\|, \ \forall B \in \mathcal{L}(V, V).$$

There is much to say concerning the choice of $u^0$, tol. and $\varepsilon_0$; a detailed discussion of these important issues can be found in, e.g., [7, Chapter 3] (see also the references therein).

### 2.3 Conjugate gradient solution of (LVP): (II) The general case

Suppose that $a(\cdot, \cdot)$ is no longer symmetric. In order to solve (LVP) by a conjugate gradient algorithm we may proceed as follows:

(i) Observe that (from the Riesz representation theorem) (LVP) is equivalent to

$$Au = l, \tag{2.13}$$

where $A$ (resp., $l$) is the unique isomorphism from $V$ onto $V$ (resp., the unique element of $V$) such that $(Av, w) = a(v, w)$ for all $v, w \in V$ (resp., $(l, v) = L(v)$ for all $v \in V$).

(ii) Denote by $A^t$ the transposed of $A$ (defined for example by $(A^t v, w) = a(w, v)$, $v, w \in V$); since $A^t$ is also an isomorphism, problem (2.13) is equivalent to (its *normal equation*):

$$A^t A u = A^t l. \tag{2.14}$$

A variational formulation of (2.14) is given by:

$$\left.\begin{array}{l} u \in V, \\ (Au, Av) = (l, Av), \quad \forall v \in V. \end{array}\right\} \tag{2.15}$$

Problem (2.15) is clearly a linear variational problem in space $V$. Since

1. $|(l, Av)| \leq \|l\| \, \|A\| \, \|v\|, \;\; \forall v \in V$,
2. $|(Av, Aw)| \leq \|A\|^2 \, \|v\| \, \|w\|, \;\; \forall v, w \in V$,
3. $(Av, Av) = \|Av\|^2 \geq \|A^{-1}\|^{-2}\|v\|^2, \;\; \forall v \in V$,
4. the bilinear functional $(v, w) \to (Av, Aw)$ is *symmetric*,

problem (2.15) can be solved by the conjugate gradient algorithm (2.1)–(2.9). We obtain then

$$u^0 \text{ is given in } V; \tag{2.16}$$

solve

$$\left.\begin{aligned} &g^0 \in V, \\ &(g^0, v) = (Au^0 - l, Av), \;\; \forall v \in V; \end{aligned}\right\} \tag{2.17}$$

if $\dfrac{\|g^0\|}{\max(\|u^0\|, \varepsilon_0)} \leq$ tol. take $u = u^0$; otherwise, set

$$w^0 = g^0. \tag{2.18}$$

For $n \geq 0$, $u^n$, $g^n$ and $w^n$ being known, with $g^n$ and $w^n$ different from 0, solve:

$$\left.\begin{aligned} &\bar{g}^n \in V, \\ &(\bar{g}^n, v) = (Aw^n, Av), \;\; \forall v \in V, \end{aligned}\right\} \tag{2.19}$$

and compute

$$\rho^n = \frac{\|g^n\|^2}{(\bar{g}^n, w^n)}, \tag{2.20}$$

$$u^{n+1} = u^n - \rho^n w^n, \tag{2.21}$$

$$g^{n+1} = g^n - \rho^n \bar{g}^n. \tag{2.22}$$

If $\dfrac{\|g^{n+1}\|}{\|g^0\|} \leq$ tol. take $u = u^{n+1}$; otherwise compute

$$\gamma^n = \frac{\|g^{n+1}\|^2}{\|g^n\|^2} \tag{2.23}$$

and update $w^n$ via

$$w^{n+1} = g^{n+1} + \gamma^n w^n. \tag{2.24}$$

Set $n = n + 1$ and return to (2.19).

Algorithm (2.16)–(2.24) has a major drawback for it involves operator $A$ and vector $l$, explicitly. A simple way to remove this difficulty is to observe that an equivalent formulation of the above algorithm is given by

$$u^0 \text{ is given in } V; \tag{2.25}$$

solve

$$\left.\begin{aligned} &y^0 \in V, \\ &(y^0, v) = a(u^0, v) - L(v), \;\; \forall v \in V; \end{aligned}\right\} \tag{2.26}$$

and then

$$\left.\begin{aligned} &g^0 \in V, \\ &(g^0, v) = a(v, y^0), \quad \forall v \in V; \end{aligned}\right\} \tag{2.27}$$

if $\dfrac{\|g^0\|}{\max(\|u^0\|, \varepsilon_0)} \leq$ tol. take $u = u^0$; otherwise, set

$$w^0 = g^0. \tag{2.28}$$

For $n \geq 0$, $u^n$, $g^n$ and $w^n$ being known, with $g^n$ and $w^n$ different from 0, solve:

$$\left.\begin{aligned} &\bar{y}^n \in V, \\ &(\bar{y}^n, v) = a(w^n, v), \;\; \forall v \in V, \end{aligned}\right\} \tag{2.29}$$

and

$$\left.\begin{aligned} &\bar{g}^n \in V, \\ &(\bar{g}^n, v) = a(v, \bar{y}^n), \;\; \forall v \in V, \end{aligned}\right\} \tag{2.30}$$

and compute

$$\rho^n = \frac{\|g^n\|^2}{(\bar{g}^n, w^n)}, \tag{2.31}$$

$$u^{n+1} = u^n - \rho^n w^n, \tag{2.32}$$

$$g^{n+1} = g^n - \rho^n \bar{g}^n. \tag{2.33}$$

if $\dfrac{\|g^{n+1}\|}{\|g^0\|} \leq$ tol. take $u = u^{n+1}$; otherwise compute

$$\gamma^n = \frac{\|g^{n+1}\|^2}{\|g^n\|^2} \tag{2.34}$$

and update $w^n$ via

$$w^{n+1} = g^{n+1} + \gamma^n w^n. \tag{2.35}$$

Set $n = n + 1$ and return to (2.29).

We observe that algorithm (2.25)–(2.35) does not require the explicit knowledge of $A$, $A^t$, $A^t A$ and $l$. On the other hand, each iteration requires the solution of two linear variational problems associated to the scalar product $(\cdot, \cdot)$.

**Remark 2.1.** Problems (2.14) and (2.15) correspond, respectively, to the operator and variational formulations of the following least squares problem:

$$\left.\begin{aligned} &u \in V, \\ &j(u) \leq j(v), \;\; \forall v \in V, \end{aligned}\right\} \tag{LSP}$$

with $j(v) = \frac{1}{2}\|Av - l\|^2$. The least squares/conjugate gradient solution of linear and nonlinear variational problems in Hilbert spaces has been discussed in, e.g., [7, Chapter 3], [10] and [11] (see also the references therein); the three above references contain many applications, to fluid mechanics problems in particular.

# 3. Application to the boundary control of the wave equation

## 3.1 Formulation of the control problem

Let $\Omega$ be a bounded domain of $\mathbb{R}^d$, with $d \geq 1$; we denote by $\Gamma$ the boundary of $\Omega$. We suppose that a physical phenomenon is taking place in $\Omega$, this phenomenon being modeled by the following *wave equation*:

$$\frac{1}{c^2}\frac{\partial^2 y}{\partial t^2} - \Delta y = 0 \text{ in } \Omega \times (0,T), \tag{WE}$$

where, in (WE):

1. $c > 0$ is the speed of propagation.
2. The Laplace operator $\Delta$ is (classically) defined by

$$\Delta = \sum_{i=1}^{d} \frac{\partial^2}{\partial x_i^2},$$

$x = \{x_i\}_{i=1}^d$ being the generic point of $\mathbb{R}^d$.

3. $0 < T < \infty$.

From now on, we assume that $c$ is independent of $t$ and $x$. Let us denote by $\varphi(t)$ the function $x \to \varphi(x,t)$, $\Gamma_0$ being a subset of $\Gamma$; our goal here is to find a function $v \in L^2(\Gamma_0 \times (0,T))$ such that

$$\left.\begin{aligned} &\frac{1}{c^2}\frac{\partial^2 y}{\partial t^2} - \Delta y = 0 \text{ in } \Omega \times (0,T),\\ &y = v \text{ on } \Gamma_0 \times (0,T), \quad y = 0 \text{ on } (\Gamma \setminus \Gamma_0) \times (0,T),\\ &y(0) = y_0, \quad \frac{\partial y}{\partial t}(0) = y_1,\\ &y(T) = z_0, \quad \frac{\partial y}{\partial t}(T) = z_1. \end{aligned}\right\} \tag{3.1}$$

In other words, starting from $\{y_0, y_1\}$, can we find a boundary control à la Dirichlet ($v \in L^2(\Gamma_0 \times (0,T))$) driving the system to $\{z_0, z_1\}$? The answer is yes, for all $\{y_0, y_1\}$ and $\{z_0, z_1\}$ belonging to $L^2(\Omega) \times H^{-1}(\Omega)$, if $\Gamma_0$ and $T$ are sufficiently large (see [12], [13] for details and further references); here, $H^{-1}(\Omega)$ is the dual space of $H_0^1(\Omega)$. In order to "force" uniqueness we will look for the unique control of minimal norm in $L^2(\Gamma_0 \times (0,T))$; we shall denote by $u$ this control. In order to have the existence of an optimal control even if problem (3.1) has no solution, we are going to use penalty to relax the final conditions $y(T) = z_0$, $\frac{\partial y}{\partial t}(T) = z_1$. This leads to the following well-posed optimal control problem (with $\varepsilon > 0$):

$$\left.\begin{array}{l}\text{Find } u_\varepsilon \in L^2(\Gamma_0 \times (0,T)) \text{ such that} \\ J_\varepsilon(u_\varepsilon) \le J_\varepsilon(v), \quad \forall v \in L^2(\Gamma_0 \times (0,T)),\end{array}\right\} \tag{OCP-$eps$}$$

with

$$\begin{aligned} J_\varepsilon(v) =& \frac{1}{2}\int_{\Gamma_0 \times (0,T)} |v|^2 \,\mathrm{d}\Gamma \,\mathrm{d}t \\ &+ \frac{1}{2\varepsilon}\left[\int_\Omega |y(T) - z_0|^2 \,\mathrm{d}x + \left\|\frac{\partial y}{\partial t}(T) - z_1\right\|_{-1}^2\right]; \end{aligned} \tag{3.2}$$

in (3.2):

- $\mathrm{d}x = \mathrm{d}x_1 \ldots \mathrm{d}x_d$.
- The $H^{-1}(\Omega)$-norm $\|\cdot\|_{-1}$ is defined by: $\|f\|_{-1} = \left(\int_\Omega |\nabla\varphi|^2 \,\mathrm{d}x\right)^{\frac{1}{2}}$ for $f \in H^{-1}(\Omega)$, function $\varphi$ being the unique solution in $H_0^1(\Omega)$ of the Dirichlet problem $-\Delta\varphi = f$ in $\Omega$, $\varphi = 0$ on $\Gamma$.

While (3.1) is an exact boundary controllability problem, problem (OCP-*eps*) is an approximate boundary controllability one. If (3.1) has solutions in $L^2(\Gamma_0 \times (0,T))$, then $\lim_{\varepsilon \to 0} u_\varepsilon = u$ in $L^2(\Gamma_0 \times (0,T))$, $u$ being the necessarily unique-solution of minimal norm of (3.1). The conjugate gradient solution of problem (OCP-*eps*) will be discussed in Section 3.3.

### 3.2 Optimality conditions for (OCP-*eps*)

Applying the methods discussed in, e.g., [6] and [14], we can prove that the unique solution of (OCP-*eps*) is characterized by the following optimality system (we suppose from now on that $c = 1$):

$$u_\varepsilon = \frac{\partial p_\varepsilon}{\partial n}\Big|_{\Gamma_0 \times (0,T)}, \tag{3.3}$$

$$\left.\begin{array}{l} \dfrac{\partial^2 y_\varepsilon}{\partial t^2} - \Delta y_\varepsilon = 0 \text{ in } \Omega \times (0,T), \\ y_\varepsilon = u_\varepsilon \text{ on } \Gamma_0 \times (0,T), \quad y_\varepsilon = 0 \text{ on } (\Gamma \setminus \Gamma_0) \times (0,T), \\ y_\varepsilon(0) = y_0, \quad \dfrac{\partial y_\varepsilon}{\partial t}(0) = y_1, \end{array}\right\} \tag{3.4}$$

$$\left.\begin{array}{l} \dfrac{\partial^2 p_\varepsilon}{\partial t^2} - \Delta p_\varepsilon = 0 \text{ in } \Omega \times (0,T), \\ p_\varepsilon = 0 \text{ on } \Gamma \times (0,T), \\ -\Delta p_\varepsilon(T) = \dfrac{1}{\varepsilon}\left[\dfrac{\partial y_\varepsilon}{\partial t}(T) - z_1\right] \text{ in } \Omega, \quad p_\varepsilon(T) = 0 \text{ on } \Gamma, \\ \dfrac{\partial p_\varepsilon}{\partial t}(T) = -\dfrac{1}{\varepsilon}[y_\varepsilon(T) - z_0]. \end{array}\right\} \tag{3.5}$$

The last two relations in (3.5) can be rewritten as

$$\left.\begin{aligned} &-\varepsilon\Delta p_\varepsilon(T) - \frac{\partial y_\varepsilon}{\partial t}(T) = -z_1 \text{ in } \Omega, \quad p_\varepsilon(T) = 0 \text{ on } \Gamma, \\ &\varepsilon\frac{\partial p_\varepsilon}{\partial t}(T) + y_\varepsilon(T) = z_0. \end{aligned}\right\} \tag{3.6}$$

Denote by $E$ the product space $H_0^1(\Omega) \times L^2(\Omega)$ and by $E'$ its dual space $H^{-1}(\Omega) \times L^2(\Omega)$. Next, we define $\Lambda \in \mathcal{L}(E, E')$ by:

Consider $\mathbf{f} = \{f_0, f_1\} \in E$, then

$$\Lambda\mathbf{f} = \left\{-\frac{\partial y}{\partial t}(T), y(T)\right\}, \tag{3.7}$$

with $y$ obtained from $\mathbf{f}$ via the solution of the two wave problems below

$$\left.\begin{aligned} &\frac{\partial^2 p}{\partial t^2} - \Delta p = 0 \text{ in } \Omega \times (0,T), \\ &p = 0 \text{ on } \Gamma \times (0,T), \\ &p(T) = f_0, \quad \frac{\partial p}{\partial t}(T) = f_1, \end{aligned}\right\} \tag{3.8}$$

and

$$\left.\begin{aligned} &\frac{\partial^2 y}{\partial t^2} - \Delta y = 0 \text{ in } \Omega \times (0,T), \\ &y = \frac{\partial p}{\partial n} \text{ on } \Gamma_0 \times (0,T), \quad y = 0 \text{ on } (\Gamma \setminus \Gamma_0) \times (0,T), \\ &y(0) = 0, \quad \frac{\partial y}{\partial t}(0) = 0. \end{aligned}\right\} \tag{3.9}$$

We have then the following

**Theorem 3.1.** *Operator $\Lambda$ defined by relations (3.7)–(3.9) is continuous, self-adjoint and positive semidefinite.*

Proof. A rigorous proof of the above results is not trivial, relying on the facts that:

1. $\mathbf{f} \in E$ implies the existence of $\partial p/\partial n$ on $\Gamma \times (0,T)$, this function belonging to $L^2(\Gamma \times (0,T))$ and the (linear) mapping $\mathbf{f} \to \partial p/\partial n$ being continuous from $E$ into $L^2(\Gamma \times (0,T))$.
2. Problem (3.9) has a solution such that $\{\frac{\partial y}{\partial t}, y\} \in C^0([0,T]; E')$, the above pair being a continuous function of the Dirichlet data on $\Gamma_0 \times (0,T)$.
3. Integration by parts is still possible, via appropriate duality pairings.

Such a proof can be found in, e.g., [12], [13]. The following proof is formal in the sense that it assumes that all the functions we are dealing with are smooth enough so that we can use "ordinary" integration by parts and Green's formula. We have then (with obvious notation):

$$\langle \Lambda \mathbf{f}, \mathbf{f}' \rangle_{\sigma(E',E)} = \int_\Omega \left[ -\frac{\partial y}{\partial t}(T) f_0' + y(T) f_1' \right] \mathrm{d}x$$
$$= \int_\Omega \left[ -\frac{\partial y}{\partial t}(T) p'(T) + y(T) \frac{\partial p'}{\partial t}(T) \right] \mathrm{d}x = \int_{\Omega \times (0,T)} \left[ \frac{\partial^2 p'}{\partial t^2} y - \frac{\partial^2 y}{\partial t^2} p' \right] \mathrm{d}x \, \mathrm{d}t$$
$$= \int_{\Omega \times (0,T)} [\Delta p' y - \Delta y p'] \, \mathrm{d}x \, \mathrm{d}t = \int_{\Gamma_0 \times (0,T)} \left[ \frac{\partial p'}{\partial n} y - \frac{\partial y}{\partial n} p' \right] \mathrm{d}\Gamma \, \mathrm{d}t$$
$$= \int_{\Gamma_0 \times (0,T)} \frac{\partial p'}{\partial n} \frac{\partial p}{\partial n} \, \mathrm{d}\Gamma \, \mathrm{d}t, \quad \forall \mathbf{f}, \mathbf{f}' \in E.$$

The above relations clearly imply that the operator $\Lambda$ is self-adjoint and positive semi-definite. If appropriate conditions on $\Gamma_0$ and $T$ hold (see, e.g., [12], [13] for details) then operator $\Lambda$ is an $E$-elliptic isomorphism from $E$ onto $E'$ (the $E$-ellipticity property meaning the existence of $\alpha > 0$ such that $\langle \Lambda \mathbf{f}, \mathbf{f} \rangle_{\sigma(E',E)} \geq \alpha \|\mathbf{f}\|_E^2$ for all $\mathbf{f} \in E$). □

Our goal, now is to show that $\mathbf{e}_\varepsilon = \{e_{0\varepsilon}, e_{1\varepsilon}\}$, with $e_{0\varepsilon} = p_\varepsilon(T)$ and $e_{1\varepsilon} = \frac{\partial p_\varepsilon}{\partial t}(T)$, is solution of a functional equation of the following kind:

$$\varepsilon \begin{pmatrix} -\Delta & 0 \\ 0 & I \end{pmatrix} \mathbf{e}_\varepsilon + \Lambda \mathbf{e}_\varepsilon = \boldsymbol{\beta}, \tag{3.10}$$

where, in (3.10), the vector-valued function $\boldsymbol{\beta} \in E'$ and can be "easily" computed from the data $y_0$, $y_1$ and $z_0$, $z_1$. Let us compute $Y_0$ via the solution of:

$$\left.\begin{aligned} &\frac{\partial^2 Y_0}{\partial t^2} - \Delta Y_0 = 0 \ \text{ in } \ \Omega \times (0,T), \\ &Y_0 = 0 \ \text{ on } \ \Gamma \times (0,T), \\ &Y_0(0) = y_0, \quad \frac{\partial Y_0}{\partial t}(0) = y_1; \end{aligned}\right\} \tag{3.11}$$

we have then $\left\{ \frac{\partial Y_0}{\partial t}(T), Y_0(T) \right\} \in E'$. Next define $\bar{y}_\varepsilon$ by $\bar{y}_\varepsilon = y_\varepsilon - Y_0$; we have then, by subtraction between (3.4) and (3.11):

$$\left.\begin{aligned} &\frac{\partial^2 \bar{y}_\varepsilon}{\partial t^2} - \Delta \bar{y}_\varepsilon = 0 \ \text{ in } \ \Omega \times (0,T), \\ &\bar{y}_\varepsilon = \frac{\partial p_\varepsilon}{\partial n} \ \text{ on } \ \Gamma_0 \times (0,T), \quad \bar{y}_\varepsilon = 0 \ \text{ on } \ (\Gamma \setminus \Gamma_0) \times (0,T) \\ &\bar{y}_\varepsilon(0) = 0, \quad \frac{\partial \bar{y}_\varepsilon}{\partial t}(0) = 0. \end{aligned}\right\} \tag{3.12}$$

From (3.12) and from the definition of operator $\Lambda$, we clearly have (3.10) with

$$\boldsymbol{\beta} = \left\{ \frac{\partial Y_0}{\partial t}(T) - z_1, z_0 - Y_0(T) \right\}. \tag{3.13}$$

Since $\varepsilon > 0$, problem (3.10) has a unique solution $\mathbf{e}_\varepsilon$; from this solution we obtain $p_\varepsilon$ via the solution of a backward wave equation and then the optimal control $u_\varepsilon = \frac{\partial p_\varepsilon}{\partial n}|_{\Gamma_0 \times (0,T)}$. A variational formulation of problem (3.10) is given by:

$$\left.\begin{array}{l} \mathbf{e}_\varepsilon (= \{e_{0\varepsilon}, e_{1\varepsilon}\}) \in E, \\ \varepsilon \int_\Omega [\nabla e_{0\varepsilon} \cdot \nabla f_0 + e_{1\varepsilon} f_1] \, dx + \langle \Lambda \mathbf{e}_\varepsilon, \mathbf{f} \rangle_{\sigma(E',E)} = \langle \boldsymbol{\beta}, \mathbf{f} \rangle_{\sigma(E',E)}, \\ \forall \mathbf{f} = \{f_0, f_1\} \in E. \end{array} \right\} \quad (3.14)$$

The bilinear functional occurring in (3.14) is continuous and symmetric; we also have (from Theorem 3.1)

$$\varepsilon \int_\Omega [\nabla f_0 \cdot \nabla f_0 + f_1 f_1] \, dx + \langle \Lambda \mathbf{f}, \mathbf{f} \rangle_{\sigma(E',E)} \geq \varepsilon \|\mathbf{f}\|_E^2, \quad \forall \mathbf{f} \in E,$$

implying that we can solve problem (3.14) (and therefore, the optimal control problem (OCP-*eps*) by the conjugate gradient algorithm (2.1)–(2.9) described in Section 2.2. As we shall see, applying the above algorithm to the solution of problem (3.14) does not require the explicit knowledge of the operator $\Lambda$ and function $\boldsymbol{\beta}$.

## 3.3 Conjugate gradient solution of (OCP-*eps*)

Applying (2.1)–(2.9) to the solution of (OCP-*eps*) via (3.14) leads to the following algorithm (where all subscripts $\varepsilon$ have been omitted):

**Step 0: Initialization**

$$\mathbf{e}^0 (= \{e_0^0, e_1^0\}) \text{ is given in } E; \quad (3.15)$$

solve first the backward wave problem

$$\left.\begin{array}{l} \dfrac{\partial^2 p^0}{\partial t^2} - \Delta p^0 = 0 \text{ in } \Omega \times (0,T), \\ p^0 = 0 \text{ on } \Gamma \times (0,T), \\ p^0(T) = e_0^0, \quad \dfrac{\partial p^0}{\partial t}(T) = e_1^0, \end{array} \right\} \quad (3.16)$$

and then the following forward one

$$\left.\begin{array}{l} \dfrac{\partial^2 y^0}{\partial t^2} - \Delta y^0 = 0 \text{ in } \Omega \times (0,T), \\ y^0 = \dfrac{\partial p^0}{\partial n} \text{ on } \Gamma_0 \times (0,T), \quad y^0 = 0 \text{ on } (\Gamma \setminus \Gamma_0) \times (0,T), \\ y^0(0) = y_0, \quad \dfrac{\partial y^0}{\partial t}(0) = y_1. \end{array} \right\} \quad (3.17)$$

Next, compute $\mathbf{g}^0 (= \{g_0^0, g_1^0\}) \in E$ via the solution of the Dirichlet problem

$$\left.\begin{aligned} -\Delta g_0^0 &= -\varepsilon \Delta e_0^0 + z_1 - \frac{\partial y^0}{\partial t}(T) \text{ in } \Omega, \\ g_0^0 &= 0 \text{ on } \Gamma, \end{aligned}\right\} \tag{3.18.1}$$

and

$$g_1^0 = \varepsilon e_1^0 + y^0(T) - z_0. \tag{3.18.2}$$

If $\dfrac{\|\mathbf{g}^0\|_E}{\max(\varepsilon_0, \|\mathbf{e}^0\|_E)} \leq$ tol., take $\mathbf{e} = \mathbf{e}^0$; otherwise set

$$\mathbf{w}^0 (= \{w_0^0, w_1^0\}) = \mathbf{g}^0. \tag{3.19}$$

For $n \geq 0$, $\mathbf{e}^n$, $\mathbf{g}^n$ and $\mathbf{w}^n$ being known, with $\mathbf{g}^n$ and $\mathbf{w}^n$ both different from $\mathbf{0}$, compute $\mathbf{e}^{n+1}$, $\mathbf{g}^{n+1}$ and, if necessary, $\mathbf{w}^{n+1}$, as follows:

**Step 1: Steepest descent**
Solve the following backward and forward wave equations:

$$\left.\begin{aligned} &\frac{\partial^2 \bar{p}^n}{\partial t^2} - \Delta \bar{p}^n = 0 \text{ in } \Omega \times (0,T), \\ &\bar{p}^n = 0 \text{ on } \Gamma \times (0,T), \\ &\bar{p}^n(T) = w_0^n, \quad \frac{\partial \bar{p}^n}{\partial t}(T) = w_1^n, \end{aligned}\right\} \tag{3.20}$$

$$\left.\begin{aligned} &\frac{\partial^2 \bar{y}^n}{\partial t^2} - \Delta \bar{y}^n = 0 \text{ in } \Omega \times (0,T), \\ &\bar{y}^n = \frac{\partial \bar{p}^n}{\partial n} \text{ on } \Gamma_0 \times (0,T), \quad \bar{y}^n = 0 \text{ on } (\Gamma \setminus \Gamma_0) \times (0,T), \\ &\bar{y}^n(0) = 0, \quad \frac{\partial y^0}{\partial t}(0) = 0. \end{aligned}\right\} \tag{3.21}$$

Next, compute $\bar{\mathbf{g}}^n (= \{\bar{g}_0^n, \bar{g}_1^n\})$ via

$$\left.\begin{aligned} -\Delta \bar{g}_0^n &= -\varepsilon \Delta w_0^n - \frac{\partial \bar{y}^n}{\partial t}(T) \text{ in } \Omega, \\ \bar{g}_0^n &= 0 \text{ on } \Gamma, \end{aligned}\right\} \tag{3.22.1}$$

and

$$\bar{g}_1^n = \varepsilon w_1^n + \bar{y}^n(T), \tag{3.22.2}$$

and then,

$$\rho_n = \frac{\int_\Omega [|\nabla g_0^n|^2 + |g_1^n|^2] \, dx}{\int_\Omega [\nabla \bar{g}_0^n \cdot \nabla w_0^n + \bar{g}_1^n w_1^n] \, dx}, \tag{3.23}$$

$$\mathbf{e}^{n+1} = \mathbf{e}^n - \rho_n \mathbf{w}^n, \tag{3.24}$$

$$\mathbf{g}^{n+1} = \mathbf{g}^n - \rho_n \bar{\mathbf{g}}^n. \tag{3.25}$$

**Step 2: Testing the convergence and construction of the new descent direction**

If

$$\frac{\int_\Omega [|\nabla g_0^{n+1}|^2 + |g_1^{n+1}|^2]\,dx}{\int_\Omega [|\nabla g_0^0|^2 + |g_1^0|^2]\,dx} \leq (\text{tol.})^2 \text{ take } \mathbf{e}_\varepsilon = \mathbf{e}^{n+1}; \text{ otherwise compute}$$

$$\gamma_n = \frac{\int_\Omega [|\nabla g_0^{n+1}|^2 + |g_1^{n+1}|^2]\,dx}{\int_\Omega [|\nabla g_0^n|^2 + |g_1^n|^2]\,dx} \tag{3.26}$$

and

$$\mathbf{w}^{n+1} = \mathbf{g}^{n+1} + \gamma_n \mathbf{w}^n. \tag{3.27}$$

Set $n = n + 1$ and return to (3.22).

The above algorithm deserves many comments and observations. We shall limit ourselves to the following remarks:

**Remark 3.1.** Suppose that $\Gamma_0$ and $T$ are such that the operator $\Lambda$ is $E$-elliptic. Then, problems (OCP-*eps*) and (3.10), (3.14) are well-posed even if $\varepsilon = 0$. To solve the above problems if $\varepsilon = 0$ we can use the conjugate gradient algorithm obtained by taking $\varepsilon = 0$ in (3.15)–(3.27).

**Remark 3.2.** The *finite element approximation* of problem (3.10), (3.14), and the related implementation of algorithm (3.15)–(3.27), lead to delicate approximation issues (reminiscent of those encountered when approximating the Stokes problems discussed in the following Section 4). A systematic way to obtain stable and converging approximations of the above problems, is to approximate space $E$ using a finite element mesh twice coarser than the one used to space-approximate the forward and backward wave equations encountered in the definition of operator $\Lambda$ and in the description of algorithm (3.15)–(3.27). On these meshes, we shall use the same type of finite element spaces (consisting of piecewise linear continuous functions, for example). For a detailed discussion on these approximation issues (including the approximation of $\frac{\partial p}{\partial n}|_{\Gamma_0 \times (0,T)}$) see [4], [6] and [15]. Evidences of the fact that the two-mesh approach is a good one are:

(i) If piecewise linear finite element spaces are used, the corresponding approximations of $\mathbf{e}(= \mathbf{e}_\varepsilon|_\varepsilon = 0)$ are of optimal order in $E$.

(ii) The number of iterations necessary for the convergence of the discrete analogues of algorithm (3.15)–(3.27) is essentially independent of the value of the space-time discretization parameters (for a given value of tol.).

Details on implementation, validation, and the results of numerical experiments can be found in the three above references (see also [16] for a thorough discussion of a finite difference implementation of the above methodology).

**Remark 3.3.** The idea of investigating the *controllability* of systems modeled by partial differential equations via the properties of operators such $\Lambda$, has been systematically investigated by J. L. Lions via his celebrated *Hilbert Uniqueness Method* (HUM) (see, e.g., [12], [13], [17], [6] and the references therein). In the present section we have shown (or tried to show) that HUM, beside being a tool for theoretical investigations, leads also to computational methods.

# 4. On the conjugate gradient solution of the generalized Stokes problem

## 4.1 Generalities. Problem formulation

Another area, where conjugate gradient has provided efficient solution methods, is *Computational Fluid Dynamics* (CFD). Among those CFD related problems, where conjugate gradient has made a significant difference, we shall single out the *generalized Stokes problem* to be discussed below. The starting point of our discussion will be the *Navier-Stokes equations* modeling *incompressible viscous flow*. In the case of a Newtonian fluid, these equations read as follows:

$$\rho\left[\frac{\partial \mathbf{u}}{\partial t} + (\mathbf{u}\cdot\nabla)\mathbf{u}\right] - \mu\Delta\mathbf{u} + \nabla p = \mathbf{f} \text{ in } \Omega\times(0,T), \tag{4.1}$$

$$\nabla\cdot\mathbf{u} = 0 \text{ in } \Omega\times(0,T), \tag{4.2}$$

$$\mathbf{u}(0) = \mathbf{u}_0, \text{ with } \nabla\cdot\mathbf{u}_0 = 0; \tag{4.3}$$

in (4.1)–(4.3):

- $\Omega$ is a bounded domain of $\mathbb{R}^d$ (with $d = 2$ or 3) and $T \in (0,\infty]$; we denote by $\Gamma$ the boundary of $\Omega$.
- $\mathbf{u} = \{u_i\}_{i=1}^d$ (resp., $p$) is the fluid *velocity* (resp., *pressure*).
- $\rho$ and $\mu$ are the fluid *density* and *viscosity*, respectively; we assume that they are both positive constants.
- $\mathbf{f}$ is the density of external forces.

System (4.1)–(4.3) has to be completed by appropriate boundary conditions; for simplicity we shall consider only *Dirichlet boundary conditions*, namely:

$$\left.\begin{array}{l}\mathbf{u} = \mathbf{g} \text{ on } \Gamma\times(0,T),\\ \text{with (necessarily) } \int_\Gamma \mathbf{g}(t)\cdot\mathbf{n}\,d\Gamma = 0 \text{ for } t\in(0,T),\end{array}\right\} \tag{4.4}$$

$\mathbf{n}$ being the outward unit normal vector at $\Gamma$; for other boundary conditions see, e.g., [7]. In order to *time-discretize* the Navier-Stokes system (4.1)–(4.4), we can use (among many other possibilities) the following scheme, of the *operator-splitting* type:

$$\mathbf{u}^0 = \mathbf{u}_0; \tag{4.5}$$

then, for $n \geq 0$, assuming that $\mathbf{u}^n$ is known, we compute $\mathbf{u}^{n+1/2}$ and $\mathbf{u}^{n+1}$ as the respective solutions of

$$\left.\begin{aligned}&\rho\frac{\mathbf{u}^{n+1/2}-\mathbf{u}^n}{\tau}-\mu\Delta\mathbf{u}^{n+1/2}+\nabla p^{n+1}=\mathbf{f}^{n+1} \text{ in } \Omega,\\&\nabla\cdot\mathbf{u}^{n+1/2}=0 \text{ in } \Omega,\\&\mathbf{u}^{n+1/2}=\mathbf{g}^{n+1} \text{ on } \Gamma,\end{aligned}\right\}\tag{4.6}$$

and

$$\left.\begin{aligned}&\frac{\partial\mathbf{u}}{\partial t}+(\mathbf{u}^{n+1/2}\cdot\nabla)\mathbf{u}=0 \text{ in } \Omega\times(t^n,t^{n+1}),\\&\mathbf{u}(t^n)=\mathbf{u}^{n+1/2},\\&\mathbf{u}=\mathbf{g}^{n+1} \text{ on } \Gamma_-^{n+1}\times(t^n,t^{n+1}),\\&\mathbf{u}^{n+1}=\mathbf{u}(t^{n+1}),\end{aligned}\right\}\tag{4.7}$$

with $\tau > 0$ a time-discretization step, $t^n = n\tau$, and $\Gamma_-^{n+1} = \{x \in \Gamma \mid \mathbf{n}(x)\cdot\mathbf{g}^{n+1}(x) < 0\}$.

The solution of the linear transport system (4.7) is addressed in, e.g., [7, Chapter 6]; the solution of the generalized Stokes problem (4.6) will be discussed in the following paragraphs.

### 4.2 Formulation of the generalized Stokes problem as a linear variational problem in $L^2(\Omega)$

Problem (4.6) is a particular case of

$$\left.\begin{aligned}&\alpha\mathbf{u}-\nu\Delta\mathbf{u}+\nabla p=\mathbf{f} \text{ in } \Omega,\\&\nabla\cdot\mathbf{u}=0 \text{ in } \Omega,\\&\mathbf{u}\ =\mathbf{g} \text{ on } \Gamma, \text{ with } \textstyle\int_\Gamma \mathbf{g}\cdot\mathbf{n}\,d\Gamma=0,\end{aligned}\right\}\tag{4.8}$$

with $\alpha$ and $\nu$ two positive constants. Suppose that $\mathbf{f} \in (H^{-1}(\Omega))^d$ and $\mathbf{g} \in (H^{1/2}(\Gamma))^d$; it follows then from, e.g., [18] and [7] that problem (4.8) has a unique solution in $\mathbf{V_g}\times(L^2(\Omega)/\mathbb{R})$, with $\mathbf{V_g} = \{\mathbf{v} \in (H^1(\Omega))^d \mid \mathbf{v} = \mathbf{g} \text{ on } \Gamma\}$. Actually, in what follows, $p$ will be the unique pressure solution of (4.8) so that $p \in L_0^2(\Omega)$, with $L_0^2(\Omega) = \{q \in L^2(\Omega) \mid \int_\Omega q\,dx = 0\}$. Next, define $\mathbf{u}_0 \in \mathbf{V_g}$ and $A \in \mathcal{L}(L^2(\Omega), L_0^2(\Omega))$ by:

(i)

$$\alpha\mathbf{u}_0-\nu\Delta\mathbf{u}_0=\mathbf{f} \text{ in } \Omega, \quad \mathbf{u}_0=\mathbf{g} \text{ on } \Gamma,\tag{4.9}$$

(ii)

$$\forall\mathbf{q}\in L^2(\Omega), \quad Aq=\nabla\cdot\mathbf{u}_q,\tag{4.10}$$

where, in (4.10), $\mathbf{u}_q$ is the unique solution in $(H_0^1(\Omega))^d$ of the following Dirichlet system:

$$\alpha\mathbf{u}_q-\nu\Delta\mathbf{u}_q=-\nabla q \text{ in } \Omega, \quad \mathbf{u}_q=\mathbf{0} \text{ on } \Gamma.\tag{4.11}$$

We have then the following theorem.

**Theorem 4.1.** *The operator $A$ is a self-adjoint, $L_0^2(\Omega)$-elliptic isomorphism of $L_0^2(\Omega)$ onto itself.*

Proof. We clearly have $A \in \mathcal{L}(L^2(\Omega), L_0^2(\Omega))$. To prove the other properties, it is sufficient to prove that the bilinear functional $\{q, q'\} \to \int_\Omega (Aq)q' \, dx$ is symmetric and $L_0^2(\Omega)$-elliptic. Concerning the symmetry, we have (from relations (4.10), (4.11) and from Green's formula):

$$\begin{aligned} \int_\Omega (Aq)q' \, dx &= \int_\Omega \nabla \cdot \mathbf{u}_q q' \, dx = -\int_\Omega \mathbf{u}_q \cdot \nabla q' \, dx \\ &= \int_\Omega \mathbf{u}_q \cdot (\alpha \mathbf{u}_{q'} - \nu \Delta \mathbf{u}_{q'}) \, dx \\ &= \int_\Omega [\mathbf{u}_q \mathbf{u}_{q'} + \nu \nabla \mathbf{u}_q : \nabla \mathbf{u}_{q'}] \, dx, \quad \forall q, q' \in L^2(\Omega), \end{aligned} \tag{4.12}$$

with $\mathbf{S} : \mathbf{T} = \sum_{1 \le i,j \le d} s_{ij} t_{ij}$ if $\mathbf{S} = (s_{ij})$ and $\mathbf{T} = (t_{ij})$. Relation (4.12) clearly implies the symmetry of the above bilinear functional (together with its positive definiteness over $L_0^2(\Omega)$). Concerning now its ellipticity property, it follows from (4.11), (4.12), and from the fact that operator $v \to \alpha \mathbf{v} - \nu \Delta \mathbf{v}$ is an isomorphism from $(H_0^1(\Omega))^d$ onto $(H^{-1}(\Omega))^d$, that there exists a positive constant $c$ so that

$$\int_\Omega (Aq)q \, dx \ge c \|\nabla q\|_{-1}^2, \quad \forall q \in L^2(\Omega); \tag{4.13}$$

the $\| \cdot \|_{-1}$ norm has been defined in Section 3.1 (for real-valued functions only but the generalization to vector-valued ones is obvious). Since (see, e.g., refs. [18]–[20]), the mapping $q \to \|\nabla q\|_{-1}$ defines over the space $L_0^2(\Omega)$ a norm equivalent to the $L^2(\Omega)$-norm, the ellipticity property we are looking for follows from relation (4.13). □

**Remark 4.1.** A more detailed proof of Theorem 4.1 can be found in, e.g., [7, Chapter 4]. Actually, the third and fourth integrals in (4.12) should be replaced by appropriate duality brackets.

Concerning now the conjugate gradient solution of problem (4.8), the key result is provided by the following assertion.

**Theorem 4.2.** *Let $p$ be the unique pressure solution of problem* (4.8) *belonging to $L_0^2(\Omega)$. We have then*

$$Ap = -\nabla \cdot \mathbf{u}_0. \tag{4.14}$$

Proof. Define $\bar{\mathbf{u}}$ by $\bar{\mathbf{u}} = \mathbf{u} - \mathbf{u}_0$. By subtraction between (4.8) and (4.9) we obtain

$$\left.\begin{aligned} &\alpha \bar{\mathbf{u}} - \nu \Delta \bar{\mathbf{u}} = -\nabla p \text{ in } \Omega, \\ &\nabla \cdot \bar{\mathbf{u}} = -\nabla \cdot \mathbf{u}_0, \\ &\bar{\mathbf{u}} = 0 \text{ on } \Gamma. \end{aligned}\right\} \tag{4.15}$$

It follows from the definition of operator $A$ (see relations (4.10) and (4.11)) that (4.15) implies (4.14). □

A *variational formulation* of problem (4.14) is given by

$$\left.\begin{array}{l} p \in L_0^2(\Omega), \\ \int_\Omega (Ap)q\,dx = -\int_\Omega \nabla\cdot\mathbf{u}_0 q\,dx, \quad \forall q \in L_0^2(\Omega). \end{array}\right\} \tag{4.16}$$

From the continuity, symmetry and $L_0^2(\Omega)$-ellipticity properties of the bilinear functional $\{q, q'\} \to \int_\Omega (Aq)q'\,dx$, problem (4.14), (4.16) can be solved by a conjugate gradient algorithm operating in $L_0^2(\Omega)$. Such an algorithm will be described in Section 4.4. Before, motivated by the construction of "good" preconditioners, we are going to investigate (in Section 4.3) some further properties of the operator $A$.

## 4.3 Some further properties of the operator $A$

We clearly have

$$A = -\nabla\cdot(\alpha I - \nu\Delta)^{-1}\nabla, \tag{4.17}$$

the operator $(\alpha I - \nu\Delta)^{-1}$ being associated to the homogeneous Dirichlet boundary conditions. Suppose that the various operator in (4.17) commute (which is not true in general). We have then

$$\left.\begin{array}{l} A \approx \nu^{-1}I \ \text{ if } \nu/\alpha \gg 1, \\ A \approx -\alpha^{-1}\Delta \ \text{ if } \alpha/\nu \gg 1. \end{array}\right\} \tag{4.18}$$

Relations (4.18) suggest that the operator $A$ "enjoys" dramatic variations as a function of the ratio $\nu/\alpha$. The idea behind our "analysis" is to find a preconditioning operator $S$ which behaves like $A$ when $\nu/\alpha$ varies from 0 to $\infty$. Following [21] (see also [7, Chapter 4]), we observe that $A^{-1}$ behaves like the operator $\nu I$ (resp., $\alpha(-\Delta)^{-1}$) for large (resp., small) values of the ratio $\nu/\alpha$; this observation leads to take for $S$ the operator defined by:

$$S^{-1} = \alpha(-\Delta)^{-1} + \nu I. \tag{4.19}$$

We have then the following theorem.

**Theorem 4.3.** *Suppose that the boundary conditions associated to* $(-\Delta)^{-1}$ *in* (4.19) *are of the homogeneous Neumann type and that*

$$\int_\Omega (-\Delta)^{-1}q\,dx = 0, \quad \forall q \in L_0^2(\Omega). \tag{4.20}$$

*Then, the operator $S$ is a self-adjoint, $L_0^2(\Omega)$-elliptic isomorphism from $L_0^2(\Omega)$ onto itself.*

Proof. To prove that the operator $S$ has the above properties it is sufficient to prove that they also hold for its inverse $S^{-1}$ or (equivalently) that the bilinear functional

$$\{q, q\} \to \int_\Omega (S^{-1}q)q'\,dx, \tag{4.21}$$

is continuous, symmetric and $L_0^2(\Omega)$-elliptic over $L_0^2(\Omega) \times L_0^2(\Omega)$. In order to show these properties, for all $q \in L_0^2(\Omega)$ define $\varphi_q$ as the unique function of $H^1(\Omega)$ (actually, of $H^2(\Omega)$ if $\Gamma$ is smooth enough and/or $\Omega$ is convex) verifying

$$\left.\begin{aligned} &-\Delta\varphi_q = q \text{ in } \Omega, \\ &\frac{\partial \varphi_q}{\partial n} = 0 \text{ on } \Gamma; \quad \int_\Omega \varphi_q \, dx = 0. \end{aligned}\right\} \tag{4.22}$$

A variational formulation of the *Neumann problem* (4.22) is given by

$$\left.\begin{aligned} &\varphi_q \in H^1(\Omega) \cap L_0^2(\Omega), \\ &\textstyle\int_\Omega \nabla\varphi_q \cdot \nabla v \, dx = \int_\Omega qv \, dx, \quad \forall v \in H^1(\Omega). \end{aligned}\right\} \tag{4.23}$$

It follows then from (4.19), (4.20) and (4.22) that

$$S^{-1}q = \alpha\varphi_q + \nu q. \tag{4.24}$$

It follows then from (4.23) and (4.24) that

$$\begin{aligned} \int_\Omega (S^{-1}q)q' \, dx &= \alpha \int_\Omega \varphi_q q' \, dx + \nu \int_\Omega qq' \, dx \\ &= \alpha \int_\Omega \nabla\varphi_q \cdot \nabla\varphi_{q'} \, dx + \nu \int_\Omega qq' \, dx, \quad \forall q, q' \in L_0^2(\Omega). \end{aligned} \tag{4.25}$$

The continuity and symmetry of the bilinear functional in (4.21) are obvious from (4.25); from this last relation we also have

$$\int_\Omega (S^{-1}q)q \, dx \geq \nu \int_\Omega |q|^2 \, dx$$

for all $q \in L_0^2(\Omega)$, which implies (since $\nu > 0$) the $L_0^2(\Omega)$-ellipticity of the above bilinear functional. □

Numerical experiments show that, indeed, the operator $S$ (in fact its discrete variants) is a quasi-optimal preconditioner concerning the conjugate gradient solution of problem (4.14), (4.16). If the generalized Stokes problem (4.8) is properly approximated the number of iterations necessary to achieve convergence is small and essentially independent of the space discretization parameters and of the ratio $\nu/\alpha$.

To conclude this paragraph we shall give another evidence of the fact that $S$ defined by (4.19), (4.20) has the potential of being a quasi-optimal preconditioner concerning the iterative solution of problem (4.14), (4.16) (other evidences can be found in, e.g., [7, Chapter 4]). Albeit operator $A$ is associated to the solution of the generalized Stokes problem, fundamentally a multidimensional problem, relation (4.17) makes sense even in one space dimension. Thus, from now on we will assume that $\Omega = (0, l)$, with $0 < l < \infty$; we have then

$$A = -\frac{\mathrm{d}}{\mathrm{d}x}\left(\alpha I - \nu \frac{\mathrm{d}^2}{\mathrm{d}x^2}\right)^{-1} \frac{\mathrm{d}}{\mathrm{d}x}, \tag{4.26}$$

the boundary conditions associated to $\left(\alpha I - \nu \frac{\mathrm{d}^2}{\mathrm{d}x^2}\right)^{-1}$ being of the homogeneous Dirichlet type. Let us define an operator $B$ from $L_0^2(0, l)$ onto itself by

$$B = \alpha\left(-\frac{\mathrm{d}^2}{\mathrm{d}x^2}\right)^{-1} + \nu I, \tag{4.27}$$

where the operator $(-\frac{\mathrm{d}^2}{\mathrm{d}x^2})^{-1}$ maps $L_0^2(0, l)$ into $H^2(0, l) \cap L_0^2(0, l)$ via the solution of the following Neumann problem

$$\left.\begin{aligned} &-\frac{\mathrm{d}^2\varphi}{\mathrm{d}x^2} = q \text{ on } (0, l),\\ &\frac{\mathrm{d}\varphi}{\mathrm{d}x}(0) = \frac{\mathrm{d}\varphi}{\mathrm{d}x}(1) = 0; \quad \int_\Omega \varphi \,\mathrm{d}x = 0. \end{aligned}\right\} \tag{4.28}$$

Let us show that $B = A^{-1}$. To prove the above relation we are going to compute $BAq$ with $q$ describing $L_0^2(0, l)$:

(i) First, we observe that

$$Aq = \frac{\mathrm{d}u_q}{\mathrm{d}x}, \tag{4.29}$$

with $u_q$ the unique solution in $H_0^1(0, l)$ of the following Dirichlet problem

$$-\nu \frac{\mathrm{d}^2 u_q}{\mathrm{d}x^2} + \alpha u_q = -\frac{\mathrm{d}q}{\mathrm{d}x} \text{ in } (0, l), \quad u_q(0) = u_q(l) = 0. \tag{4.30}$$

We observe also that $\frac{\mathrm{d}u_q}{\mathrm{d}x} \in L_0^2(0, l)$ (since $\int_\Omega \frac{\mathrm{d}u_q}{\mathrm{d}x}\,\mathrm{d}x = u_q(l) - u_q(0) = 0$).

(ii) Next, we have from (4.27)–(4.29)

$$BAq = \alpha\varphi_q + \nu \frac{\mathrm{d}u_q}{\mathrm{d}x}, \tag{4.31}$$

with $\varphi_q$ the unique solution in $H^2(0, l) \cap L_0^2(0, l)$ of the following Neumann problem

$$-\frac{\mathrm{d}^2\varphi_q}{\mathrm{d}x^2} = \frac{\mathrm{d}u_q}{\mathrm{d}x} \text{ in } (0, l), \quad \frac{\mathrm{d}\varphi_q}{\mathrm{d}x}(0) = \frac{\mathrm{d}\varphi_q}{\mathrm{d}x}(l) = 0. \tag{4.32}$$

It follows from (4.32) that

$$-\frac{\mathrm{d}\varphi_q}{\mathrm{d}x} = u_q \text{ in } (0, l). \tag{4.33}$$

Combining (4.33) with (4.30), yields $-\frac{d\varphi_q}{dx} = \alpha^{-1}\big(\nu\frac{d^2u_q}{dx^2} - \frac{dq}{dx}\big)$ in $(0,l)$ which implies in turn

$$\varphi_q = \alpha^{-1}\Big(-\nu\frac{du_q}{dx} + q\Big) \text{ in } (0,l). \tag{4.34}$$

Combining (4.34) with (4.31), we obtain $BAq = q$ for all $q \in L_0^2(0,l)$, i.e., $B = A^{-1}$. The above result, albeit one-dimensional, provides a strong incentive to use $S$ defined by (4.19), (4.20) as preconditioner for the conjugate gradient solution of problem (4.14), (4.16). The resulting algorithm will be described in the next paragraph.

**Remark 4.2.** The construction of quasi-optimal preconditioners for the conjugate gradient solution of generalized Stokes problems such as

$$\left.\begin{aligned}&\alpha\mathbf{u} - \nu\Delta\mathbf{u} + \nabla p = \mathbf{f} \text{ in } \Omega,\\ &\nabla\cdot\mathbf{u} = 0 \text{ in } \Omega,\\ &\mathbf{u} = \mathbf{g}_0 \text{ on } \Gamma_0 \text{ and } \nu\frac{\partial\mathbf{u}}{\partial n} - \mathbf{n}p = \mathbf{g}_1 \text{ on } \Gamma_1,\end{aligned}\right\} \tag{4.35}$$

with $\Gamma_0 \cup \Gamma_1 = \Gamma$ and $\Gamma_0 \cap \Gamma_1 = \emptyset$ in (4.35) is discussed in, e.g., [7, Chapter 4].

### 4.4 Conjugate gradient solution of problem (4.14), (4.16)

It follows from Section 4.2 that problem (4.14), (4.16) is a particular case of those linear variational problems whose conjugate gradient solution has been addressed in Section 2.2. Suppose that the Hilbert space $L_0^2(\Omega)$ is equipped with the scalar product $\{q,q'\} \to \int_\Omega (Sq)q'\,dx$, the operator $S$ being like in Section 4.3. Applying algorithm (2.1)–(2.9) to the solution of problem (4.14), (4.16) leads to:

**Step 0: Initialization**

$$p^0 \text{ is given in } L_0^2(\Omega); \tag{4.36}$$

solve

$$\left.\begin{aligned}&\alpha\mathbf{u}^0 - \nu\Delta\mathbf{u}^0 = \mathbf{f} - \nabla p^0 \text{ in } \Omega,\\ &\mathbf{u}^0 = \mathbf{g} \text{ on } \Gamma,\end{aligned}\right\} \tag{4.37}$$

and set

$$r^0 = \nabla\cdot\mathbf{u}^0. \tag{4.38}$$

If $\frac{\int_\Omega |r^0|^2\,dx}{\max\left(\int_\Omega [\alpha|\mathbf{u}^0|^2 + \nu|\nabla\mathbf{u}^0|^2\,dx, \varepsilon_0^2\right)} \le (\text{tol.})^2$, take $\{\mathbf{u},p\} = \{\mathbf{u}^0,p^0\}$; otherwise, solve

$$\left.\begin{aligned}&-\Delta\varphi^0 = r^0 \text{ in } \Omega,\\ &\frac{\partial\varphi^0}{\partial n} = 0 \text{ on } \Gamma; \quad \int_\Omega \varphi^0\,dx = 0,\end{aligned}\right\} \tag{4.39}$$

and set

$$g^0 = \alpha\varphi^0 + \nu r^0, \tag{4.40}$$

$$w^0 = g^0. \tag{4.41}$$

For $n \geq 0$, assuming that $p^n$, $r^n$, $g^n$, $w^n$ are known, with the last three functions different from 0, proceed as follows:

**Step 1: Steepest descent**

Solve

$$\left.\begin{aligned} &\alpha\bar{\mathbf{u}}^n - \nu\Delta\bar{\mathbf{u}}^n = -\nabla w^n \text{ in } \Omega, \\ &\bar{\mathbf{u}}^n = 0 \text{ on } \Gamma, \end{aligned}\right\} \tag{4.42}$$

then compute

$$\bar{r}^n = \nabla \cdot \bar{\mathbf{u}}^n, \tag{4.43}$$

and solve

$$\left.\begin{aligned} &-\Delta\bar{\varphi}^n = \bar{r}^n \text{ in } \Omega, \\ &\frac{\partial\bar{\varphi}^n}{\partial n} = 0 \text{ on } \Gamma; \quad \int_\Omega \bar{\varphi}^n \,\mathrm{d}x = 0. \end{aligned}\right\} \tag{4.44}$$

Compute

$$\bar{g}^n = \alpha\bar{\varphi}^n + \bar{r}^n, \tag{4.45}$$

$$\rho_n = \frac{\int_\Omega r^n g^n \,\mathrm{d}x}{\int_\Omega \bar{g}^n w^n \,\mathrm{d}x}, \tag{4.46}$$

$$p^{n+1} = p^n - \rho^n w^n. \tag{4.47}$$

**Step 2: Testing the convergence.**
**Construction of the new descent direction**

Compute

$$r^{n+1} = r^n - \rho_n \bar{r}^n, \tag{4.48}$$

$$g^{n+1} = g^n - \rho_n \bar{g}^n. \tag{4.49}$$

If $\frac{\int_\Omega r^{n+1} g^{n+1} \,\mathrm{d}x}{\int_\Omega r^0 g^0 \,\mathrm{d}x} \leq (\text{tol.})^2$, take $p = p^{n+1}$; otherwise, compute

$$\gamma_n = \frac{\int_\Omega r^{n+1} g^{n+1} \,\mathrm{d}x}{\int_\Omega r^n g^n \,\mathrm{d}x}, \tag{4.50}$$

and

$$w^{n+1} = g^{n+1} + \gamma_n w^n. \tag{4.51}$$

Set $n = n + 1$ and return to (4.42).

Despite its seeming complexity the above algorithm is fairly easy to implement, via an appropriate finite element approximation, for example; it requires essentially:

(i) A Neumann solver for the solution of problems (4.39) and (4.44).
(ii) A solver for problems (4.37) and (4.42).

For flow at *high Reynolds numbers*, the kinematic viscosity $\nu(=\mu/\rho)$ is usually small, while the fast dynamics of these flow requires a small time step $\tau$, implying that $\alpha(\approx 1/\tau)$ is large. From these properties of coefficients $\alpha$ and $\nu$, the discrete analogues of problems (4.37) and (4.42) are, in general, better conditioned than those of problems (4.39) and (4.44), implying that they are easier to solve iteratively (for more details on these issues, and the treatment of other boundary conditions than Dirichlet's in (4.8), see, e.g., [7, Chapter 5]). It is worth noticing that if the finite element methods of Bercovier-Pironneau or Hood-Taylor are used to approximate the generalized Stokes problem (4.8), the finite element spaces used to approximate the pressure $p$ are of smaller dimension than those used to approximate the velocity $\mathbf{u}$, namely 8 (resp., 24) times smaller if the flow region $\Omega$ is two-dimensional (resp., three-dimensional); incidentally, the above observation gives a strong incentive to use "non-equal" interpolation techniques to approximate Stokes problems such (4.8); see, again, [7, Chapter 4] for more details, comments, and further references.

**Acknowledgements.** The authors would like to thank J. Cahouet, J.P. Chabard, E.J. Dean, J.W. He, C.H. Li, T.W. Pan, J. Periaux, O. Pironneau, and E. Zuazua for their contributions to the various matters discussed in this article.

## References

[1] M.R. Hestenes, E.L. Stiefel, *Methods of conjugate gradients for solving linear systems*, J. Res. National Bureau of Standards, Section B **49** (1952), 409–436.

[2] G.H. Golub, D.P. O'Leary, *Some history of the conjugate gradient and Lanczos algorithms: 1948–1976*, SIAM Rev. **31** (1989), 50–102.

[3] J. Daniel, *The Approximate Minimization of Functionals*, Prentice-Hall, Englewood Cliffs, 1970.

[4] R. Glowinski, *Ensuring well-posedness by analogy: Stokes problem and boundary control for the wave equation*, J. Comput. Phys. **103** (1992), 189–221.

[5] R. Glowinski, J.L. Lions, *Exact and approximate controllability for distributed parameter systems. Part 1*, in: Acta Numerica 1994, Cambridge University Press, Cambridge, 1994, pp. 269–378.

[6] R. Glowinski, J.L. Lions, *Exact and approximate controllability for distributed parameter systems. Part 2*, in: Acta Numerica 1995, Cambridge University Press, Cambridge, 1995, pp. 159–333.

[7] R. Glowinski, *Finite element methods for incompressible viscous flow*, in: Handbook of Numerical Analysis IX (P. G. Ciarlet and J.L. Lions, eds.), North-Holland, Amsterdam, 2003, pp. 3–1176.

[8] P. G. Ciarlet, *The Finite Element Method for Elliptic Problems*, North-Holland, Amsterdam, 1978, reprinted as Vol. 40, SIAM Classics in Applied Mathematics, SIAM, Philadelphia, 2002.

[9] P. G. Ciarlet, *Basic error estimates for elliptic problems*, in: Handbook of Numerical Analysis II (P. G. Ciarlet and J.L. Lions, eds.), North-Holland, Amsterdam, 1991, pp. 17–351.

[10] R. Glowinski, *Numerical Methods for Nonlinear Variational Problems*, Springer-Verlag, New York, 1984.

[11] R. Glowinski, H.B. Keller, and L. Reinhart, *Continuation-conjugate gradient methods for the least-squares solution of nonlinear boundary value problems*, SIAM J. Sci. Stat. Comp. **4** (1985), 793–832.

[12] J.L. Lions, *Controllability, stabilization and perturbations for distributed systems*, SIAM Review **30** (1988), 1–68.

[13] J.L. Lions, *Controlabilité Exacte, Stabilisation et Perturbation des Systèmes Distribués*, Masson, Paris, 1988.

[14] J.L. Lions, *Optimal Control of Systems Governed by Partial Differential Equations*, Springer-Verlag, New York, 1971.

[15] R. Glowinski, C.H. Li, *On the numerical implementation of the Hilbert uniqueness method for the exact controllability of the wave equation*, C. R. Acad. Sci. Paris, Ser. I, **311** (1990), 135–142.

[16] M. Asch, G. Lebeau, *Geometrical aspects of exact boundary controllability for the wave equation – a numerical study, ESAIM: Control, Optimization and Calculus of Variations* **3** (1998), 163–212.

[17] J.L. Lions, J.E. Lagnese, *Modeling, Analysis and Control of Thin Plates*, Masson, Paris, 1988.

[18] V. Girault, P.A. Raviart, *Finite Element Methods for Navier-Stokes Equations: Theory and Algorithms*, Springer-Verlag, Berlin, 1986.

[19] J. Nečas, *Equations aux Dérivées Partielles*, Presses de l' Université de Montréal, Montréal, 1965.

[20] L. Tartar, *Topics in Nonlinear Analysis, Publications Mathématiques d'Orsay*, Université Paris-Sud, Département de Mathématiques, Paris, 1978.

[21] J. Cahouet, J.P. Chabard, *Some fast 3-D solvers for the generalized Stokes problem*, Internat. J. Numer. Methods Fluids **8** (1988), 369–395.

# On the Application of Preconditioning Operators for Nonlinear Elliptic Problems

Owe Axelsson,[1] István Faragó,[2] János Karátson[3]

[1]Department of Mathematics, University of Nijmegen, 6525ED Nijmegen, Netherlands axelsson@sci.kun.nl
[2]Department of Applied Analysis, Eötvös Loránd University, P.O. Box 120, H–1518 Budapest, Hungary faragois@cs.elte.hu
[3]Department of Applied Analysis, Eötvös Loránd University, P.O. Box 120, H–1518 Budapest, Hungary karatson@cs.elte.hu

## 1 Introduction

In this paper we study the preconditioned iterative solution of nonlinear elliptic problems

$$\left.\begin{aligned} -\operatorname{div} f(x, \nabla u) &= g(x) \text{ in } \Omega, \\ u_{|\partial\Omega} &= 0 \end{aligned}\right\} \tag{1}$$

on a bounded domain $\Omega \subset \mathbb{R}^N$ with suitable ellipticity conditions. Such problems arise in many applications in physics and other fields, for instance in elasto-plasticity, magnetic potential equations and subsonic flow problems [7], [15], [16]. In these problems the nonlinearity $f$ often has the form

$$f(x, \eta) = a(|\eta|)\eta \tag{2}$$

with some positive real $C^1$ function, i.e., the corresponding operator is

$$T(u) = -\operatorname{div}(a(|\nabla u|)\, \nabla u).$$

The most frequently used numerical methods for nonlinear elliptic problems rely on some discretized form of the problem, whose solution is obtained by an iterative method. As for linear problems, preconditioning is a crucial element of the construction of the iterative method. The choice of preconditioners is often supported by Hilbert space background, which helps both the construction of methods and the study of convergence. Important examples of this are the equivalent operator approach [9] for linear equations and the Sobolev gradient technique [17], [18] for nonlinear problems.

The authors' related results (see e.g. [4], [6], [10] rely similarly on a Hilbert space framework, and give the idea of preconditioning operators.

In this paper we first give a brief summary on the idea of preconditioning operators for discretized nonlinear elliptic problems. This means that the proposed preconditioning matrices are the discretizations of suitable linear elliptic operators. In other words, the preconditioner is the projection of a preconditioning operator from the Sobolev space into the same discretization subspace as was used for the original nonlinear problem. Accordingly, the iterative sequence for the discretized problem is the similar projection of a theoretical iteration, defined in the Sobolev space, into the discretization subspace. This means that the preconditioning operator is chosen for the boundary value problem itself on the continuous level, before discretization is executed. In this way the properties of the original problem can be exploited more directly than with usual preconditioners defined for the discretized system.

Second, we present a typical application of the preconditioning operator idea, in which the derivative operator is preconditioned using a diagonal approximation of its coefficient. This is a generalization of [4]. The construction of the related preconditioners and estimates of the resulting condition number are given. The latter depends only on the coefficients, showing that the condition numbers for the discretized problems are mesh independent, which is an important feature of the preconditioning operator approach.

## 2 The general idea of preconditioning operators

The idea of preconditioning operators can be summarized in the following way, considering one-step iterations and allowing the preconditioners to vary stepwise.

Let us consider a nonlinear boundary value problem

$$F(u) = b. \tag{3}$$

The standard way of numerical solution consists of 'discretization plus iteration'. That is, one first discretizes the problem in some subspace $V_h$ and obtains a nonlinear algebraic system $F_h(u_h) = b_h$. Then one looks for positive definite matrices $\mathcal{A}_h^{(n)}$ $(n \in \mathbb{N})$ and defines the iterative sequence $\{u_h^{(n)}\}_{n\in\mathbb{N}}$ in $V_h$ with these matrices as (variable) preconditioners, as shown by Fig. 1.

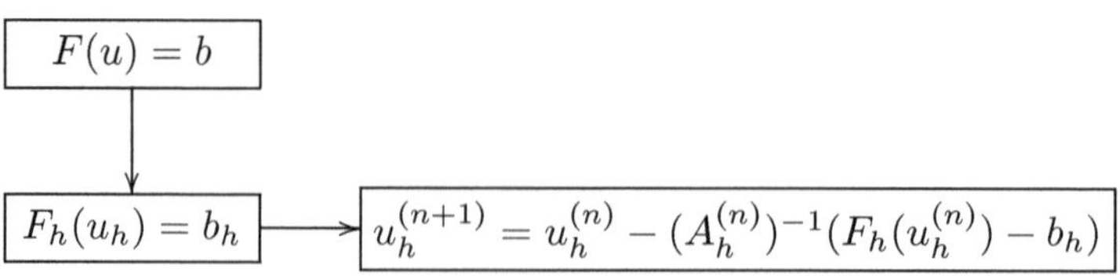


Figure 1. Discretization plus iteration.

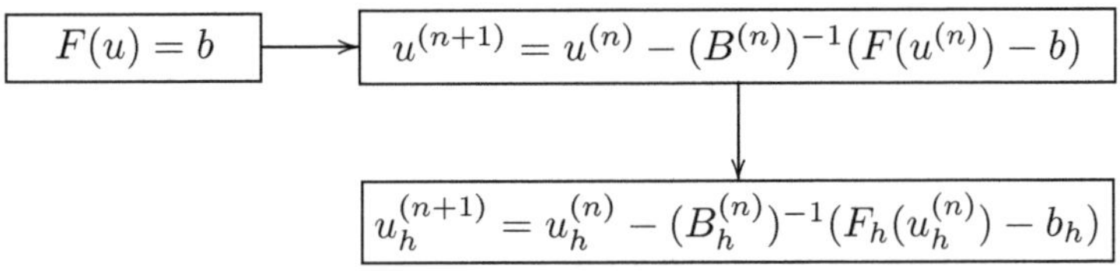


Figure 2. Iteration plus discretization: the preconditioning operator idea.

The idea of preconditioning operators means that one first defines suitable linear elliptic differential operators $B^{(n)}$ and a sequence $\{u^{(n)}\}_{n\in\mathbb{N}}$ with these operators as preconditioners in the corresponding Sobolev space, then one proposes the preconditioning matrices

$$\mathcal{A}_h^{(n)} = B_h^{(n)} \tag{4}$$

for the iteration in $V_h$, which means that the preconditioning matrices are obtained using the same discretization for the operators $B^{(n)}$ as was used to obtain the system $F_h(u_h) = b_h$ from problem (3). (This way of derivation is indicated in the notation $B_h^{(n)}$.) The above process is shown schematically in Fig. 2.

As a matter of fact, both approaches yield the same type of sequence running in $V_h$, and and the difference lies in the special choice of preconditioners in the second case. On the other hand, although in Fig. 2 the discretization parameter $h$ is fixed, its idea remains just the same if $h$ is redefined in each step $n$, i.e., in the setting of a multilevel or projection-iteration method [5], [8].

We note that, clearly, the preconditioning operator approach does not include important preconditioning methods which rely on different matrix techniques.

## 3 Preconditioning operators for nonlinear elliptic problems

We consider problem (1):

$$\left.\begin{aligned} -\operatorname{div} f(x,\nabla u) &= g(x) \quad \text{in } \Omega, \\ u_{|\partial\Omega} &= 0 \end{aligned}\right\} \tag{5}$$

under the following assumptions:

**Assumptions 1.**

(i) The function $f\colon \Omega \times \mathbb{R}^N \to \mathbb{R}^N$ is measurable and bounded w.r.t. the variable $x \in \Omega$ and $C^1$ w.r.t. the variable $\eta \in \mathbb{R}^N$, further, its Jacobians $\frac{\partial f(x,\eta)}{\partial \eta}$ are symmetric and their eigenvalues $\lambda$ satisfy

$$0 < \mu_1 \le \lambda \le \mu_2 < \infty \tag{6}$$

with constants $\mu_1$ and $\mu_2$ independent of $(x,\eta)$.

(ii) $g \in L^2(\Omega)$.

Let $V_h \subset H_0^1(\Omega)$ be a finite element subspace of the real Sobolev space $H_0^1(\Omega)$ with the inner product

$$\langle u, v \rangle := \int_\Omega \nabla u \cdot \nabla v. \tag{7}$$

Let $F\colon V_h \to V_h$ denote the operator defined by

$$\langle F(u), v \rangle = \int_\Omega f(x, \nabla u) \cdot \nabla v \qquad (v \in V_h) \tag{8}$$

and $b \in V_h$ the element defined by

$$\langle b, v \rangle = \int_\Omega gv \quad (v \in V_h). \tag{9}$$

Then the FEM discretization of problem (5) is written as

$$\langle F(u), v \rangle = \langle b, v \rangle \quad (v \in V_h). \tag{10}$$

Owing to the uniform ellipticity, problem (5) as well as the discretized problems (10) have a unique weak solution, and the latter tend to the former as $h \to 0$. Denote by $u^* \in V_h$ the solution of (10). (For simplicity, no index $h$ is given to $u^*$.)

The operator $F$ is Gâteaux differentiable (see, e.g., [16]), and its derivative is given by

$$\langle F'(u)v, z \rangle = \int_\Omega \frac{\partial f}{\partial \eta}(x, \nabla u)\, \nabla v \cdot \nabla z \quad (u, v, z \in V_h).$$

### 3.1 Simple and variable preconditioning

Before turning to the exact formulation of construction and convergence, we briefly summarize the studied iterative methods in this subsection. The aim of this is twofold: first, we demonstrate that the sequences in $V_h$ for the discretized problems arise directly as the projections of the similar sequences in $H_0^1(\Omega)$. Second, we can arrive from simple iterations to Newton-like methods in three steps which gradually improve the convergence, and this shows that the concept of preconditioning is able to provide a general framework to discuss simple iterations and Newton-like methods.

Let us sketch how this way of discussion proceeds. Here $V$ may stand either for $H_0^1(\Omega)$ or for $V_h$.

First, simple iterations can be accelerated by using a fixed linear preconditioning operator $B$ based on spectral equivalence:

$$m\langle Bv, v \rangle \le \langle F'(u)v, v \rangle \le M\langle Bv, v \rangle \quad (u, v \in V) \tag{11}$$

(with some constants $M \geq m > 0$), which means that

$$\mathrm{cond}(B^{-1}F'(u)) \leq \frac{M}{m} \quad (u \in V).$$

This yields that the preconditioned sequence

$$u_{n+1} = u_n - \frac{2}{M+m} B^{-1}(F(u_n) - b) \quad (n \in \mathbb{N}) \tag{12}$$

converges with ratio

$$q = \frac{M-m}{M+m}.$$

If the spectral interval $[m, M]$ in (11) is too wide (i.e., $\frac{M-m}{M+m}$ is too large), then one can use variable preconditioners $B_n$ based on spectral equivalence in the distinct steps:

$$m_n \langle B_n v, v \rangle \leq \langle F'(u_n)v, v \rangle \leq M_n \langle B_n v, v \rangle \quad (n \in \mathbb{N}, \;\; v \in V) \tag{13}$$

(with some constants $M_n \geq m_n > 0$ satisfying $0 < m \leq m_n \leq M_n \leq M$), which means that

$$\mathrm{cond}(B_n^{-1}F'(u_n)) \leq \frac{M_n}{m_n} \quad (n \in \mathbb{N}).$$

In this case the preconditioned sequence

$$u_{n+1} = u_n - \frac{2}{M_n + m_n} B_n^{-1}(F(u_n) - b) \quad (n \in \mathbb{N}) \tag{14}$$

converges with ratio estimated by

$$q = \limsup \frac{M_n - m_n}{M_n + m_n}.$$

(If the latter is 0 then superlinear convergence is achieved.) On the one hand this variable preconditioning generalizes the preconditioned simple iteration (12). On the other hand, condition (13) means that the sequence (14) defines a quasi-Newton method in which $B_n$ is an approximate derivative operator. Accordingly, the sequence (14) can only converge globally provided that a suitable damping is also used.

Finally, if one chooses $B_n = F'(u_n)$ in (13) with the corresponding spectral bounds $m_n = M_n = 1$, then one arrives at Newton's method as an extreme case of (14). This shows that (14) can be regarded as a general form of one-step iterations. We note that (in contrary to the previous cases) the operators $B_n = F'(u_n)$ are not really considered as preconditioners, since they cannot be chosen to yield suitably simple auxiliary linear equations, as was expected of $B$ and $B_n$ in (12) and (14). Hence, one generally uses some inner iteration for the auxiliary linear problems and applies preconditioning therein.

### 3.2 Construction and convergence of the iterations

The construction of the iterations (12) and (14) involves the solution of auxiliary linear elliptic problems, for which one can rely on a highly developed background of efficient solvers (see the references in Subsection 3.3). Regarding convergence, the preconditioned discretized problems have mesh independent condition numbers, since their convergence properties are asymptotically determined by the theoretical sequence in the Sobolev space.

#### 3.2.1 Simple iterations

Let $G \in L^\infty(\Omega, \mathbb{R}^{N\times N})$ be a symmetric matrix-valued function for which there exist constants $M \geq m > 0$ such that

$$m\langle G(x)\xi,\xi\rangle \leq \Big\langle \frac{\partial f(x,\eta)}{\partial \eta}\xi,\xi\Big\rangle \leq M\langle G(x)\xi,\xi\rangle \quad ((x,\eta)\in\Omega\times\mathbb{R}^N,\ \ \xi\in\mathbb{R}^N). \tag{15}$$

We introduce the self-adjoint positive linear operator

$$\langle Bz,v\rangle = \int_\Omega G(x)\nabla z\cdot\nabla v \quad (z,v\in V). \tag{16}$$

Then (15) implies

$$m\langle Bv,v\rangle \leq \langle F'(u)v,v\rangle \leq M\langle Bv,v\rangle \quad (u,v\in V). \tag{17}$$

**Theorem 1.** *For any $u_0 \in V$, the sequence*

$$u_{n+1} = u_n - \frac{2}{M+m}B^{-1}(F(u_n)-b) \quad (n\in\mathbb{N})$$

*converges linearly to $u^*$ according to the estimate*

$$\|u_n - u^*\|_B \leq \frac{1}{mp^{1/2}}\|F(u_0)-b\|\Big(\frac{M-m}{M+m}\Big)^n \quad (n\in\mathbb{N}), \tag{18}$$

*where $p>0$ is the lower bound of the operator $B$.*

The proof essentially follows from the well-known special case when $B = I$ and (17) is nothing but the ellipticity of the operator $F'(u)$ (see e.g. [12]). Namely, $B^{-1}F$ inherits the properties of $F$ in the energy space $V_B$ of the operator $B$ (i.e., with respect to the inner product $\langle u,v\rangle_B = \langle Bu,v\rangle$).

### 3.2.2 Variable preconditioning

In order to improve the bounds that arise in (15) and (17), we stepwise introduce symmetric matrix-valued functions $G_n \in L^\infty(\Omega, \mathbb{R}^{N\times N})$ for which

$$m_n\langle G_n(x)\xi,\xi\rangle \le \Big\langle \frac{\partial f}{\partial \eta}(x,\nabla u_n(x))\xi,\xi\Big\rangle \le M_n\langle G_n(x)\xi,\xi\rangle \tag{19}$$

with suitable constants $M_n \ge m_n > 0$ independent of $x \in \Omega$, $\xi \in \mathbb{R}^N$. The corresponding positive linear operator is

$$\langle B_n z, v\rangle = \int_\Omega G_n(x)\nabla z\cdot\nabla v \quad (z,v\in V), \tag{20}$$

which now satisfies

$$m_n\langle B_n v,v\rangle \le \langle F'(u_n)v,v\rangle \le M_n\langle B_n v,v\rangle \quad (v\in V).$$

The formulation of the iteration uses the norms

$$\|v\|_n = \langle F'(u_n)^{-1}v,v\rangle^{1/2} \quad (n\in\mathbb{N}), \qquad \|v\|_* = \langle F'(u^*)^{-1}v,v\rangle^{1/2}. \tag{21}$$

**Assumptions 2.**

Let the items (i)–(ii) of Assumptions 1 hold and let the Jacobians $\frac{\partial f(x,\eta)}{\partial\eta}$ be Lipschitz continuous w.r.t. $\eta$.

Since $V_h$ is finite-dimensional, the operator $F'$ inherits the Lipschitz continuity of $\partial f/\partial\eta$. Let $L$ denote the Lipschitz constant of $F'$.

**Theorem 2** (see [14]). *For arbitrary $u_0 \in V_h$ let $(u_n)$ be the sequence defined by*

$$u_{n+1} = u_n - \frac{2\tau_n}{M_n+m_n} B_n^{-1}(F(u_n)-b) \quad (n\in\mathbb{N}), \tag{22}$$

*where the following conditions hold:*

(iii) *$M_n \ge m_n > 0$, further, using notation $\omega(u_n) = L\mu_1^{-2}\|F(u_n)-b\|$, there exist constants $K>1$ and $\varepsilon>0$ such that $M_n/m_n \le 1+2/(\varepsilon + K\,\omega(u_n))$;*

(iv) *the damping parameters $\tau_n$ are defined by*

$$\tau_n = \min\Big\{1, \frac{1-Q_n}{2\rho_n}\Big\}, \tag{23}$$

*where $Q_n = \frac{M_n-m_n}{M_n+m_n}(1+\omega(u_n))$ (and by the assumption (iii) $Q_n<1$), further, $\rho_n = 2LM_n^2\mu_1^{-3/2}(M_n+m_n)^{-2}\|F(u_n)-b\|_n(1+\omega(u_n))^{1/2}$, $\omega(u_n)$ is as in condition (iii) and $\|\cdot\|_n$ is defined in (21). (This value of $\tau_n$ ensures optimal contractivity in the $n$-th step in the $\|\cdot\|_*$-norm.)*

*Then we have*

$$\|u_n - u^*\| \le \mu_1^{-1} \|F(u_n) - b\| \to 0,$$

*namely,*

$$\limsup \frac{\|F(u_{n+1}) - b\|_*}{\|F(u_n) - b\|_*} \le \limsup \frac{M_n - m_n}{M_n + m_n} < 1. \tag{24}$$

*Moreover, if in addition we assume* $M_n/m_n \le 1 + c_1 \|F(u_n) - b\|^\gamma$ $(n \in \mathbb{N})$ *with some constants* $c_1 > 0$ *and* $0 < \gamma \le 1$, *then*

$$\|F(u_{n+1}) - b\|_* \le d_1 \|F(u_n) - b\|_*^{1+\gamma} \quad (n \in \mathbb{N}) \tag{25}$$

*with some constant* $d_1 > 0$.

Owing to the equivalence of the norms $\|\cdot\|$ and $\|\cdot\|_*$, the orders of convergence corresponding to the estimate (25) can be formulated with the original norm:

**Corollary 1** (Rate of convergence in the original norm). *Let*

$$M_n/m_n \le 1 + c_1 \|F(u_n - b)\|^\gamma$$

*with some constants* $c_1 > 0$, $0 < \gamma \le 1$. *Then we have*

$$\|F(u_{n+1}) - b\| \le d_1 \|F(u_n) - b\|^{1+\gamma} \quad (n \in \mathbb{N}),$$

*and consequently*

$$\|u_n - u^*\| \le \mu_1^{-1} \|F(u_n) - b\| \le \text{const} \cdot \rho^{(1+\gamma)^n}$$

*with some constant* $0 < \rho < 1$.

### 3.2.3 Inner-outer iterations

Similarly to the previous paragraph, we impose the following

**Assumptions 3.**

Let the items (i)–(ii) of Assumptions 1 hold and let the Jacobians $\frac{\partial f(x,\eta)}{\partial \eta}$ be Lipschitz continuous w.r.t. $\eta$. Let $L$ denote the Lipschitz constant of $F'$.

The iterative sequence $(u_n) \subset V_h$ is constructed by the following inner-outer iteration. Let $u_0 \in V_h$ be arbitrary.

(a) The outer iteration defines the sequence

$$u_{n+1} = u_n + \tau_n p_n \quad (n \in \mathbb{N}), \tag{26}$$

where $p_n \in V_h$ is the numerical solution of the auxiliary linear problem

$$\langle F'(u_n) p_n, v \rangle = -\langle F(u_n) - b, v \rangle \quad (v \in V_h) \tag{27}$$

or

$$\int_\Omega \frac{\partial f}{\partial \eta}(x, \nabla u_n)\nabla p_n \cdot \nabla v \;= -\int_\Omega (f(x, \nabla u_n) \cdot \nabla v \;- gv) \quad (v \in V_h),$$

further,

$$\delta_n > 0 \text{ is some prescribed constant satisfying } 0 < \delta_n \le \delta_0 < 1, \qquad (28)$$

$$\tau_n = \min\Big\{1, \frac{(1-\delta_n)}{(1+\delta_n)} \frac{\mu_1}{L\|p_n\|}\Big\} \in (0,1].$$

(b) To determine $p_n$ in (27), the inner iteration defines a sequence

$$(p_n^{(k)}) \subset V_h \quad (k \in \mathbb{N})$$

using a preconditioned conjugate gradient method. Namely, we can choose constants $M_n \ge m_n > 0$ and a symmetric matrix-valued function $G_n \in L^\infty(\Omega, \mathbb{R}^{N\times N})$, satisfying $\sigma(G(x)) \subset [\mu_1, \mu_2]$ $(x \in \Omega)$ with $\mu_1, \mu_2$ from (6), such that (19) holds, and let $B_n\colon V_h \to V_h$ be the corresponding linear operator defined by (20).

Then we consider the preconditioned form of (27):

$$B_n^{-1}F'(u_n)p_n = -B_n^{-1}(F(u_n) - b), \qquad (29)$$

and construct the sequence $(p_n^{(k)})_{k\in\mathbb{N}}$ by the standard conjugate gradient method for the equation (29) (see, e.g., [2]). For convenience we let $p_n^{(0)} = 0$.

Finally, $p_n := p_n^{(k_n)} \in V_h$ is defined with the smallest index $k_n$ for which the relative error estimate

$$\|F'(u_n)p_n^{(k_n)} + (F(u_n) - b)\|_{B_n^{-1}} \le \rho_n \|F(u_n) - b\|_{B_n^{-1}} \qquad (30)$$

holds with $\rho_n = (\mu_1/\mu_2)^{1/2}\delta_n$ and $\delta_n > 0$ defined in (28).

**Theorem 3.** *Let Assumptions 3 be satisfied. Then the following convergence results hold:*

(1) *We have*

$$\operatorname{cond}(B_n^{-1}F'(u_n)) \le \frac{M_n}{m_n} \qquad (31)$$

*and, accordingly, the inner iteration satisfies*

$$\|F'(u_n)p_n^{(k)} + (F(u_n) - b)\|_{B_n^{-1}} \le \Big(\frac{\sqrt{M_n} - \sqrt{m_n}}{\sqrt{M_n} + \sqrt{m_n}}\Big)^k \|F(u_n) - b\|_{B_n^{-1}} \quad (k \in \mathbb{N}).$$

*Therefore, the number of inner iterations for the n-th outer step is at most $k_n \in \mathbb{N}$ determined by the inequality*

$$\left(\frac{\sqrt{M_n} - \sqrt{m_n}}{\sqrt{M_n} + \sqrt{m_n}}\right)^{k_n} \leq \rho_n \tag{32}$$

*(where $\rho_n = (\mu_1/\mu_2)^{1/2}\delta_n$).*

(2) *The outer iteration $(u_n)$ satisfies*

$$\|u_n - u^*\| \leq \mu_1^{-1}\|F(u_n) - b\|,$$

*which converges to 0 monotonically with speed depending on the sequence $(\delta_n)$ up to locally quadratic order. Namely, if $\delta_n \equiv \delta_0 < 1$, then the convergence is linear. Further, if*

$$\delta_n \leq const \cdot \|F(u_n) - b\|^\gamma$$

*with some constant $0 < \gamma \leq 1$, then the convergence is locally of order $1 + \gamma$:*

$$\|F(u_{n+1}) - b\| \leq c_1\|F(u_n) - b\|^{1+\gamma} \quad (n \geq n_0)$$

*with some index $n_0 \in \mathbb{N}$ and constant $c_1 > 0$, yielding also the convergence estimate of weak order $1 + \gamma$*

$$\|F(u_n) - b\| \leq d_1 q^{(1+\gamma)^n} \quad (n \in \mathbb{N})$$

*with suitable constants $0 < q < 1$, $d_1 > 0$.*

The construction of the iteration and the proof of parts (1) and (2) of the theorem are based on [2] and [1], respectively.

**Remark 1.** The estimate (30) in the algorithm can be checked as follows: denoting the residuals corresponding to (29) by

$$r_n^{(k)} := B_n^{-1}F'(u_n)p_n^{(k)} + B_n^{-1}(F(u_n) - b) \quad (k = 0, 1, \ldots, k_n), \tag{33}$$

inequality (30) is equivalent to the computable estimate

$$\|r_n^{(k_n)}\|_{B_n} \leq \rho_n\|r_n^{(0)}\|_{B_n}. \tag{34}$$

### 3.3 Advantages of the preconditioning operator idea

The advantages of the preconditioning operator idea appear in both areas that are involved in the requirements of good preconditioners, i.e., easy construction and solvability, and convenient conditioning estimates. Namely:

- Defining the auxiliary systems as discretizations of suitable linear elliptic operators, one can rely on a highly developed background for their solution. Efficient standard solvers including fast direct methods are summarized, e.g., in [3], [13], [19].

- In addition, the construction of a Sobolev space preconditioner is straightforward from the underlying preconditioning operator, and is achieved without studying the actual form of the discretized system. In this way the properties of the original problem can be exploited more directly than with usual preconditioners defined for the discretized system. For instance, this way can help to handle certain difficulties such as discontinuous coefficients or sharp gradients. (See, e.g., [4], [17].)
- The preconditioned discretized problems have mesh independent condition numbers, since their convergence properties are asymptotically determined by the theoretical sequence in the Sobolev space. Hence, the conditioning properties exploit the fact that the preconditioning operators are chosen for the boundary value problem itself on the continuous level. Moreover, one can obtain a priori bounds for the condition numbers by carrying out analytic estimates, without using the discretized systems. (See, e.g., [4].)

## 4 Diagonal coefficient preconditioners

In this section we discuss preconditioning operators with diagonal (i.e., scalar-valued) coefficients. The motivation for this kind of preconditioner is that the Jacobian of the coefficient is typically not diagonal, and hence, $F'(u_n)$ is an elliptic operator with a general structure.

We study the scalar nonlinearity (2)

$$f(x,\eta) = a(|\eta|)\eta$$

with some real $C^1$ function $a\colon \mathbb{R}^+ \to \mathbb{R}^+$ satisfying the ellipticity condition

$$0 < \mu_1 \le a(r) \le a(r) + a'(r)r \le \mu_2 \quad (r \ge 0), \tag{35}$$

and consider the operator

$$T(u) = -\operatorname{div}(a(|\nabla u|)\nabla u),$$

which then satisfies Assumption 1.

The suggested preconditioner will depend on the step $n$, i.e., it will be described in the setting of Theorems 2 and 3. (We note that the resulting iteration is obviously simpler than the exact Newton iteration if the proposed operator is used in the quasi-Newton setting of Theorem 2. If it is used in an inner iteration in the setting of Theorem 3, then the overall cost of the iteration is determined by the actual coefficient.)

### 4.1 The preconditioning operator

The preconditioning operator is defined as follows. Let

$$c(r) := a(r) + a'(r)r \quad (r \geq 0).$$

We choose a real function $d\colon \mathbb{R}^+ \to \mathbb{R}^+$ such that

$$a(r) \leq d(r) \leq c(r) \quad (r \geq 0), \tag{36}$$

and let

$$\langle B_n z, v\rangle = \int_\Omega d(|\nabla u_n|)\nabla z \cdot \nabla v \quad (z, v \in H_0^1(\Omega)). \tag{37}$$

This preconditioning operator corresponds to the diagonal coefficient matrix

$$G_n(x) \equiv d(|\nabla u_n(x)|) \cdot I,$$

where $I$ is the identity matrix. This preconditioner uses a simplified operator compared to $F'(u_n)$, since the Jacobian of $f(x,\eta) = a(|\eta|)\eta$ w.r.t. $\eta$ is not diagonal:

$$\frac{\partial f}{\partial \eta}(x, \nabla u_n) = a(|\nabla u_n|)I \ + \frac{a'(|\nabla u_n|)}{|\nabla u_n|}(\nabla u_n \cdot \nabla u_n^t),$$

where $(\nabla u_n \cdot \nabla u_n^t)$ is the diadic product matrix of $\nabla u_n$, hence, $F'(u_n)$ is an elliptic operator with a general structure.

The diagonal factor will also appear in the discretized form (38) below.

### 4.2 Construction of the preconditioning matrix

The corresponding preconditioning matrix w.r.t. a basis $v_1, \ldots, v_k$ of $V_h$ is given by

$$((B_n)_h)_{i,j} = \int_\Omega d(|\nabla u_n|)\nabla v_i \cdot \nabla v_j \quad (i, j = 1, \ldots, k).$$

That is, the auxiliary equations are the discretizations of linear elliptic problems of the type

$$\text{Find } z \in V_h\colon \int_\Omega d(|\nabla u_n|)\nabla z \cdot \nabla v \ = \int_\Omega rv \quad (v \in V_h).$$

By virtue of [7], the matrix $(B_n)_h$ has the product form

$$(B_n)_h = \mathcal{Z}\mathcal{W}_n\mathcal{Z}^t, \tag{38}$$

where the matrices $\mathcal{Z}$ and $\mathcal{Z}^t$ correspond to the discretization of $-\operatorname{div}$ and $\nabla$, respectively, i.e. they are independent of $n$ and hence, need not be updated.

Since the operator $B_n$ has the scalar-valued coefficient $d_n$, the weight matrix $\mathcal{W}_n$ is diagonal.

The solution of the linear algebraic systems containing the preconditioner $(B_n)_h$ relies on efficient elliptic solvers, see, e.g., Subsection 3.3.

## 4.3 Conditioning

We determine the condition number of the operator $B_n^{-1}F'(u_n)$ in $H_0^1(\Omega)$ based on spectral equivalence.

A simple calculation yields

$$\begin{aligned} a(|\nabla u_n(x)|)|\xi|^2 &\le \frac{\partial f}{\partial \eta}(x, \nabla u_n(x))\xi \cdot \xi \\ &\le (a(|\nabla u_n(x)|) + a'(|\nabla u_n(x)|)|\nabla u_n(x)|)|\xi|^2, \end{aligned} \tag{39}$$

which implies that

$$\begin{aligned} \left(\min_{r\ge 0} \frac{a(r)}{d(r)}\right)\langle B_n v, v\rangle &\le \int_\Omega a(|\nabla u_n|)|\nabla v_n|^2 \le \langle F'(u_n)v, v\rangle \\ &\le \int_\Omega c(|\nabla u_n|)|\nabla v_n|^2 \le \left(\max_{r\ge 0} \frac{c(r)}{d(r)}\right)\langle B_n v, v\rangle \end{aligned}$$

for all $v \in V_h$. Hence, we obtain the estimate

$$\operatorname{cond}(B_n^{-1}F'(u_n)) \le \max_{r\ge 0} \frac{c(r)}{d(r)} \max_{r\ge 0} \frac{d(r)}{a(r)}. \tag{40}$$

It is clear from the obtained expression that $\operatorname{cond}(B_n^{-1}F'(u_n))$ is independent of the actual subspace $V_h$.

Two possible choices of the function $d$ are as follows:

- First, let us observe that

  $$\max_{r\ge 0} \frac{c(r)}{d(r)} \max_{r\ge 0} \frac{d(r)}{a(r)} \ge \max_{r\ge 0} \frac{c(r)}{a(r)}$$

  and the latter is achieved when

  $$d(r) = (a(r)c(r))^{1/2}. \tag{41}$$

  Hence, this $d$ is the optimal choice concerning the possible condition numbers.
- Second, a simpler structure of the preconditioner is obtained if $d$ is chosen to be a piecewise constant function; then $d(|\nabla u_n|)$ is a piecewise constant weight function, similarly as introduced in [4].

  We note that the preconditioner using (41) can be regarded as a limiting case of the piecewise constant coefficient preconditioners as the decomposition is refined, and the achieved condition number is smaller than by any piecewise constant coefficient.

**Acknowledgement.** The second author was supported by the Hungarian Research Fund OTKA under grant No. T031807 and T043765; the third author was supported by the Hungarian Research Fund OTKA under grant No. F034840 and T031807.

## References

[1] O. Axelsson, *On global convergence of iterative methods*, Iterative solution of nonlinear systems of equations. Lecture Notes in Math. 953, Springer, 1982, pp. 1–19.

[2] O. Axelsson, *Iterative Solution Methods*, Cambridge Univ. Press, 1994.

[3] O. Axelsson, V. A. Barker, *Finite Element Solution of Boundary Value Problems*, Academic Press, 1984.

[4] O. Axelsson, I. Faragó, and J. Karátson, *Sobolev space preconditioning for Newton's method using domain decomposition*, Numer. Lin. Alg. Appl. **9** (2002), 585–598.

[5] O. Axelsson, I. Kaporin, *Minimum residual adaptive multilevel finite element procedure for the solution of nonlinear stationary problems*, SIAM J. Numer. Anal. **35** (1998), 1213–1229 (electronic).

[6] O. Axelsson, J. Karátson, *Double Sobolev gradient preconditioning for elliptic problems*, Report 0016, Numer. Methods Partial Differential Equations (to appear).

[7] O. Axelsson, J. Maubach, *On the updating and assembly of the Hessian matrix in finite element methods*, Comput. Methods Appl. Mech. Engrg. **71** (1988), 41–67.

[8] R. Blaheta, *Multilevel Newton methods for nonlinear problems with applications to elasticity*, Copernicus 940820, Technical report.

[9] V. Faber, T. Manteuffel, and S. V. Parter, *On the theory of equivalent operators and application to the numerical solution of uniformly elliptic partial differential equations*, Adv. in Appl. Math. **11** (1990), 109–163.

[10] I. Faragó, J. Karátson, *The gradient-finite element method for elliptic problems*, Comp. Math. Appl. **42** (2001), 1043–1053.

[11] I. Faragó, J. Karátson, *Numerical Solution of Nonlinear Elliptic Problems Via Preconditioning Operators. Theory and Applications*, NOVA Science Publishers, New York, 2002.

[12] H. Gajewski, K. Gröger, and K. Zacharias, *Nichtlineare Operatorgleichungen und Operatordifferentialgleichungen*, Akademie-Verlag, Berlin, 1974.

[13] W. Hackbusch, *Elliptic Differential Equations. Theory and Numerical Treatment. Springer Series in Computational Mathematics 18*, Springer, Berlin, 1992.

[14] J. Karátson, I. Faragó, *Variable preconditioning via inexact Newton methods for nonlinear problems in Hilbert space*, SIAM J. Numer. Anal. **41** (2003), 1242–1262.

[15] M. Křížek, P. Neittaanmäki, *Mathematical and Numerical Modelling in Electrical Engineering: Theory and Applications*, Kluwer Academic Publishers, 1996.

[16] S. G. Mikhlin, *The Numerical Performance of Variational Methods*, Walters-Noordhoff, 1971.

[17] J. W. Neuberger, *Sobolev Gradients and Differential Equation*, Lecture Notes in Mathematics No. 1670, Springer, 1997.

[18] W. B. Richardson Jr., *Sobolev gradient preconditioning for PDE applications*, Iterative Methods in Scientific Computation IV (D. R. Kincaid, A. C. Elster, eds.), IMACS, New Jersey, 1999, pp. 223–234.

[19] T. Rossi, J. Toivanen, *A parallel fast direct solver for block tridiagonal systems with separable matrices of arbitrary dimension*, SIAM J. Sci. Comput. **20** (1999), 1778–1796 (electronic).

# On a Conjugate Gradient/Newton/Penalty Method for the Solution of Obstacle Problems. Application to the Solution of an Eikonal System with Dirichlet Boundary Conditions

Bernard Dacorogna,[1] Roland Glowinski,[2] Yuri Kuznetsov,[3] Tsorng-Whay Pan[4]

[1]EPFL-DMA, CH–1015 Lausanne, Switzerland Bernard.Dacorogna@epfl.ch
[2]Department of Mathematics, University of Houston, Houston, TX 77204–3008, USA roland@math.uh.edu
[3]Department of Mathematics, University of Houston, Houston, TX 77204–3008, USA kuz@math.uh.edu
[4]Department of Mathematics, University of Houston, Houston, TX 77204–3008, USA pan@math.uh.edu

*Dedicated to M. Hestenes, C. Lanczos, E. Stieffel, and to all those individuals who made conjugate gradient the universally used method it is today.*

*Dedicated to G. H. Golub on his 70th birthday.*

## 1 Introduction and historical perspective

As the dedication suggests, many scientists in several countries have contributed significantly during these last fifty years to make *conjugate gradient* the efficient and versatile method it is today. However, this article is not limited to conjugate gradient and from that point of view, it owes a lot to *M. Hestenes* for his major contributions, not only to conjugate gradient, but also to *calculus of variations* and *augmented Lagrangian methods.* Indeed, in this article after investigating the solution of *obstacle problems* for *elliptic* and *parabolic* linear *operators* by a methodology combining *penalty, Newton's method* and *conjugate gradient algorithms,* we shall apply the above methodology to the search for *non-negative solutions* to an *Eikonal system* with Dirichlet boundary conditions, after transforming it into a (non-convex) problem from *Calculus of Variations;* in order to solve this non-convex minimization problem we shall apply an *operator-splitting method,* and as shown in, e.g., [1], [2] operator splitting methods have close links with augmented Lagrangian methods and therefore with M. Hestenes contributions to Applied and Computational Mathematics. Since, in this article, we make a systematic use of *conjugate gradient algorithms,* and of *cyclic reduction methods* for the *fast solution of discrete linear elliptic problems,* a special credit has to be given to *G. H. Golub* for his outstanding contributions to these topics (among many other contributions, which include, in fact, complementarity methods for the solution of discrete obstacle problems). We deeply regret he could not attend the Jyväskylä Conference on "50 Years of Conjugate Gradients".

To conclude this section on a more personal note, we shall mention that the initial intention of the second author was to write a survey on some conjugate gradient success stories of the past thirty years, particularly in Fluid Mechanics and in Control; however, after the first author of this article brought to the attention of his colleagues a (challenging) *system of Eikonal equations* with Dirichlet boundary conditions, modeling *anisotropic microstructural phenomena from Material Science,* it was decided to report on this new problem, where *conjugate gradient algorithms for the fast solution of obstacle problems* play a fundamental role.

## 2 An obstacle problem for the Laplace operator: Motivation and generalizations

In Section 9 of this article, we will show that *non-negative* solutions to the following *system* of *Eikonal equations,* with *Dirichlet boundary conditions,*

$$u \in H_0^1(\Omega), \quad \left|\frac{\partial u}{\partial x_i}\right| = 1 \quad a.e., \ \forall i = 1, \ldots, d, \tag{2.1}$$

can be obtained via an algorithm (of the operator-splitting type) requiring at each iteration (in fact time step, and among other things) the solution of an *obstacle problem* of the following type:

$$\int_\Omega \nabla u \cdot \nabla(v-u)\,\mathrm{d}x \geq \int_\Omega \mathbf{p} \cdot \nabla(v-u)\,\mathrm{d}x \quad \forall v \in K^+; \quad u \in K^+. \tag{2.2}$$

In (2.1), (2.2):

- $\Omega$ is a bounded domain of $\mathbb{R}^d$ $(d \geq 1)$ with boundary $\Gamma$ and $\mathrm{d}x = \mathrm{d}x_1 \ldots \mathrm{d}x_d$;
- $\mathbf{p}$ is a given vector-valued function of $(L^2(\Omega))^d$;
- $K^+ = \{v \mid v \in H_0^1(\Omega), \ v \geq 0, \ \text{a.e. on } \Omega\}$.

Let us mention immediately that in the context of problem (2.1) the over-relaxation methods with projection, and Uzawa type algorithms, discussed in, e.g., [3] and [4] perform pretty poorly proving too slow. Looking for alternatives it was decided to investigate an approach combining (exterior) *penalty* and *Newton's method.* Actually, with this combination one has been able to simulate (see [5], [6] for details) the vibrations of strings and beams, when their motion is constrained by *obstacles*; however, the above were obstacle problems in *one space dimension,* where the linear systems resulting from the combination penalty/Newton (after appropriate finite element discretizations) could be solved easily by *direct methods* taking advantage of the sparsity and band structure of the corresponding matrices. Since some of the finite element meshes (or finite difference grids) associated to the solution of problem (2.1) involve more than $10^6$ points, it is clear that, as today, direct methods are not a feasible option as components of the penalty/Newton solution of problem (2.2) (for most practitioners at least). It will be shown in this article that replacing the above direct methods by a well-chosen *conjugate gradient algorithm,* leads to a *fast converging* iterative method, the only requirement being the access to a *fast solver* for *linear second order elliptic boundary value problems.* Since vector $\mathbf{p}$ in (2.2) is provided by the solution of an initial value problem coupled – in some sense – to (2.2), we shall take as model problem the *parabolic variational inequality* below (where $0 < T \leq +\infty$):

$$u(0) = u_0 \ (\in K), \tag{2.3}$$

$$\left.\begin{aligned} &\text{for } t \in (0,T), \ \mathbf{u}(t) \in K \text{ and}, \forall v \in K, \\ &\Big\langle \frac{\partial}{\partial t} u(t), v - u(t) \Big\rangle + \int_\Omega \nabla u(t) \cdot \nabla (v - u(t))\, \mathrm{d}x \\ &\qquad \geq \langle f(t), v - u(t) \rangle; \end{aligned}\right\} \tag{2.4}$$

in (2.3), (2.4), $f \in L^2(0,T; H^{-1}(\Omega))$, $\langle \cdot, \cdot \rangle$ denotes the duality pairing between $H^{-1}(\Omega)$ and $H_0^1(\Omega)$, and the convex $K$ is defined by

$$K = \{ v \mid v \in H_0^1(\Omega), \quad v \geq \psi \text{ a.e. on } \Omega \}, \tag{2.5}$$

with $\psi \in C^0(\overline{\Omega}) \cap H^1(\Omega)$ and $\psi \leq 0$ on the boundary $\Gamma$ of $\Omega$. It follows from, e.g., [7] that the obstacle problem (2.3), (2.4) has a unique solution in $L^2(0,T; H_0^1(\Omega)) \cap C^0([0,T]; L^2(\Omega))$.

## 3 Backward Euler time discretization of (2.3) and (2.4)

Let $\Delta t$ ($> 0$) be a time-discretization step. The *backward Euler time discretization* of problem (2.3), (2.4) leads to

$$u^0 = u_0; \tag{3.1}$$

then, for $n \geq 0$, $u^n$ being known, we obtain $u^{n+1}$ from the solution of

$$\left.\begin{aligned} &u^{n+1} \in K; \ \forall v \in K \\ &\int_\Omega \frac{u^{n+1} - u^n}{\Delta t} (v - u^{n+1})\, \mathrm{d}x + \int_\Omega \nabla u^{n+1} \cdot \nabla (v - u^{n+1})\, \mathrm{d}x \\ &\quad \geq \langle f^{n+1}, v - u^{n+1} \rangle. \end{aligned}\right\} \tag{3.2}$$

The *elliptic variational inequality problem* (3.2) has a unique solution. Problem (3.2) is, $\forall n \geq 0$, of the following type

$$\int_\Omega u(v-u)\,\mathrm{d}x + \mu \int_\Omega \nabla u \cdot \nabla(v-u)\,\mathrm{d}x \geq \langle f, v-u\rangle \quad \forall v \in K; \;\; u \in K \quad (3.3)$$

with $\mu$ a *positive constant* and $f \in H^{-1}(\Omega)$.

## 4 Penalty approximation of (3.3)

Let $\varepsilon$ be a *positive parameter* and denote $\sup(0, -v)$ by $v^-$. We approximate problem (3.3) by

$$\left.\begin{array}{l} \displaystyle\int_\Omega u^\varepsilon v\,\mathrm{d}x + \mu \int_\Omega \nabla u^\varepsilon \cdot \nabla v\,\mathrm{d}x - \varepsilon^{-1} \int_\Omega ((u^\varepsilon - \psi)^-)^2 v\,\mathrm{d}x = \langle f, v\rangle, \\ \forall v \in H_0^1(\Omega); \;\; u^\varepsilon \in H_0^1(\Omega). \end{array}\right\} \quad (4.1)$$

Problem (4.1) is obtained by *penalization* of the constraint $v-\psi \geq 0$ a.e. in $\Omega$, via the introduction of the functional $(\varepsilon^{-1}/3)\int_\Omega ((v-\psi)^-)^3\,\mathrm{d}x$ and its differentiation with respect to $v$. Indeed, (4.1) is the *Euler-Lagrange equation* of the following problem of the *Calculus of Variations*:

$$J_\varepsilon(u^\varepsilon) \leq J_\varepsilon(v) \quad \forall v \in H_0^1(\Omega); \quad u^\varepsilon \in H_0^1(\Omega), \quad (4.2)$$

with

$$J_\varepsilon(v) = \frac{1}{2}\int_\Omega v^2\,\mathrm{d}x + \frac{\mu}{2}\int_\Omega |\nabla v|^2\,\mathrm{d}x - \langle f, v\rangle + \frac{\varepsilon^{-1}}{3}\int_\Omega ((v-\psi)^-)^3\,\mathrm{d}x.$$

Using *Convex Analysis* methods (see, e.g., [8], [9]) we can easily show that problem (4.1), (4.2) has a unique solution; moreover $\lim_{\varepsilon \to 0^+} \|u^\varepsilon - u\|_{H^1(\Omega)} = 0$, with $u$ the solution of problem (3.3).

## 5 Newton's method for the solution of (4.1) and (4.2)

Let us drop the superscript $\varepsilon$; problem (4.1) can also be written as:

$$F_\varepsilon(u) = 0, \quad (5.1)$$

where $F_\varepsilon \colon H_0^1(\Omega) \to H^{-1}(\Omega)$ is defined by

$$F_\varepsilon(v) = v - \mu\Delta v - \varepsilon^{-1}((v-\psi)^-)^2 - f, \;\; \forall v \in H_0^1(\Omega). \quad (5.2)$$

Applying Newton's method to the solution of problem (5.1) leads (with obvious notation) to:

$$u^0 \text{is given in } H_0^1(\Omega) \quad (u^0 \in K \textit{ is recommended}); \tag{5.3}$$

*for $m \geq 0$, $u^m$ being known, compute $u^{m+1}$ via*

$$u^{m+1} = u^m - F_\varepsilon'(u^m)^{-1} F_\varepsilon(u^m), \tag{5.4}$$

*where*

$$\left.\begin{aligned} &\forall v, w \in H_0^1(\Omega), \ \ \langle F_\varepsilon'(u^m)v, w\rangle = \\ &\int_\Omega vw \, dx + \mu \int_\Omega \nabla v \cdot \nabla \, dx + 2\varepsilon^{-1} \int_\Omega (u^m - \psi)^- vw \, dx. \end{aligned}\right\} \tag{5.5}$$

It follows from (5.2), (5.5) that (5.4) is equivalent to the following (well-posed) *linear variational problem:*

$$\left.\begin{aligned} &\overline{u}^m \in H_0^1(\Omega); \ \ \forall v \in H_0^1(\Omega), \\ &\int_\Omega \overline{u}^m v \, dx + \mu \int_\Omega \nabla \overline{u}^m \cdot \nabla v \, dx + 2\varepsilon^{-1} \int_\Omega (u^m - \psi)^- \overline{u}^m v \, dx \\ &= \int_\Omega u^m v \, dx + \mu \int_\Omega \nabla u^m \cdot \nabla v \, dx - \varepsilon^{-1} \\ &\times \int_\Omega ((u^m - \psi)^-)^2 v \, dx - \langle f, v\rangle, \end{aligned}\right\} \tag{5.6}$$

with $\overline{u}^m = u^m - u^{m+1}$. The *finite element implementation* of Newton's algorithm (5.3), (5.6), and the *conjugate gradient solution* of the discrete analogues of problem (5.6) will be discussed in the following Sections 6 and 7.

## 6 Finite element implementation of the Newton's algorithm (5.3), (5.6)

From now on, we suppose that $\Omega$ is a bounded polygonal domain of $\mathbb{R}^2$; we introduce next a finite element triangulation $\mathcal{T}_h$ of $\overline{\Omega}$ (with $h$ the length of the largest edge(s) of the triangles of $\mathcal{T}_h$) and approximate $H_0^1(\Omega)$, $K$, (3.3), (4.1), and the Newton's algorithm (5.3), (5.6) by, respectively,

$$V_{0h} = \{v_h \mid v_h \in C^0(\overline{\Omega}), \quad v_h|_T \in P_1, \ \forall T \in \mathcal{T}_h, \ v_h = 0 \text{ on } \Gamma\} \tag{6.1}$$

(with $P_1$ the space of the polynomials in two variables of degree $\leq 1$),

$$K_h = \{v_h \mid v_h \in V_{0h}, \quad v_h(P) \geq \psi(P), \quad \forall P \in \Sigma_{0h}\} \tag{6.2}$$

(with $\Sigma_{0h} = \{P \mid P \text{ is a vertex of } \mathcal{T}_h, \ P \notin \Gamma\}$),

$$\left.\begin{aligned} &u_h \in K_h; \quad \forall v_h \in K_h \\ &\int_\Omega u_h(v_h - u_h) \, dx + \mu \int_\Omega \nabla u_h \cdot \nabla(v_h - u_h) \, dx \geq \langle f, v_h - u_h\rangle, \end{aligned}\right\} \tag{6.3}$$

$$\left.\begin{aligned} &\int_\Omega u_h^\varepsilon v_h \,\mathrm{d}x + \mu \int_\Omega \nabla u_h^\varepsilon \cdot \nabla v_h \,\mathrm{d}x - (\varepsilon^{-1}/3) \\ &\quad \times \textstyle\sum_{P\in\Sigma_{0h}} A_P((u_h^\varepsilon(P) - \psi(P))^-)^2 v_h(P) \\ &= \langle f, v\rangle \quad \forall v_h \in V_{0h};\ u_h^\varepsilon \in V_{0h}, \end{aligned}\right\} \tag{6.4}$$

and (dropping some of the subscripts $h$):

$$u^0 \ \textit{is given in } V_{0h} \ (\textit{in } K_h \ \textit{if possible}); \tag{6.5}$$

for $m \geq 0$, $u^m$ being known, denote $u^m - u^{m+1}$ by $\overline{u}^m$ and solve

$$\left.\begin{aligned} &\overline{u}^m \in V_{0h}; \quad \forall v \in V_{0h} \\ &\int_\Omega \overline{u}^m v \,\mathrm{d}x + \mu \int_\Omega \nabla \overline{u}^m \cdot \nabla v \,\mathrm{d}x + \frac{2\varepsilon^{-1}}{3} \\ &\quad \times \textstyle\sum_{P\in\Sigma_{0h}} A_P(u^m(P) - \psi(P))^- \overline{u}^m(P) v(P) \\ &= \int_\Omega u^m v \,\mathrm{d}x + \mu \int_\Omega \nabla u^m \cdot \nabla v \,\mathrm{d}x - \frac{\varepsilon^{-1}}{3} \\ &\quad \times \sum_{P\in\Sigma_{0h}} A_P((u^m(P) - \psi(P))^-)^2 v(P) - \langle f, v\rangle; \end{aligned}\right\} \tag{6.6}$$

in (6.4), (6.6), $A_P$ is the measure of the polygonal union of the triangles of $\mathcal{T}_h$ which have $P$ as a common vertex (the penalty related terms in (6.4) and (6.6) have been obtained by approximating $\int_\Omega ((u^\varepsilon - \psi)^-)^2 v \,\mathrm{d}x$, and other similar integrals, by the *trapezoidal rule*, in order to "diagonalize" the matrices associated to the penalty treatment of the condition $u \geq \psi$ on $\Omega$; actually, the trapezoidal rule can also be employed to approximate the various $L^2(\Omega)$-scalar products encountered in (6.3), (6.4) and (6.6)).

## 7 Conjugate gradient solution of problem (6.6)

Problem (6.6) is equivalent to a linear system of the following form

$$AX + \varepsilon^{-1} DX = b, \tag{7.1}$$

where, in (7.1), $A$ is an $N \times N$ matrix, symmetric and positive definite, $D$ is a diagonal $N \times N$ matrix, positive semi-definite, and $b \in \mathbb{R}^N$. Define $Y$ by

$$Y = \varepsilon^{-1} D^{1/2} X; \tag{7.2}$$

system (7.1) can be reformulated as

$$\varepsilon Y + D^{1/2} A^{-1} D^{1/2} Y = D^{1/2} A^{-1} b. \tag{7.3}$$

Matrix $\varepsilon I + D^{1/2} A^{-1} D^{1/2}$ being *symmetric* and *positive definite* it makes sense to attempt solving problem (7.3) by a *conjugate gradient algorithm.*

*Without preconditioning*, such an algorithm reads as follows (where $V \cdot W = \sum_{i=1}^{N} V_i W_i$, $\forall V = \{V_i\}_{i=1}^{N}$, $W = \{W_i\}_{i=1}^{N} \in \mathbb{R}^N$):

$$Y^0 \text{ is given in } \mathbb{R}^N \quad (Y^0 = 0, \text{ for example}); \tag{7.4}$$

*solve*

$$Ar^0 = D^{1/2}Y^0 - b, \tag{7.5}$$

*and set*

$$g^0 = \varepsilon Y^0 + D^{1/2}r^0, \tag{7.6}$$

$$w^0 = g^0. \tag{7.7}$$

*For $k \geq 0$, assuming that $Y^k$, $g^k$, $w^k$ are known, solve*

$$A\overline{r}^k = D^{1/2}w^k, \tag{7.8}$$

*and set*

$$\overline{g}^k = \varepsilon w^k + D^{1/2}\overline{r}^k. \tag{7.9}$$

*Compute*

$$\rho_k = g^k \cdot g^k / \overline{g}^k \cdot w^k, \tag{7.10}$$

$$Y^{k+1} = Y^k - \rho_k w^k, \tag{7.11}$$

$$g^{k+1} = g^k - \rho_k \overline{g}^k. \tag{7.12}$$

*If $g^{k+1} \cdot g^{k+1}/g^0 \cdot g^0 \leq \eta$, take $X = A^{-1}(D^{1/2}Y^{k+1} - b)$; else, compute*

$$\gamma_k = g^{k+1} \cdot g^{k+1}/g^k \cdot g^k, \tag{7.13}$$

$$w^{k+1} = g^{k+1} + \gamma_k w^k. \tag{7.14}$$

*Do $k = k+1$ and return to (7.8).*

It follows from (7.5) and (7.8) that each iteration of algorithm (7.4)–(7.14) requires the solution of a linear system associated to matrix $A$, i.e., in the context of the discrete obstacle problem (6.3), of a linear system associated to a discrete analogue of operator $I - \mu\Delta$ with Dirichlet boundary conditions, a classical problem indeed. Concerning the *speed of convergence* of algorithm (7.4)–(7.14), it can be shown that, in the neighborhood of the solution of problem (6.4), the *condition number* $\nu$ of the corresponding matrix $\varepsilon I + D^{1/2}A^{-1}D^{1/2}$ is $O(\varepsilon^{-1/2})$, implying (from the classical relation $\|Y^k - Y\| \leq C\left(\frac{\sqrt{\nu}-1}{\sqrt{\nu}+1}\right)^k \|Y^0 - Y\|$, $\forall k \geq 1$) that, in fact, the speed of convergence of (7.4)–(7.14) is "controlled" by $\varepsilon^{-1/4}$, a not too small number, even if $\varepsilon$ is small (of the order of $10^{-4}$, for example). From this observation, we can expect a reasonably fast convergence of algorithm (7.4)–(7.14); the numerical experiments to be reported in Section 8 confirm this prediction.

**Remark 7.1.** The methodology discussed in Sections 4 to 7 can be easily modified to handle problem (2.2). Actually, it will be quite successful (as we shall see in Section 13) at finding non-negative solutions to problem (2.1), a much more complicated obstacle problem indeed.

# 8 Application to the solution of a time dependent obstacle problem: Numerical experiments and results

In order to validate the methodology discussed in Sections 2 to 7, we consider the variant of problem (2.3), (2.4) with $\Omega = (0,1) \times (0,1)$, $f = C$, and $K = \{v \mid v \in H_0^1(\Omega),\ v(x) \leq \delta(x,\Gamma),\ \text{a.e., on } \Omega\}$; here $\delta(x,\Gamma) =$ distance from $x$ to the boundary $\Gamma$ of $\Omega$. The above problem is well documented and related to the elasto-plastic torsion of a infinitely long cylinder of cross section $\Omega$, $C$ being a torsion angle per unit length (see, e.g., [3], [4] for details). To approximate the solution of this problem we have used a finite element approximation like the one discussed in Section 6, with $V_{0h}$ and $K_h$ defined from a uniform triangulation $\mathcal{T}_h$ of $\Omega$, allowing the use of fast elliptic solvers in algorithm (7.4)–(7.14).

Table 1

| $\varepsilon$ | $N_{\text{Newton}}$ | $N_{\text{CG}}$ | $L^2$-error |
|---|---|---|---|
| 0.0032 | 4 | 3 | $8.9604339 \times 10^{-2}$ |
| 0.0016 | 5 | 3 | $6.9899789 \times 10^{-2}$ |
| 0.0008 | 5 | 4 | $5.3565620 \times 10^{-2}$ |
| 0.0004 | 5 | 4 | $4.0485722 \times 10^{-2}$ |
| 0.0002 | 6 | 4 | $3.0261847 \times 10^{-2}$ |
| 0.0001 | 6 | 5 | $2.2398988 \times 10^{-2}$ |
| 0.00005 | 7 | 5 | $1.6440081 \times 10^{-2}$ |
| 0.000025 | 9 | 6 | $1.1985428 \times 10^{-2}$ |
| 0.0000125 | 11 | 6 | $8.6906453 \times 10^{-3}$ |

All calculations have been done with $\Delta t = 2.5 \times 10^{-4}$, $h = 1/256$, and $C = 10$. The Newton's iterations have been stopped when $\sum_{i,j} |\bar{u}_{m,ij}| \leq 10^{-4}$. In the conjugate gradient algorithm we have taken $10^{-6}$ for $\eta$ in the stopping criterion. Finally concerning the time discretization scheme itself, we consider that a steady state has been reached when $\sum_{i,j} |u_{ij}^{n+1} - u_{ij}^n| \leq 10^{-4}$. In Table 1, we have shown the maximal numbers of iterations in the Newton's iteration and the conjugate gradient algorithm and the $L^2$-error (compared with the solution obtained with same parameters by the relaxation method discussed in [4]) with different choices of the penalty parameter $\varepsilon$. It is clear from Table 1 that both the Newton's and conjugate gradient methods have fast convergence properties. There is no doubt that the fact that we initialize Newton's method with the solution obtained at the previous time step is an important factor of this good convergence property. However, when applying a similar initialization strategy for SOR projection when solving problem (2.1) the convergence is quite slow while the penalty/Newton/conjugate gradient approach performs very efficiently: the good initial guess does not explain everything. Our final comment is that the $L^2$-error in Table 1 behaves essentially like $\sqrt{\varepsilon}$ which is a result we were expecting.

## 9 A system of Eikonal equations with Dirichlet conditions: Related variational problems

Motivated by the analysis of nonlinear models from *μ-Mechanics* and *Material Sciences*, the first author of this article has been lead to investigate (see [10] for details) the properties of the solution of the following nonlinear boundary value problem:

$$\text{Find } u \in H_0^1(\Omega) \text{ such that } \left|\frac{\partial u}{\partial x_i}\right| = 1 \text{ a.e. on } \Omega, \quad \forall i = 1, \ldots, d, \tag{9.1}$$

where, in (9.1), $\Omega$ is a bounded domain of $\mathbb{R}^d$ $(d \geq 1)$ with the boundary $\Gamma$; problem (9.1) can be viewed as a boundary value problem for a *system of Eikonal equations*. Since problem (9.1) has an infinity of solutions, we are going to focus on those solutions which *maximize* (or "nearly" maximize) functional $v \to \int_\Omega v\,dx$. This leads us to consider the following problem of the *Calculus of Variations*:

$$u \in E; \quad \int_\Omega u\,dx \geq \int_\Omega v\,dx \quad \forall v \in E, \tag{9.2}$$

with $E = \{v \mid v \in H_0^1(\Omega), \left|\frac{\partial v}{\partial x_i}\right| = 1 \text{ a.e.}, \forall i = 1, \ldots, d\}$. If $d = 2$ and $\Omega = \{x \mid x = (x_1, x_2),\ |x_1 \pm x_2| < 1\}$, the unique solution of problem (9.2) is given by $u(x) = 1 - |x_1| - |x_2|$ for all $x \in \Omega$; on the other hand, problem (2.1) has no solution in general. We can easily show that if $u$ is a solution of (9.2), then $u(x) \geq 0$ on $\overline{\Omega}$, moreover, since $|\nabla v|^2 = d$ a.e. in $\Omega$ if $v \in E$, problem (9.2) is equivalent to:

$$u \in E^+; \quad J(u) \leq J(v), \quad \forall v \in E^+, \tag{9.3}$$

where

$$E^+ = \{v \mid v \in E, \quad v \geq 0 \text{ on } \overline{\Omega}\}, \tag{9.4}$$

and

$$J(v) = \frac{1}{2}\int_\Omega |\nabla v|^2\,dx - C\int_\Omega v\,dx, \tag{9.5}$$

where, in (9.5), $C$ is an (arbitrary) *positive constant*. There are, in principle, several (if not many) ways to solve *numerically* problem (9.2), (9.3); motivated by some earlier work on the numerical solution of a *Ginzburg-Landau* equation (see refs. [11], [12]), we are going to treat the constraints $\left|\frac{\partial v}{\partial x_i}\right| = 1$, $\forall i = 1, \ldots, d$, by a (exterior) *penalty* method. Moreover, in order to "control" mesh-related oscillations, we are going to bound $\|\Delta v\|_{L^2(\Omega)}$. This leads us to approximate the functional $J(\cdot)$ in (9.3), (9.5) by $J^\varepsilon(\cdot)$ defined as follows:

$$J^\varepsilon(v) = (\varepsilon_1/2)\int_\Omega |\Delta v|^2\,\mathrm{d}x + J(v) + (\varepsilon_2^{-1}/4)\sum_{i=1}^{d}\int_\Omega\left(\left|\frac{\partial v}{\partial x_i}\right|^2 - 1\right)^2 \mathrm{d}x; \quad (9.6)$$

in (9.6), $\varepsilon = \{\varepsilon_1, \varepsilon_2\}$, $\varepsilon_1$ and $\varepsilon_2$ being both *positive* and *"small"*. We then approximate (9.2), (9.3) by

$$u^\varepsilon \in K^+; \quad J^\varepsilon(u^\varepsilon) \le J^\varepsilon(v), \quad \forall v \in H^2(\Omega) \cap K^+, \tag{9.7}$$

with

$$K^+ = \{v \mid v \in H_0^1(\Omega), \quad v \ge 0 \text{ a.e. on } \Omega\}. \tag{9.8}$$

Proving that problem (9.7) (a "beautiful" problem from Calculus of Variations, indeed) has at least a solution is an (almost) elementary exercise. The *numerical solution* of the (*obstacle*) problem (9.7) is the main objective of the remainder of this article.

**Remark 9.1.** *Augmented Lagrangian methods* (closely related to those discussed in, e.g., [1]–[4], [13]) can be applied to the solution of problem (9.7). Preliminary results look promising and will be reported elsewhere.

**Remark 9.2.** Suppose that one looks for a solution of problem (9.1) as close as possible (in $H_0^1(\Omega)$) of $\varphi_d \in H_0^1(\Omega)$; in that case, we should take in (9.5), (9.6), the functional $J(\cdot)$ defined by $J(v) = (1/2)\int_\Omega |\nabla v|^2\,\mathrm{d}x - \int_\Omega \nabla\varphi_d \cdot \nabla v\,\mathrm{d}x$.

## 10 An equivalent variational formulation of problem (9.7) and an associated boundary value problem

Let us denote $(L^2(\Omega))^d$ by $\Lambda$ and $\nabla u^\varepsilon$ by $\mathbf{p}^\varepsilon$; problem (9.7) is clearly *equivalent* to

$$\mathbf{p}^\varepsilon \in \Lambda; \quad j^\varepsilon(\mathbf{p}^\varepsilon) \le j^\varepsilon(\mathbf{q}), \ \forall \mathbf{q} \in \Lambda, \tag{10.1}$$

with $\mathbf{q} = \{q_i\}_{i=1}^d$, and, for all $\mathbf{q} \in \Lambda$

$$\begin{aligned} j^\varepsilon(\mathbf{q}) = (1/2)\int_\Omega |\mathbf{q}|^2\,\mathrm{d}x - C\int_\Omega \nabla\varphi_1 \cdot \mathbf{q}\,\mathrm{d}x \\ + (\varepsilon_2^{-1}/4)\sum_{i=1}^{d}\int_\Omega (|q_i|^2 - 1)^2\,\mathrm{d}x + I_+(\mathbf{q}); \end{aligned} \tag{10.2}$$

in (10.2), $\varphi_1$ is the unique solution in $H_0^1(\Omega)$ of the Dirichlet problem $-\Delta\varphi_1 = 1$ in $\Omega$, $\varphi_1 = 0$ on $\Gamma$, and $I_+(\cdot)$ is the functional defined as follows:

$$I_+(\mathbf{q}) = \begin{cases} (\varepsilon_1/2)\displaystyle\int_\Omega |\nabla\cdot\mathbf{q}|^2\,dx & \text{if } \mathbf{q}\in\nabla(H^2(\Omega)\cap K^+), \\ +\infty & \text{elsewhere.} \end{cases} \tag{10.3}$$

It is easy to show that the functional $I_+(\cdot)$ is *convex, proper*, and *lower semi-continuous* over the space $\Lambda$. The *variational formulation* of the (formal) *Euler-Lagrange equation* associated to problem (10.1) reads as follows:

$$\left.\begin{aligned} &\textstyle\int_\Omega \mathbf{p}^\varepsilon\cdot\mathbf{q}\,dx + \varepsilon_2^{-1}\sum_{i=1}^d \int_\Omega (|p_i^\varepsilon|^2-1)p_i^\varepsilon q_i\,dx + \langle\partial I_+(\mathbf{p}^\varepsilon),\mathbf{q}\rangle \\ &= C\textstyle\int_\Omega \nabla\varphi_1\cdot\mathbf{q}\,dx, \quad \forall\mathbf{q}\in\Lambda; \quad \mathbf{p}^\varepsilon\in\Lambda, \end{aligned}\right\} \tag{10.4}$$

with $\partial I_+(\cdot)$ the *sub-differential* of $I_+(\cdot)$. We associate to equation (10.4) the following *initial value problem*

$$\left.\begin{aligned} &\mathbf{p}^\varepsilon(0) = \mathbf{p}_0, \\ &\textstyle\int_\Omega (\partial\mathbf{p}^\varepsilon/\partial t)\cdot\mathbf{q}\,dx + \int_\Omega \mathbf{p}^\varepsilon\cdot\mathbf{q}\,dx \\ &\quad +\varepsilon_2^{-1}\textstyle\sum_{i=1}^d \int_\Omega (|p_i^\varepsilon|^2-1)p_i^\varepsilon q_i\,dx + \langle\partial I_+(\mathbf{p}^\varepsilon),\mathbf{q}\rangle \\ &= C\textstyle\int_\Omega \nabla\varphi_1\cdot\mathbf{q}\,dx, \quad \forall\mathbf{q}\in\Lambda,\ t>0. \end{aligned}\right\} \tag{10.5}$$

From now on, our goal is to "capture" the *steady state solutions* of (10.5); to achieve this objective, we shall time-discretize (10.5) by an *operator-splitting scheme à la Marchuk-Yanenko* (simply because it is the simplest splitting-scheme we can think of; more sophisticated splitting-schemes will be discussed elsewhere).

## 11 Time discretization of problem (10.5) by operator splitting

Let $\Delta t$ ($>0$) be a time-discretization step; dropping the superscript $\varepsilon$, we time-discretize (10.5) by the following *operator-splitting scheme* (of the *Marchuk-Yanenko* type):

$$\mathbf{p}^0 = \mathbf{p}_0 \quad (=\mathbf{0},\ \text{or } \nabla\varphi_1, \quad \text{for example}), \tag{11.0}$$

then, for $n\geq 0$, assuming $\mathbf{p}^n$ known, solve successfully:

$$\left.\begin{aligned} &(\Delta t)^{-1}\int_\Omega (\mathbf{p}^{n+1/2}-\mathbf{p}^n)\cdot\mathbf{q}\,dx + \int_\Omega \mathbf{p}^{n+1/2}\cdot\mathbf{q}\,dx \\ &\quad +\varepsilon_2^{-1}\textstyle\sum_{i=1}^d \displaystyle\int_\Omega (|p_i^{n+1/2}|^2-1)p_i^{n+1/2}q_i\,dx = C\int_\Omega \nabla\varphi_1\cdot\mathbf{q}\,dx, \\ &\forall\mathbf{q}\in\Lambda;\ \mathbf{p}^{n+1/2}\in\Lambda, \end{aligned}\right\} \tag{11.1}$$

$$\left.\begin{aligned}&(\Delta t)^{-1}\int_\Omega(\mathbf{p}^{n+1}-\mathbf{p}^{n+1/2})\cdot\mathbf{q}\,dx+\langle\partial I_+(\mathbf{p}^{n+1}),\mathbf{q}\rangle=0,\\&\forall\mathbf{q}\in\Lambda;\ \ \mathbf{p}^{n+1}\in\Lambda.\end{aligned}\right\}\tag{11.2}$$

Problem (11.1) has a *unique* solution "as soon" as $\Delta t \leq \varepsilon_2$; moreover, (11.1) can be solved "pointwise", since for all $i = 1,\ldots,d$, and a.e. on $\Omega$, $p_i^{n+1/2}(x)$ is the solution of a cubic equation of the following type

$$(1-\Delta t\varepsilon_2^{-1}+\Delta t)z+\Delta t\varepsilon_2^{-1}z^3=\text{RHS}.\tag{11.3}$$

If $\varepsilon_2 \geq \Delta t$, problem (11.3) has a unique solution which can be easily solved by *Newton's method.* On the other hand, the solution of problem (11.2) is given by

$$\mathbf{p}^{n+1}=\nabla u^{n+1},\tag{11.4}$$

function $u^{n+1}$ being the solution of the following *well-posed elliptic variational inequality*

$$\left.\begin{aligned}&\int_\Omega\nabla u^{n+1}\cdot\nabla(v-u^{n+1})\,dx+\Delta t\varepsilon_1\int_\Omega\Delta u^{n+1}\Delta(v-u^{n+1})\,dx\\&\geq\int_\Omega\mathbf{p}^{n+1/2}\cdot\nabla(v-u^{n+1})\,dx,\ \ \forall v\in H^2(\Omega)\cap K^+;\\&\qquad u^{n+1}\in H^2(\Omega)\cap K^+.\end{aligned}\right\}\tag{11.5}$$

In order to facilitate the numerical solution of problem (11.5), we perform a (variational) crime by *"approximately" factoring* the above problem as follows (other approximate factorizations are possible):

$$\left.\begin{aligned}&\omega^{n+1}\in K^+;\ \ \forall v\in K^+,\\&\int_\Omega\nabla\omega^{n+1}\cdot\nabla(v-\omega^{n+1})\,dx\geq\int_\Omega\mathbf{p}^{n+1/2}\cdot\nabla(v-\omega^{n+1})\,dx,\end{aligned}\right\}\tag{11.6}$$

$$u^{n+1}-\Delta t\varepsilon_1\Delta u^{n+1}=\omega^{n+1}\ \text{ in }\Omega,\ \ u^{n+1}=0\ \text{ on }\Gamma.\tag{11.7}$$

From the *non-negativity* of $\omega^{n+1}$, the *maximum principle for second order elliptic operators implies* (via (11.7)) the *non-negativity* of $u^{n+1}$.

**Remark 11.1.** There is no basic difficulty in solving problem (11.5) numerically, but the approximate factorization (11.6), (11.7) of the above problem produces smooth non-negative solutions, as (11.5) does, while leading to faster algorithms. Indeed, $u^{n+1} \in H^3(\Omega) \cap K^+$ if $\Gamma$ is smooth enough.

## 12 Finite element implementation of scheme (11.0), (11.1), (11.6), (11.7), (11.4)

Since any solution of problem (9.1) is *continuous* and *piecewise affine*, it makes sense to compute these solutions via *globally continuous, piecewise affine finite element approximations.* From now on, we shall assume that $\Omega$ is a bounded *polygonal* domain of $\mathbb{R}^2$ (the non-polygonal case is almost as easy to treat). Let $\mathcal{T}_h$ be a *finite element triangulation* of $\overline{\Omega}$, like those considered in Section 6. We approximate $L^2(\Omega)$ (and $H^1(\Omega)$), $H_0^1(\Omega)$, $K^+$, and $\Lambda$ by:

$$V_h = \{v_h \mid v_h \in C^0(\overline{\Omega}), \quad v_h|_T \in P_1, \ \forall T \in \mathcal{T}_h\}, \tag{12.1}$$

$$V_{0h} = V_h \cap H_0^1(\Omega) = \{v_h \mid v_h \in V_h, \quad v_h = 0 \ \text{on} \ \Gamma\}, \tag{12.2}$$

$$K_h^+ = \{v_h \mid v_h \in V_{0h}, \ v_h(P) \geq 0, \ \forall P \text{ vertex of } \mathcal{T}_h\}, \tag{12.3}$$

and

$$\Lambda_h = \{\mathbf{q}_h \mid \mathbf{q}_h \in \Lambda, \quad \mathbf{q}_h|_T \in \mathbb{R}^2, \ \forall T \in \mathcal{T}_h\}, \tag{12.4}$$

respectively, with $P_1$ the space of linear polynomials in $x_1$, $x_2$. We clearly have $\nabla V_h \subset \Lambda_h$ and $\dim(\Lambda_h) = 2\,\mathrm{Card}(\mathcal{T}_h)$. A *finite element implementation* of scheme (11.0), (11.1), (11.6), (11.7), (11.4), corresponding to (12.1)–(12.4), reads as follows (with $m(T) = \int_T \mathrm{d}x$, and other obvious notation):

$$\mathbf{p}_h^0 = \mathbf{p}_{0h}, \quad \textit{with } \mathbf{p}_{0h} \textit{ given in } \Lambda_h; \tag{12.5}$$

*for* $n \geq 0$, $\mathbf{p}_h^n$ *being known, compute* $\mathbf{p}_h^{n+1/2}$, $\omega_h^{n+1}$, $u_h^{n+1}$, *and* $\mathbf{p}_h^{n+1}$ *as follows:*

$$\left.\begin{aligned} &(p_{iT}^{n+1/2} - p_{iT}^n)/\Delta t + p_{iT}^{n+1/2} + \varepsilon_2^{-1}(|p_{iT}^{n+1/2}|^2 - 1)p_{iT}^{n+1/2} \\ &\qquad = [C/m(T)] \int_T \frac{\partial \varphi_1}{\partial x_i}\,\mathrm{d}x, \ \forall i = 1,2, \ \forall T \in \mathcal{T}_h, \end{aligned}\right\} \tag{12.6}$$

$$\left.\begin{aligned} &\omega_h^{n+1} \in K_h^+; \ \forall v_h \in K_h^+, \\ &\int_\Omega \nabla \omega_h^{n+1} \cdot \nabla(v_h - \omega_h^{n+1})\,\mathrm{d}x \geq \int_\Omega \mathbf{p}_h^{n+1/2} \cdot \nabla(v_h - \omega_h^{n+1})\,\mathrm{d}x, \end{aligned}\right\} \tag{12.7}$$

$$\left.\begin{aligned} &\mathbf{p}_h^{n+1} = \nabla u_h^{n+1}, \quad \textit{with } u_h^{n+1} \in V_{0h}, \textit{ and } \forall v_h \in V_{0h}, \\ &\int_\Omega (u_h^{n+1} v_h + \Delta t \varepsilon_1 \nabla u_h^{n+1} \cdot \nabla v_h)\,\mathrm{d}x = \int_\Omega \omega_h^{n+1} v_h\,\mathrm{d}x. \end{aligned}\right\} \tag{12.8}$$

The cubic equations in (12.6) are particular cases of problem (11.3); they have a unique solution if $\Delta t \leq \varepsilon_2$, and can be easily solved by Newton's method. The *discrete obstacle problem* (12.7) is a pretty classical one, in principle; however, in the context of (the difficult) problem (9.7), it turns out that the solution methods based on *over-relaxation with projection* and *Uzawa algorithms with projections* (see, e.g., refs [3], [4] for details) have troubles to achieve fast convergence, forcing us to look for more efficient alternatives. Such an alternative is provided by the *penalty/Newton/conjugate gradient method* discussed in Sections 4 to 8. For the test cases to be considered in

Section 13, the above method never failed and converged, always, in few iterations, even for small values of its penalty parameter. Finally, problem (12.8) is a fairly standard one (see, e.g., [4, Appendix 1] and the references therein), and does not deserve further comments, therefore, other than: a sufficient condition to have a *discrete maximum principle* is to have the angles of $\mathcal{T}_h$ not exceeding $\frac{\pi}{2}$ (a classical result indeed; see, e.g., [14]). To summarize:

***We have reduced the approximate solution of problem (9.2) to that of a sequence of simple cubic equations and of, essentially, discrete linear elliptic equations.***

**Remark 12.1.** Let us denote by $\{u_h^\varepsilon\}_h^\varepsilon$ the family of approximate solutions provided by algorithm (12.5)–(12.8). The role of the *biharmonic regularization (à la Tychonoff)* in problems (12.7), (12.8) is to enhance the *compactness* properties of $\{u_h^\varepsilon\}_h^\varepsilon$ (and both things being related) control spatial oscillations of short wave length (of order $h$, typically).

**Remark 12.2.** As can be expected, *biharmonic regularizations* are not new; they have been used in, e.g., [15] for the solution of boundary control problems for the wave equation, and in [16] for the solutions of problems such as (9.1), precisely (by methods quite different from those discussed in this article). Indeed, on the basis of *boundary layer thickness considerations* (similar to those in [15]) we can expect that the optimal value of $\Delta t\varepsilon_1$ in (12.8) has to be of order $h^2$; the numerical experiments to be reported in Section 13 will validate this prediction, showing that a near optimal value of $\Delta t\varepsilon_1$ is $h^2/36$, for all $h$'s sufficiently small.

**Remark 12.3.** With minor modifications, the methodology discussed above can be applied to those variants of (9.1), where the boundary condition $u = 0$ on $\Gamma$ is replaced by $u = g$, with $g$ a *Lipschitz continuous function* such that $|g(x) - g(y)| \leq |x - y|$, $\forall x, y \in \Gamma$, and $\max_{\{x,y\}\in\Gamma}[g(x) - g(y)]$ not too large. Actually, for validation purposes, some of the test problems to be considered in Section 13 will involve non-zero Dirichlet data (i.e., $g \neq 0$) and will correspond to situations, where a unique maximal (or minimal) solution to problem (9.1) exists.

## 13 Numerical Experiments

For the test problems discussed below we took $\Omega = (0,1) \times (0,1)$ and used a uniform triangulation of $\Omega$. From a computational point of view, this implies that the finite element approximations reduce to finite difference ones, allowing, for example, the use of *cyclic reduction based* on fast Poisson and Helmholtz solvers for the solution of the discrete elliptic problems encountered at each iteration of the numerical procedure discussed in the above sections. The main goal of the first and second test problems is to validate the numerical methodology, since both of them have closed form solutions.

The third test problem is the one of interest, since the boundary conditions being incompatible with the constraints $|\partial u/\partial x_1| = |\partial u/\partial x_2| = 1$ a.e., we can expect a non-smooth behavior of the solutions (*fractal*, in fact) near the boundary.

*First two test problems*: They are defined by

$$\left|\frac{\partial u}{\partial x_1}\right| = \left|\frac{\partial u}{\partial x_2}\right| = 1 \quad \text{a.e. in } \Omega, \quad u = g \text{ on } \Gamma, \tag{13.1}$$

with

$$\left.\begin{aligned} g(x_1,0) = g(x_1,1) = \min(x_1, 1-x_1), 0 \le x_1 \le 1, \\ g(0,x_2) = g(1,x_2) = \min(x_2, 1-x_2), 0 \le x_2 \le 1. \end{aligned}\right\}$$

Here, the maximal (which turns out to be also a viscosity) solution of problem (13.1) is given by

$$u_{\max}(x_1,x_2) = 1 - \left|\frac{1}{2} - x_1\right| - \left|\frac{1}{2} - x_2\right| = \min(x_1, 1-x_1) + \min(x_2, 1-x_2).$$

The minimal solution is given by

$$u_{\min}(x_1,x_2) = \begin{cases} |x_1 - x_2|, \;\; 0 \le x_1, \;\; x_2 \le \frac{1}{2}, \text{ and } \frac{1}{2} \le x_1, \;\; x_2 \le 1, \\ |1 - x_1 - x_2|, \;\; \frac{1}{2} \le x_1 \le 1, \;\; 0 \le x_2 \le \frac{1}{2}, \\ \qquad \text{and } 0 \le x_1 \le \frac{1}{2}, \;\; \frac{1}{2} \le x_2 \le 1. \end{cases}$$

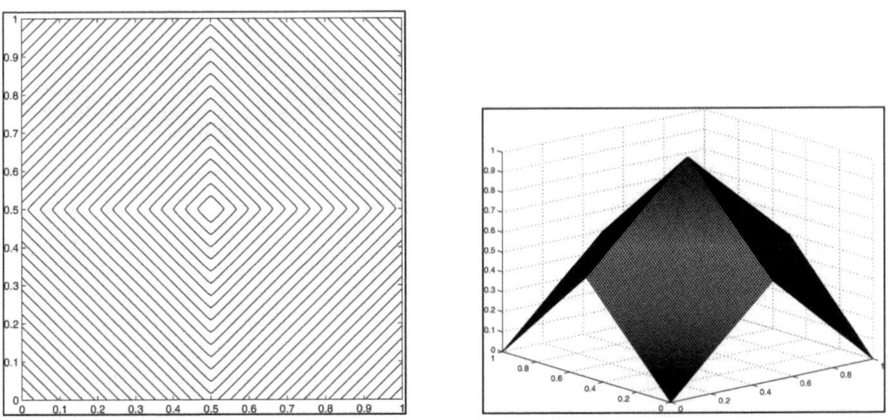


Figure 1. Contours (*left*) and graph (*right*) of the computed maximal solution of problem (13.1) ($h = 1/128$).

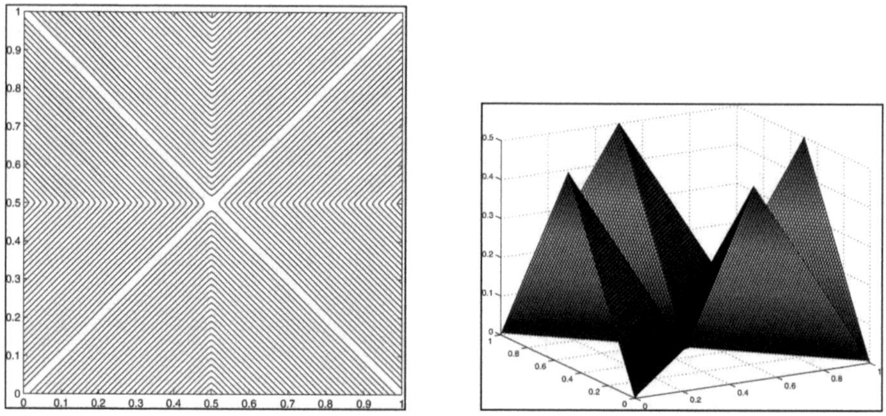

Figure 2. Contours (*left*) and graph (*right*) of the computed minimal solution of problem (13.1) ($h = 1/128$).

The computed solution $u_h$ obtained with $h = 1/128$, $\Delta t \varepsilon_1 = h^2/36$, $\varepsilon_2 = 0.001$, $\Delta t = 0.0001$, and $C = 10$ for the maximal case (resp., $C = -10$ for the minimal case) has been visualized in Fig. 1 (resp., Fig. 2); it coincides quite well with the exact solution, since approximation errors are $\|u_h - u\|_{0,\Omega} = 3.268 \times 10^{-4}$ and $\|u_h - u\|_{\infty,\Omega} = 5.086 \times 10^{-3}$ (resp., $\|u_h - u\|_{0,\Omega} = 7.133 \times 10^{-4}$ and $\|u_h - u\|_{\infty,\Omega} = 5.715 \times 10^{-3}$).

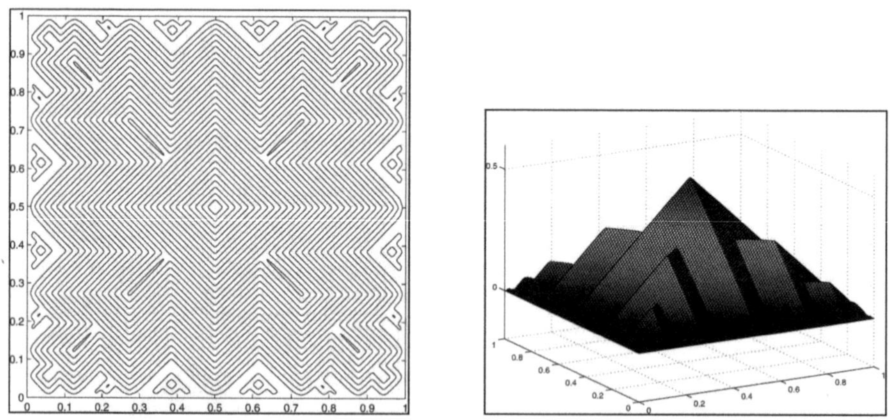

Figure 3. Contours (*left*) and graph (*right*) of the computed maximal solution of problem (2.1), (9.1) ($h = 1/512$).

*Third test problem*: In this case we look for a nearly maximal solution of problem (2.1), (9.1), recalling that in this case maximal solutions do not exist. We have visualized in Figs. 3 and 4 the graphs and contours of the discrete solutions computed with $h = 1/512$ and $1/1024$. The other parameters are

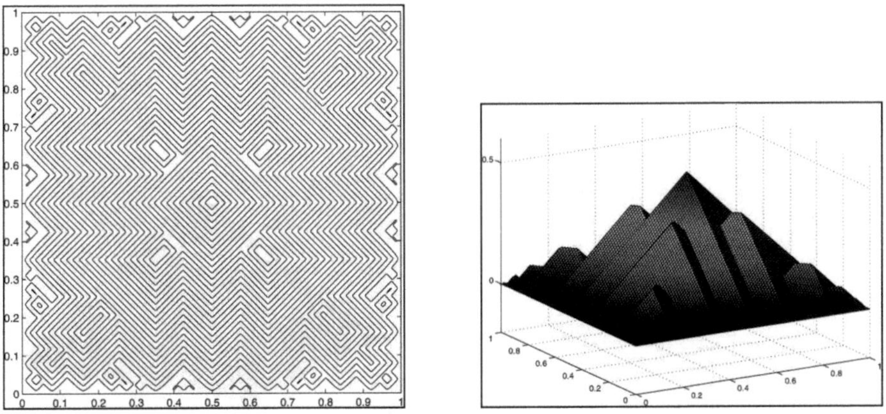


Figure 4. Contours (*left*) and graph (*right*) of the computed maximal solution of problem (2.1), (9.1) ($h = 1/1024$).

$\Delta t \varepsilon_1 = h^2/9$, $\varepsilon_2 = 0.001$, $C = 10$ and $\Delta t = 0.0001$. These figures show that as one refines the mesh new structures appear clearly showing (as expected) a fractal behavior near the boundary (and particularly in the corners). Incidentally, the $L^1$-norms corresponding to $h = 1/512$ and $h = 1/1024$ are 0.142263 and 0.143316, respectively, while the maximal values (reached at $x_1 = x_2 = 1/2$) are 0.52457 and 0.51494 (they should be equal to 0.5, but realize that we are solving a highly non-smooth problem).

To further validate the methodology discussed in Sections 9 to 12, we have applied it to the solution of a *genuine Eikonal equation*, namely

$$|\nabla u| = 1 \text{ a.e. in } \Omega, \quad u = g \text{ on } \Gamma. \tag{13.2}$$

If one looks for the maximal solution (the *distance function* to the boundary if $g = 0$ in (13.2)), the same methodology used for problem (2.1), (9.1) can be used except that we replace $\sum_{i=1}^{d} \int_\Omega \left( \left| \frac{\partial v}{\partial x_i} \right|^2 - 1 \right)^2 \mathrm{d}x$ in (9.6) by $\int_\Omega (|\nabla v|^2 - 1)^2 \, \mathrm{d}x$, which is in fact simpler. If one looks for a minimal solution, we will take $C < 0$ in (9.5). The resulting Eikonal solver has been validated via the following test problems (both with $\Omega = (0,1) \times (0,1)$):

*First test problem:* We take $g = 0$ in (13.2), implying that the maximal solution is given by $u(x) = \delta(x, \Gamma)$, where $\delta(x, \Gamma)$ is the distance from $x$ to the boundary $\Gamma$. This solution has been captured very quickly with $h = 1/256$, $\Delta t \varepsilon_1 = h^2/49$, $\varepsilon_2 = 0.001$, $C = 10$, and $\Delta t$ =0.0001. The approximation errors are $\|u_h - u\|_{0,\Omega} = 1.638 \times 10^{-3}$ and $\|u_h - u\|_{\infty,\Omega} = 9.811 \times 10^{-3}$. Also for this solution, we have $u(1/2, 1/2) = 1/2$ and $\int_\Omega u \, \mathrm{d}x = 1/6$ ($= 0.16666\ldots$), the computed values being 0.50789761 and 0.16737264. The graph and contours of the computed solution are shown in Fig. 5.

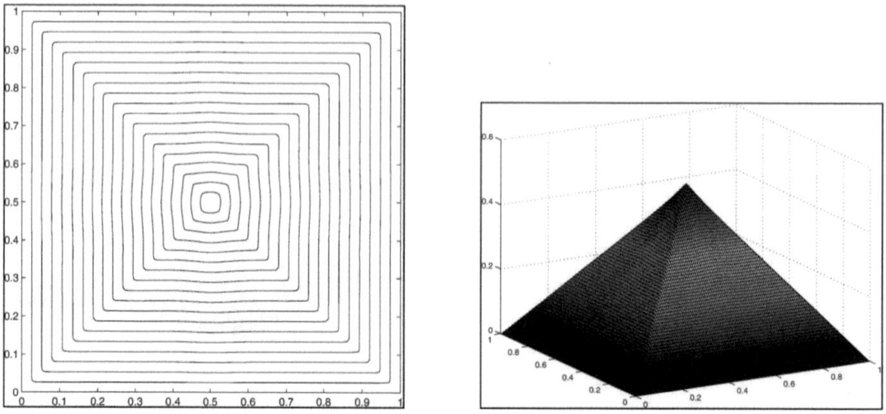

Figure 5. Contours (*left*) and graph (*right*) of the computed maximal solution of problem (13.2) with $g = 0$ ($h = 1/256$).

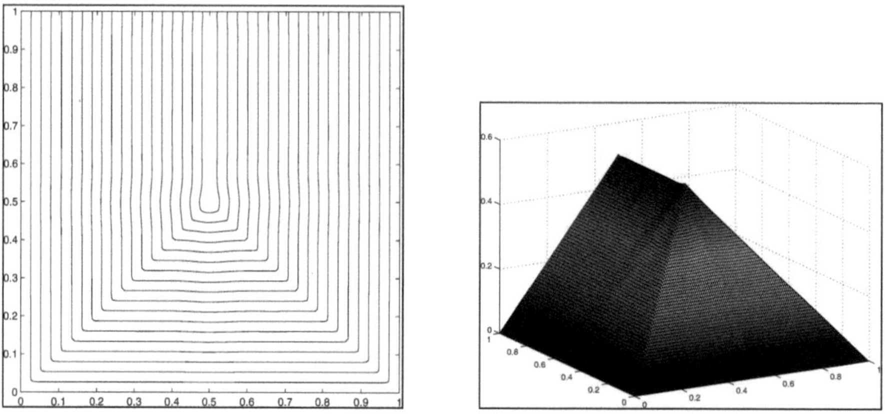

Figure 6. Contours (*left*) and graph (*right*) of the computed maximal solution of problem (13.2) with $g$ given in (13.3) ($h = 1/256$).

*Second test problem*: Here $g$ in (13.2) is given by

$$\left.\begin{aligned} g(x_1, 0) &= 0, 0 \leq x_1 \leq 1, \\ g(x_1, 1) &= \min(x_1, 1 - x_1), 0 \leq x_1 \leq 1, \\ g(0, x_2) = g(1, x_2) &= 0, 0 \leq x_2 \leq 1, \end{aligned}\right\} \tag{13.3}$$

implying that the maximal solution of (13.2) is given by

$$u(x_1, x_2) = \begin{cases} x_2, & x_1 \geq x_2 \text{ and } 1 - x_1 \geq x_2, \\ x_1, & x_1 \leq x_2 \text{ and } x_1 \leq \dfrac{1}{2}, \\ 1 - x_1, & 1 - x_1 \leq x_2 \text{ and } x_1 \geq \dfrac{1}{2}. \end{cases}$$

The parameters for this test case are $\Delta t\varepsilon_1 = h^2/49$, $\varepsilon_2 = 0.001$, $C = 10$, and $\Delta t$ =0.0001. Despite the (small) overshots shown in Fig. 6, the exact solution has been captured quite accurately, since $\|u_h - u\|_{0,\Omega} = 1.1158 \times 10^{-3}$ for $h = 1/256$. The maximal value of $u(x_1, x_2)$ is 1/2, whereas the computed value is 0.50642583; also $\int_\Omega u\,\mathrm{d}x = 5/24$ ($= 0.2083333\ldots$), whereas $\int_\Omega u_h\,\mathrm{d}x = 0.2088403$. The contours of the approximated maximal solution obtained with $h = 1/256$ are shown in Fig. 6.

## 14 Final comments

For those wondering if the solutions computed via "our" *biharmonic regularization technique* are *viscosity solutions*, the answer is definitely yes. To be convinced by this strong statement, let's consider the following *Eikonal equation*

$$|\nabla\psi|^2 = f\ (\geq 0)\ \text{a.e. in}\ \Omega, \quad \psi = g\ \text{on}\ \Gamma. \tag{14.1}$$

We associate to (14.1) the following penalized/regularized problem:

$$\psi_\varepsilon \in V_g;\ \ J^\varepsilon(\psi_\varepsilon) \leq J^\varepsilon(\varphi), \quad \forall \varphi \in V_g, \tag{14.2}$$

where $V_g = \{\varphi \mid \varphi \in H^2(\Omega),\ \ \varphi = g\ \text{on}\ \Omega\}$, and

$$\begin{aligned} J^\varepsilon(\varphi) = \frac{\varepsilon_1}{2}\int_\Omega |D^2\varphi|^2\,\mathrm{d}x + \frac{1}{2}\int_\Omega |\nabla\varphi|^2\,\mathrm{d}x - C\int_\Omega \varphi\,\mathrm{d}x \\ + \frac{\varepsilon_2^{-1}}{4}\int_\Omega (|\nabla\varphi|^2 - f)^2\,\mathrm{d}x, \end{aligned} \tag{14.3}$$

with

$$D^2\varphi = (\partial^2\varphi/\partial x_i\partial x_j)_{1\leq i,\,j\leq d}.$$

The Euler-Lagrange equation associated to (14.3) reads as follows:

$$\left.\begin{aligned} &\psi_\varepsilon \in V_g;\ \forall\varphi \in V_0\ \ (= H^2(\Omega) \cap H_0^1(\Omega)), \\ &\varepsilon_1\int_\Omega D^2\psi_\varepsilon : D^2\varphi\,\mathrm{d}x + \int_\Omega \nabla\psi_\varepsilon\cdot\nabla\varphi\,\mathrm{d}x \\ &+\varepsilon_2^{-1}\int_\Omega (|\nabla\psi_\varepsilon|^2 - f)\nabla\psi_\varepsilon\cdot\nabla\varphi\,\mathrm{d}x = C\int_\Omega \varphi\,\mathrm{d}x. \end{aligned}\right\} \tag{14.4}$$

Suppose now that $d = 2$ and that $\Omega$ is simply connected. Introduce then $\mathbf{u}_\varepsilon = \{\frac{\partial\psi_\varepsilon}{\partial x_2}, -\frac{\partial\psi_\varepsilon}{\partial x_1}\}$ and $\mathbf{v} = \{\frac{\partial\varphi}{\partial x_2}, -\frac{\partial\varphi}{\partial x_1}\}$; we clearly have $\mathbf{u}_\varepsilon \in (H^1(\Omega))^2$, $\mathbf{v} \in (H^1(\Omega))^2$, $\nabla\cdot\mathbf{u}_\varepsilon = \nabla\cdot\mathbf{v} = 0$, and $\mathbf{u}_\varepsilon\cdot\mathbf{n} = \mathrm{d}g/\mathrm{d}s$, $\mathbf{v}\cdot\mathbf{n} = 0$ on $\Gamma$, where

$s$ is a curvilinear coordinate on $\Gamma$, $\Gamma$ being oriented counterclockwise. Problem (14.4) takes then the equivalent (kind of "velocity-pressure") following formulation:

$$\left.\begin{aligned}&\mathbf{u}_\varepsilon \in W_g,\\&\varepsilon_1 \int_\Omega \nabla \mathbf{u}_\varepsilon : \nabla \mathbf{v}\,dx + \int_\Omega \mathbf{u}_\varepsilon \cdot \mathbf{v}\,dx + \varepsilon_2^{-1} \int_\Omega (|\mathbf{u}_\varepsilon|^2 - f)\mathbf{u}_\varepsilon \cdot \mathbf{v}\,dx\\&\quad + \int_\Omega p\nabla \cdot \mathbf{v}\,dx = C \int_\Omega \mathbf{u}_1 \cdot \mathbf{v}\,dx, \quad \forall \mathbf{v} \in W_0,\\&\nabla \cdot \mathbf{u}_\varepsilon = 0 \text{ in } \Omega,\end{aligned}\right\} \tag{14.5}$$

where $W_0 = \{\mathbf{v} \mid \mathbf{v} \in (H^1(\Omega))^2,\ \mathbf{v}\cdot\mathbf{n} = 0 \text{ on } \Gamma\}$, $W_g = \{\mathbf{v} \mid \mathbf{v} \in (H^1(\Omega))^2,\ \mathbf{v}\cdot\mathbf{n} = dg/ds \text{ on } \Gamma\}$, and $\mathbf{u}_1 = \{\frac{\partial\varphi_1}{\partial x_2}, -\frac{\partial\varphi_1}{\partial x_1}\}$ with $-\Delta\varphi_1 = 1$ in $\Omega$, $\varphi = 0$ on $\Gamma$.

Problem (14.5) is clearly a *Stokes* variant of a *Ginzburg-Landau* equation, $\varepsilon_1$ playing here the role of a *viscosity coefficient.* We have shown thus that our *biharmonic regularization procedure* is indeed a *viscosity solution method* in the *sense of Stokes* (of course, once $\mathbf{u}_\varepsilon$ is known, we obtain $\psi_\varepsilon$ via the solution of $-\Delta\psi_\varepsilon = \frac{\partial u_{2\varepsilon}}{\partial x_1} - \frac{\partial u_{1\varepsilon}}{\partial x_2}$ in $\Omega$, $\psi_\varepsilon = g$ on $\Gamma$); the case where $\Omega$ has holes is a little more complicated to handle, but can be treated essentially in a similar way.

Our very last comments will be about fractals and geometry: in the September 2002 issue of SIAM NEWS, Ph. J. Davis wrote "Since 1977, when Mandelbrot published his *Fractals: Form, Chance and Dimension*, geometry, to a very large constituency, has meant fractals and their production by iteration." There is no doubt (in our mind, at least; see Figs. 3 and 4) that this article is another contribution to fractal geometry.

**Acknowledgment.** The authors would like to thank Professors P. G. Ciarlet and P. Neittaanmäki (and the City University of Hong-Kong and University of Jyväskylä) for a decisive role in bringing this article into existence. They also thank Professor Jiwen He for helpful comments and suggestions, and P. J. Muscarello for processing this article. The supports of the National Science Foundation (Grant DMS-0209066) and of the LACSI/DOE program are also acknowledged.

## References

[1] M. Fortin, R. Glowinski, *Augmented Lagrangian Methods: Applications to the Numerical Solution of Boundary-Value Problems*, North-Holland, Amsterdam, 1983.

[2] R. Glowinski, P. LeTallec, *Augmented Lagrangians and Operator Splitting Methods in Nonlinear Mechanics*, SIAM, Philadelphia, 1989.

[3] R. Glowinski, J.-L. Lions, and R. Tremolières, *Numerical Analysis of Variational Inequalities*, North-Holland, Amsterdam, 1981.

[4] R. Glowinski, *Numerical Methods for Nonlinear Variational Problems*, Springer-Verlag, New York, 1984.

[5] E. J. Dean, R. Glowinski, Y, M. Kuo, and M. G. Nasser, *On the discretization of some second order in time differential equations. Applications to nonlinear wave problems*, Computational Techniques in Identification and Control of Flexible Flight Structures (A. V. Balakrishnan, ed.), Optimization Software, Inc., Los Angeles, 1990, pp. 199–246.

[6] H. Carlsson, R. Glowinski, *Vibrations of Euler-Bernoulli beams with pointwise obstacles*, Advances in Kinetic Theory and Continuum Mechanics (R. Gatignol and Soubbaramayer, eds.), Springer-Verlag, Berlin, 1991, pp. 261–275.

[7] G. Duvaut, J.-L. Lions, *Inequalities in Mechanics and Physics*, Springer-Verlag, Berlin, 1976.

[8] J.-J. Lions, *Contrôle Optimal des Systèmes Gouvernés par des Équations aux Dérivées Partielles*, Dunod-Gauthier Villars, Paris, 1968.

[9] I. Ekeland, R. Temam, *Convex Analysis and Variational Problems*, North-Holland, Amsterdam, 1976.

[10] B. Dacorogna, P. Marcellini, *Implicit Partial Differential Equations*, Birkhauser, Basel, 1999.

[11] R. Glowinski, T.-W. Pan, and J. Periaux, *A Lagrange multiplier/fictitious domain method for the Dirichlet problem. Generalization to some flow problems*, Japan J. Indust. Appl. Math. 12 (1995), 87–108.

[12] R. Glowinski, *Finite Element Methods for Incompressible Viscous Flow*, Handbook of Numerical Analysis, Vol. IX (P. G. Ciarlet and J.-L. Lions, eds.), North-Holland, Amsterdam, 2003.

[13] R. Glowinski, M. Holmstrom, *Constrained motion problems with applications by nonlinear programming methods*, Surveys on Math. for Industry **5** (1995), 75–108.

[14] P. G. Ciarlet, *The Finite Element Method for Elliptic Problems*, North-Holland, Amsterdam, 1978.

[15] R. Glowinski, C. H. Li, and J.-L. Lions, *A numerical approach to the exact boundary controllability of the wave equation (I) Dirichlet controls: Description of the numerical methods*, Japan J. Appl. Math. **7** (1990), 1–76.

[16] P.-A. Gremaud, N. R. Ide, *Computation of nonclassical solutions to Hamilton-Jacobi problems*, SIAM J. Sci. Comput. **21** (1999), 502–521.

# Part V

# Advanced Applications: From Image Processing to Computational Physics

# The Use of Bilinear Rectangular Elements in Reconstruction of Panoramic Images

Pavel Křížek

Center for Machine Perception, Faculty of Electrical Engineering, Czech Technical University, Karlovo náměstí 13, CZ–121 35 Prague 2, Czech Republic
krizekp1@feld.cvut.cz

## 1 Introduction

The aim of this paper is to develop an algorithm for mosaicing a sequence of panoramic images of the pipeline inner surface obtained by an inspection robot that is equipped by an omnidirectional camera. This camera is a combination of a regular camera and an axially symmetric hyperbolic mirror. The construction of the robot (see Fig. 1) enabled us to take and store a sequence of images of the inner pipeline surface under the viewing angle 25°–106° from the camera optical axis and 360° around this axis. Since the axes of pipeline and camera were not identical, each picture was considerably deformed.

In Section 3 we describe how to get nondeformed rectangular panoramic images by means of bilinear elements and an appropriate nonlinear transform. Mosaicing particular panoramic images by a correlation function in Section 4, we acquire a global image of the whole inner surface of the pipeline, which is unwarped to the plane (see Fig. 2).

## 2 Acquiring panoramic images

It is evident that the whole camera viewing angle was nonuniformly illuminated (cf. Fig. 1). Therefore, we have developed a procedure to correct this nonuniformity. It was also necessary to correct the hyperbolic surface of the axially symmetric mirror. Deformations of the mirror were approximated by higher-degree polynomials whose coefficients were established by means of the least squares method (see [5]). The resulting system of linear algebraic equations with a symmetric and positive definite matrix was solved by the standard conjugate gradient method which converges in a finite number of iterations for (see [1]). All calculations were performed in double precision arithmetic. A very good agreement between the measured data and the associated polynomial approximation is presented in detail in [3].

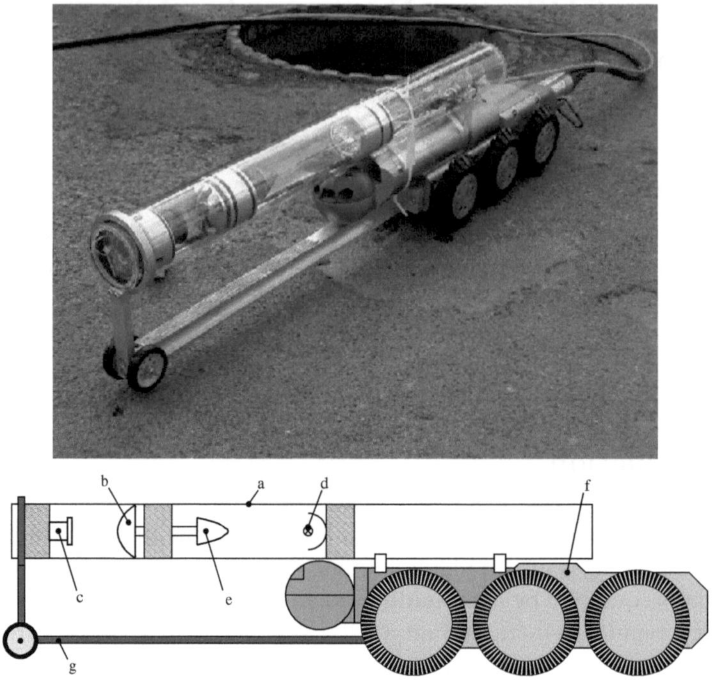


Figure 1. Photo and the associated schematic illustration of the inspection robot; a – glass tube, b – panoramic hyperbolic mirror, c– camera, d – light bulb, e – additional hyperbolic mirror, f – prospecting tractor of the British Company Pearpoint, g – chassis.

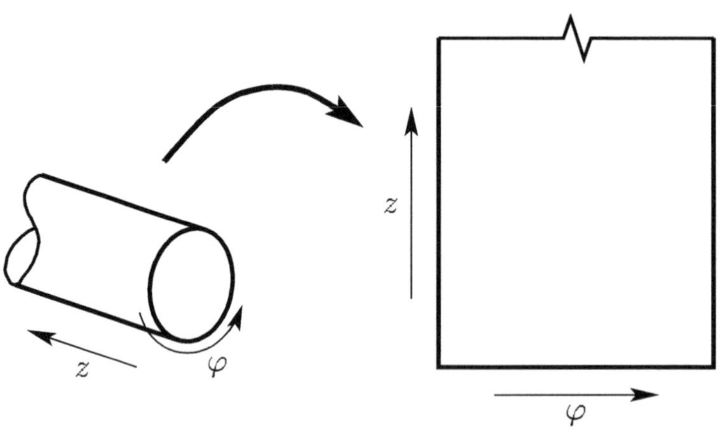


Figure 2. Unwarping the pipeline inner surface of constant radius into the plane by the standard cylindrical coordinates.

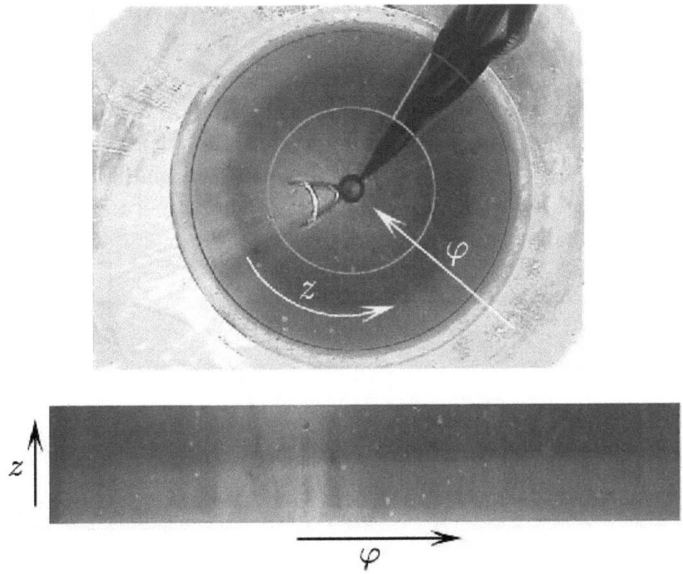


Figure 3. Unwarping the image of the mirror into the coordinates $(\varphi, z)$.

In Fig. 3 we observe a typical image of the omnidirectional camera and the associated unwarped panoramic image in the case that the axis of the pipeline and the optical axis of the camera coincide.

However, in practice the axis of the pipeline is usually different from the optical axis of camera. Therefore, the used camera was calibrated in a fibreglass test pipeline whose inner surface was covered by a uniform black and white checker pattern. Due to camera misalignment we see in Fig. 4 that the black and white squares were considerably deformed.

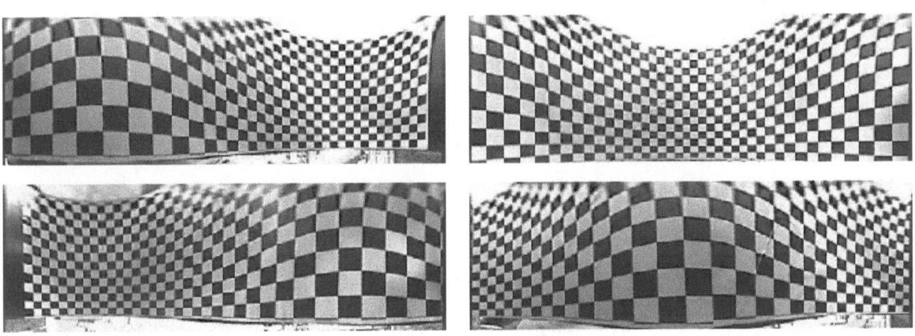

Figure 4. Examples of panoramic images of the inner pipeline surface covered by a uniform black and white checker pattern. Large nonlinear deformations are due to camera misalignment.

## 3 Correction of a parallel misalignment of the camera

The orientation of the axes of the Cartesian coordinate system $(x, y, z)$ and the position of the camera in the pipeline are sketched in Fig. 5. Assume now that the axis of the pipeline is parallel with the optical axis of the camera. The absolute value $|d|$ stands for their distance (we shall assume that $d > 0$ if the axis of the pipeline is above the axis of the camera, otherwise $d \leq 0$).

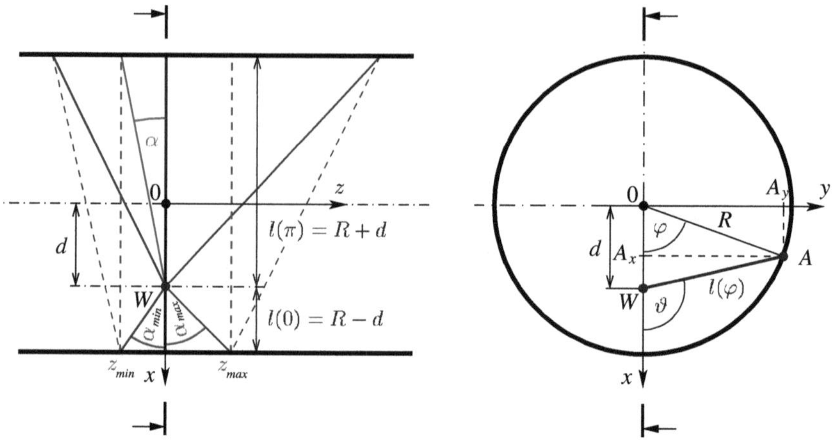


Figure 5. Schematic illustration of pipeline a) along, and b) across the axis of symmetry. The position of the camera is denoted by $W$, the distance between the axis of the pipeline and optical axis of the camera is $|d|$, $\alpha_{min}$ and $\alpha_{\max}$ determine the viewing angle of the camera, and $R$ stands for radius of the pipeline.

Fig. 6 shows the corresponding cylindrical coordinates, the position of the coordinate $(u, v)$ system in the image of the camera with the origin at the centre of the mirror, and also a rectangular domain in coordinates $(\varphi, z)$, where the unknown image intensity function is searched.

We shall seek the coordinates $u$ and $v$ in the image of the mirror (see Fig. 6b)) which correspond to a given angle $\varphi$ and the height $z$ in the unwarped panoramic image (see Fig. 6c)). Hence, we shall consider the transformation

$$u = u(\varphi, z), \tag{3.1}$$

$$v = v(\varphi, z). \tag{3.2}$$

Notice that the angle $\vartheta$ in Figs. 5b) and 6b) is the same. Since the searched transformations (3.1) and (3.2) depend on the angle $\varphi$, we shall first look for the angle $\vartheta$ as a function of $\varphi$, i.e.,

$$\vartheta = \vartheta(\varphi). \tag{3.3}$$

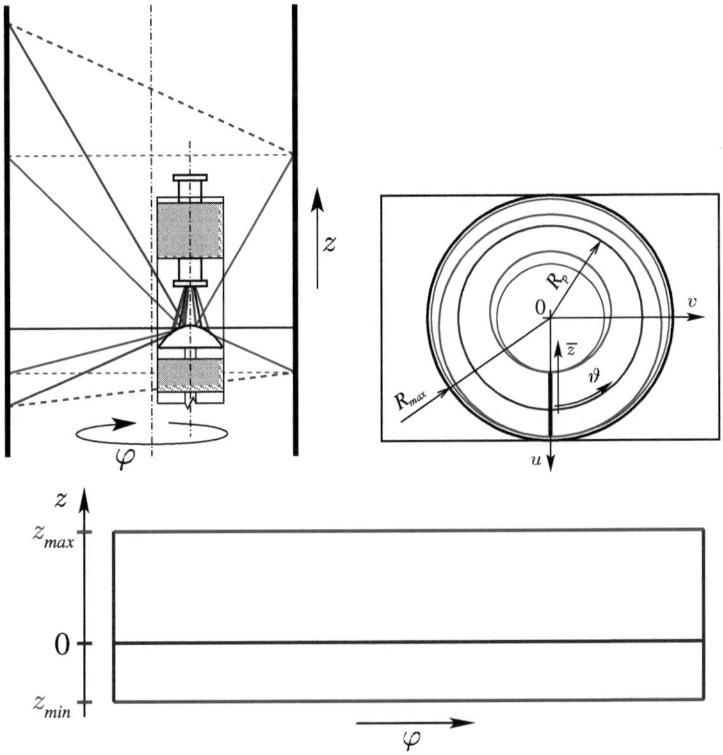


Figure 6. Schematic illustration of the pipeline a) along the axis $z$, b) the image obtained by the camera, c) the unwarped panoramic image. The coordinate system in the image of the camera is denoted by $(u, v)$, $R_{\max}$ is the radius of the mirror, $R_p$ is the radius of the circle which corresponds to the cross section of the pipeline through the point $W$, where the camera is situated, $(\varphi, z)$ and $(\vartheta, \overline{z})$ are coordinate systems in the unwarped image and image of the mirror, respectively.

Consider the pipeline with radius $R$ (see Fig. 5b)). The coordinates of the point $A = (A_x, A_y)$ satisfy the equations

$$A_x = R\cos\varphi, \tag{3.4}$$

$$A_y = R\sin\varphi. \tag{3.5}$$

The distance between the point $A$ and the point of intersection $W$ can be easily expressed by the cosine theorem,

$$l(\varphi) = \sqrt{R^2 + d^2 - 2dR\cos\varphi}. \tag{3.6}$$

Then for the angle $\vartheta$ between the axis $x$ and the line segment $WA$ (see Fig. 5b)), we obtain

$$\cos\vartheta = -\frac{d - A_x}{l(\varphi)},$$
$$\sin\vartheta = \frac{A_y}{l(\varphi)}.$$

From this, (3.4) and (3.5), we find that

$$\cos\vartheta = \frac{R\cos\varphi - d}{l(\varphi)}, \tag{3.7}$$
$$\sin\vartheta = \frac{R\sin\varphi}{l(\varphi)}. \tag{3.8}$$

The panoramic image of the inner surface of the pipeline corresponds to the strip, whose height is given by the interval $[z_{\min}, z_{\max}]$, where

$$z_{\min} = (R - |d|)\tan\alpha_{\min},$$
$$z_{\max} = (R - |d|)\tan\alpha_{\max}.$$

Concentric circles with centre at 0 in the image of the mirror (see Fig. 6b)) correspond to rays coming to the camera at the constant angle $\alpha$. For a fixed coordinate $z$ and for a given angle $\varphi$, we have to determine the angle $\alpha$ as

$$\alpha = \alpha(\varphi, z) = \arctan\frac{z}{l(\varphi)}. \tag{3.9}$$

Suppose that the image of the mirror is only linearly deformed in each radial direction. In this case the image of the pipeline in the mirror is, of course, nonlinearly deformed for $d \neq 0$. For a fixed angle $\vartheta$, the associated angle $\varphi$ is constant, and thus we have

$$\bar{z} = k\tan\alpha = k\frac{z}{l(\varphi)}, \tag{3.10}$$

where $\bar{z}$ corresponds to the radial coordinate in the image of the mirror and $k$ is a constant, which corresponds to the scale in the $z$-axis.

Now we can define a resulting nonlinear transformation between the Cartesian and polar coordinates. The angle in the image of the mirror corresponds to the value $\vartheta$ and the radius corresponds to the coordinate $\bar{z}$. From relations (3.7), (3.8), and (3.10) we find that

$$u = R_p(1 - \bar{z})\cos\vartheta, \tag{3.11}$$
$$v = R_p(1 - \bar{z})\sin\vartheta, \tag{3.12}$$

where $R_p$ is the radius of the circle (in the camera image) which determines the perpendicular view on the pipeline.

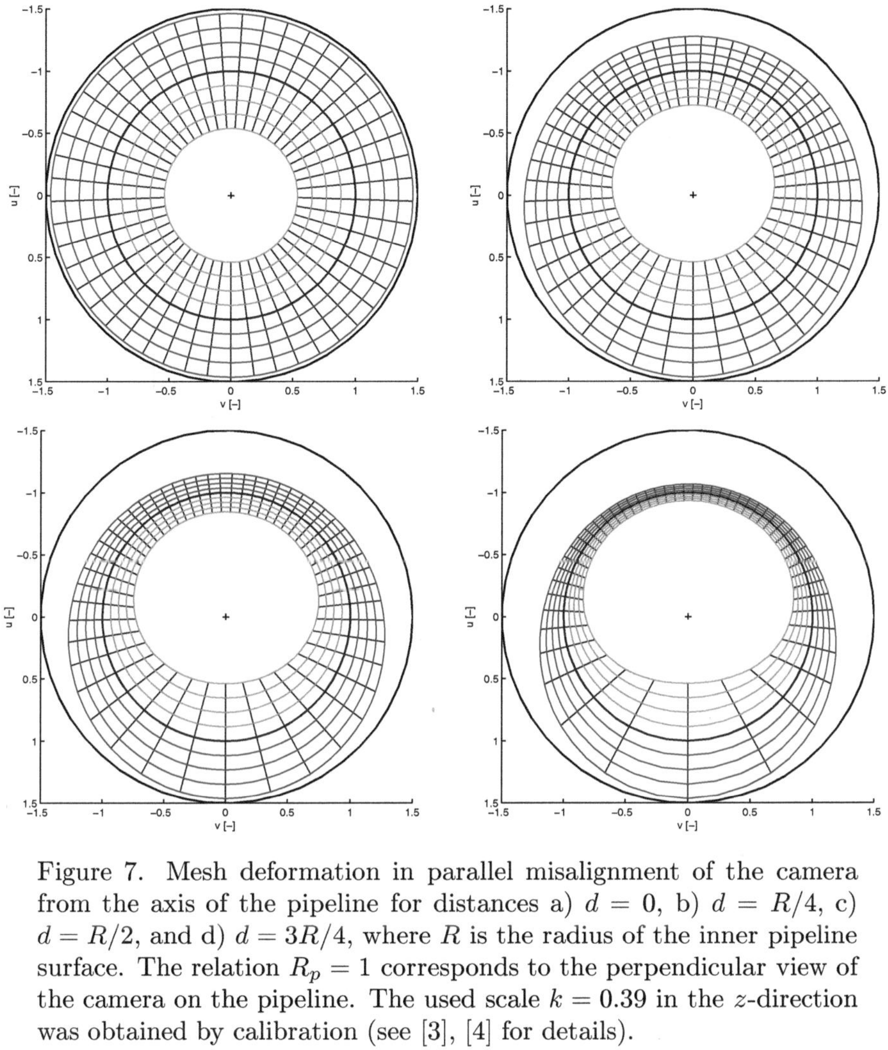


Figure 7. Mesh deformation in parallel misalignment of the camera from the axis of the pipeline for distances a) $d = 0$, b) $d = R/4$, c) $d = R/2$, and d) $d = 3R/4$, where $R$ is the radius of the inner pipeline surface. The relation $R_p = 1$ corresponds to the perpendicular view of the camera on the pipeline. The used scale $k = 0.39$ in the $z$-direction was obtained by calibration (see [3], [4] for details).

Fig. 7 shows the deformation of the regular mesh in the image of mirror after the nonlinear transformation (3.11)–(3.12), which involves a parallel misalignment of the camera.

The standard bilinear rectangular elements (see [2]) on a uniform mesh of the domain of Fig. 6b) were used to interpolate the values of the intensity function after the nonlinear transformation (3.11)–(3.12). The reason for this approach is the following: Since the coordinates $\vartheta$ and $\overline{z}$ attain only discrete values (multiples of integers) in the searched bit map, the associated coordinates $u$ and $v$ defined by (3.11)–(3.12) are, in general, not at nodal points (pixels) of Fig. 6b). Consequently, the unknown intensity function $\overline{I}$ attaining the values $\{0, 1, \ldots, 255\}$ in the rectangular domain of Fig. 6c) was calculated by the formula

$$\overline{I}(\vartheta, \overline{z}) = I(u, v),$$

where the coordinates $(u, v)$ and $(\vartheta, \overline{z})$ satisfy (3.11)–(3.12), and $I = I(u, v)$ is a continuous piecewise bilinear approximation of known intensity function rounded to the values $\{0, 1, \ldots, 255\}$ at the discrete values of $u$ and $v$ in the image of the mirror.

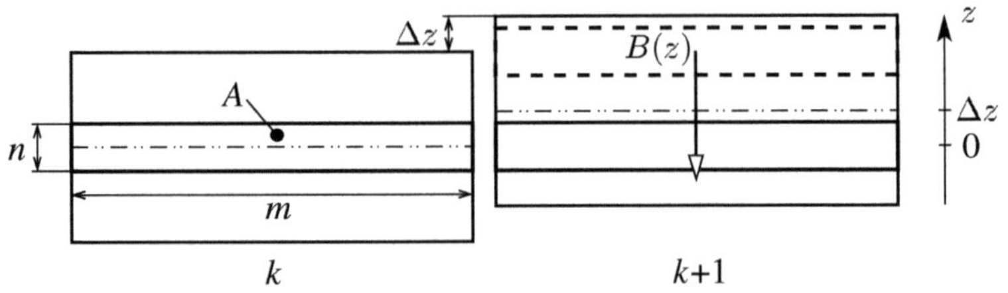


Figure 8. Searching the correct shift in the $z$-direction.

A special case of the transformation (3.11)–(3.12) for $d = 0$ corresponds to the situation when the axis of the camera is coincident with the axis of the pipeline. Formulae (3.6), (3.7), (3.8), and (3.10) then have the form

$$\begin{aligned} l(\varphi) &= \sqrt{R^2 + 0 - 0} = R, \\ \cos\vartheta &= \frac{R\cos\varphi - 0}{R} = \cos\varphi, \\ \sin\vartheta &= \frac{R\sin\varphi}{R} = \sin\varphi, \\ \overline{z} &= k\frac{z}{R} = Kz. \end{aligned}$$

Substituting this into (3.11) and (3.12), we obtain

$$\begin{aligned} u &= R_p(1 - Kz)\cos\varphi, \\ v &= R_p(1 - Kz)\sin\varphi. \end{aligned}$$

## 4 Mosaicing panoramic images

In this section we show how to obtain a global "infinite" bitmap of the whole inner surface of a given pipeline. To find a correct shift between two subsequent nondeformed panoraic images, we choose a fixed window $A$ in the first image and in the second one, we assume first of all that the window $B = B(z)$ can be moved only in the $z$-direction. We look for a $\Delta z$ for which the image $B(\Delta z)$ is most similar to the one in the window $A$ (see Fig. 8).

Similarity between these two windows can be established by the *correlation function*, which is defined by the formula

$$c(A,B) = \frac{\sum_{i=1}^{m}\sum_{j=1}^{n}\left(A_{ij}-\overline{A}\right)\left(B_{ij}-\overline{B}\right)}{\sqrt{\left(\sum_{i=1}^{m}\sum_{j=1}^{n}\left(A_{ij}-\overline{A}\right)^2\right)\left(\sum_{i=1}^{m}\sum_{j=1}^{n}\left(B_{ij}-\overline{B}\right)^2\right)}},$$

where $A = (A_{ij})$ and $B = (B_{ij})$ are images of correlation windows defined by digital data $A_{ij}, B_{ij} \in \{0, 1, \ldots, 255\}$,

$$\overline{A} = \frac{1}{mn}\sum_{i=1}^{m}\sum_{j=1}^{n} A_{ij}, \quad \overline{B} = \frac{1}{mn}\sum_{i=1}^{m}\sum_{j=1}^{n} B_{ij},$$

and $m$, $n$ are dimensions of the correlation windows.

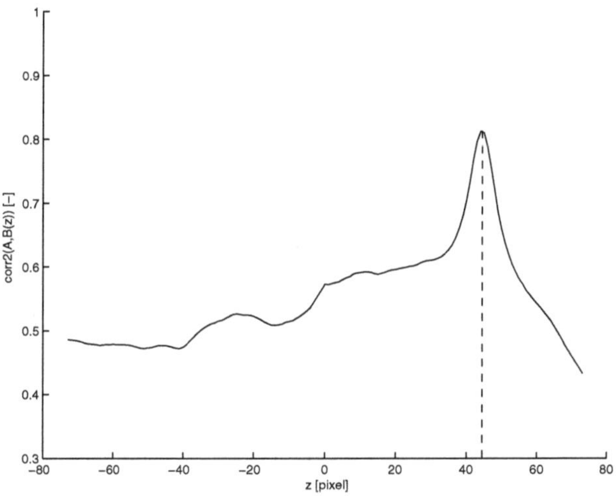


Figure 9. The graph of a correlation function $c = c(A, B(z))$.

The correct shift $\Delta z$ between two subsequent panoramic images is obtained when the correlation function attains its global maximum (see Fig. 9). This was found by sophisticated iterative methods (see [3]). Notice that the correlation function may have several local maxima.

Since the robot does not move, in general, along a straight line, we assumed that the window $B$ depends also on the angle $\varphi$, i.e., $B = B(\varphi, z)$. In this case we had to maximize the correlation function depending on two variables $\varphi$ and $z$, which takes a lot of arithmetical operations. To speed up the evaluation of $c(A, B)$, we employed pyramidal data structures introduced in [6]. The computational complexity of this procedure is described in [3]. Fig. 10 shows the graph of the associated correlation function $c = c(A, B(\varphi, z))$. Colour pictures of several reconstructed inner surfaces of the pipeline are shown in [3], which is available in web site [w1].

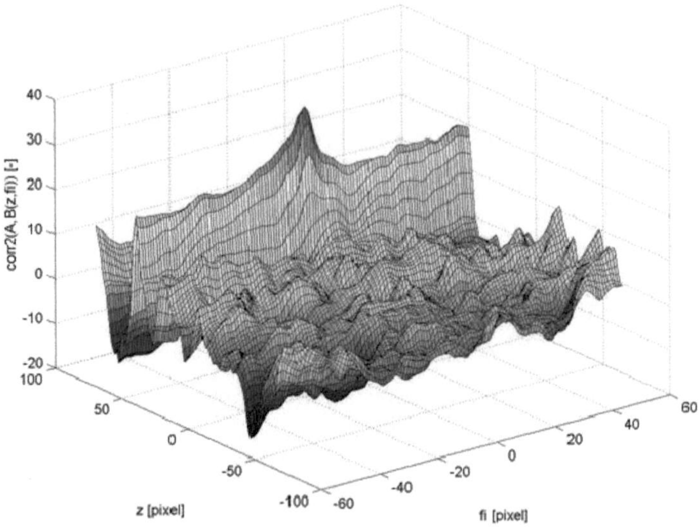


Figure 10. The graph of a continuous and piecewise bilinear correlation function $c = c(A, B(\varphi, z))$.

**Acknowledgement.** This paper was supported by grant No. 102/01/0971 of the Grant Agency of the Czech Republic. The author is indebted to Prof. Václav Hlaváč (Prague) and Prof. Lawrence Somer (Washington, D.C.) for fruitful discussions.

## References

[1] M. R. Hestenes, E. Stiefel, *Methods of conjugate gradients for solving linear systems*, J. Res. Nat. Bur. Standards **49** (1952), 409–436.

[2] M. Křížek, P. Neittaanmäki, *Finite Element Approximation of Variational Problems and Applications*, Longman Scientific & Technical, Harlow, 1990.

[3] P. Křížek, *Mosaicing of panoramic images*, Master Thesis, Faculty of Electrical Engineering, Prague, 2003, 1–62.

[4] P. Křížek, V. Smutný, *Mosaicing unwarped panoramic images into an infinite bitmap*, Research Report CTU-CMP-2002-21, Centre for Machine Perception, Czech Technical University, Prague (2002), 1–29.

[5] K. Rektorys, *Survey of Applicable Mathematics*, Kluwer, Amsterdam, 1994.

[6] M. Šonka, V. Hlaváč, R. Boyle, *Image Processing Analysis and Machine Vision*, 2nd ed., PWS, Boston, 1998.

[w1] `http://cmp.felk.cvut.cz/~krizekp1/`

# Finite Element Discretization and Iterative Solution Techniques for Multiphase Flows in Gas-Liquid Reactors

Dmitri Kuzmin,[1] Stefan Turek[2]

[1]Institute of Applied Mathematics, University of Dortmund (LS III), Vogelpothsweg 87, D–44227 Dortmund, Germany
kuzmin@math.uni-dortmund.de
[2]Institute of Applied Mathematics, University of Dortmund (LS III), Vogelpothsweg 87, D–44227 Dortmund, Germany ture@featflow.de

## 1 Introduction

Bubble columns and airlift loop reactors are widely used in industry as contacting devices in which gaseous and liquid species are brought together to engage in chemical reactions. The liquid supplied continuously or in a batch mode is agitated by bubbles fed at the bottom of the reactor. As the bubbles rise, the gaseous component is gradually absorbed into the bulk liquid where it may react with other species. The geometric simplicity of bubble columns makes them rather easy to build, operate and maintain. At the same time, the prevailing flow patterns are very complex and unpredictable, which represents a major bottleneck for the design of industrial units. In airlift loop reactors, internal parts are installed in order to enforce a stable circulation of liquid characterized by the presence of pronounced *riser* and *downcomer* zones. Shape optimization appears to be a promising way to improve the reactor performance by adjusting the geometry of internals.

In this paper, we touch upon the mathematical modeling of gas-liquid flows and propose a numerical algorithm based on the finite element method. A hierarchy of successively refined unstructured meshes is used to deal with complex domains. The velocity-pressure coupling via the incompressibility constraint is imposed in the framework of the Multilevel Pressure Schur Complement formulation originally developed for single-phase flows [29]. The instability of convective terms poses a hazard to the positivity of gas holdup and concentrations. Standard remedies like upwinding degrade the accuracy of the numerical solution and produce unacceptable results. The only way to obtain a sharp resolution of steep fronts without generating wiggles is to use a nonlinear combination of high- and low-order methods. The main contribution of this paper is a truly multidimensional generalization of high-resolution

TVD schemes to finite element discretizations on arbitrary meshes. Alternatively, convective terms can be handled by the implicit FEM-FCT methodology which is presented in detail elsewhere [11], [13], [14]. Furthermore, we discuss some other algorithmic aspects relevant for simulation of gas-liquid flows. Numerical examples are presented to give a flavor of feasible applications. However, a detailed analysis of the model and of the obtained results is beyond the scope of this paper.

## 2 Mathematical model

Detailed hydrodynamic models for dispersed two-phase flow can be classified into those of Euler-Euler and Euler-Lagrange type. The former approach implies that both phases are treated as space-sharing interpenetrating continua. The macroscopic conservation laws can be postulated using some heuristic arguments or derived mathematically by applying a suitable averaging procedure to the associated single-phase continuity and momentum equations [4], [5]. In the Euler-Lagrange formulation, only the liquid phase is considered to be continuous, while individual bubbles or bubble clusters are tracked in a Lagrangian way [15], [25]. The most detailed treatment of gas-liquid flow is provided by direct numerical simulation (DNS), which amounts to solving a formidable free boundary problem for the deformation of all bubbles and interactions between them. At present, this is feasible only for a very limited number of bubbles. Nevertheless, DNS contributes to the understanding of processes taking place at the microscopic level and constitutes a valuable tool for the derivation of input parameters and correlations for less sophisticated CFD models.

In this paper, we subscribe to the Euler-Euler approach whereby the computational cost is largely independent of the number of bubbles present in the flow field. The instantaneous distribution of phases in the reactor is determined by the local gas holdup $\alpha_G = \varepsilon$. The volume fraction of liquid equals $\alpha_L = 1 - \varepsilon$. The macroscopic continuity equations

$$\frac{\partial \tilde{\rho}_G}{\partial t} + \nabla \cdot (\tilde{\rho}_G \mathbf{u}_G) = -m_{\text{int}}, \quad \tilde{\rho}_G = \alpha_G \rho_G, \tag{1}$$

$$\frac{\partial \tilde{\rho}_L}{\partial t} + \nabla \cdot (\tilde{\rho}_L \mathbf{u}_L) = m_{\text{int}}, \quad \tilde{\rho}_L = \alpha_L \rho_L \tag{2}$$

are coupled by the mass transfer term $m_{\text{int}}$ which models the absorption of gas into the liquid. Here and below the tilde denotes multiplication by the volume fraction of the respective phase. The macroscopic momentum equations read

$$\tilde{\rho}_G \left[ \frac{\partial \mathbf{u}_G}{\partial t} + \mathbf{u}_G \cdot \nabla \mathbf{u}_G \right] = \alpha_G \nabla \cdot \mathcal{S}_G + \tilde{\rho}_G \mathbf{g} + \mathbf{f}_{\text{int}}, \tag{3}$$

$$\tilde{\rho}_L \left[ \frac{\partial \mathbf{u}_L}{\partial t} + \mathbf{u}_L \cdot \nabla \mathbf{u}_L \right] = \alpha_L \nabla \cdot \mathcal{S}_L + \tilde{\rho}_L \mathbf{g} - \mathbf{f}_{\text{int}}, \tag{4}$$

where $\mathcal{S}$ refers to the stress tensor and the coupling term $\mathbf{f}_{\text{int}}$ in both right-hand sides is responsible for the momentum exchange at the gas-liquid interface.

The above system of equations is known as the *two-fluid model*, which is widespread in the CFD community. It provides a general description of the two-phase flow but is practically worthless as long as the terms $m_{\text{int}}$, $\mathbf{f}_{\text{int}}$, $\mathcal{S}_G$, $\mathcal{S}_L$ are left unspecified. The derivation of appropriate closure relations is a challenging task due to various interactions between the two phases. In what follows, we present constituitive equations suitable for buoyancy-driven bubbly flows with mass transfer and chemical reaction.

## 2.1 Interphase mass transfer

The mass transfer rate is proportional to the difference between the concentration $c_A$ of the dissolved gas and an equilibrium concentration $c_A^*$. The latter is related to the local pressure $p$ via Henry's law $p = Hc_A^*$, where $H$ is an empirical coefficient depending on the solubility of gas. The total flux of gas into the liquid is given by

$$m_{\text{int}} = k_L a_S (c_A^* - c_A)\eta, \tag{5}$$

where $a_S$ stands for the specific interfacial area, $\eta$ denotes the molecular weight of gaseous species and $k_L$ is the liquid-side mass transfer coefficient. As a rule, the resistance to mass transfer on the gas side can be neglected.

Chemical reactions in the liquid phase can substantially accelerate the absorption process by increasing the concentration gradient. The concomitant increase in the gas flux is taken into account by applying an enhancement factor $E$ to the physical mass transfer coefficient:

$$k_L = E k_L^0, \tag{6}$$

where $k_L^0$ is a function of the Sherwood number Sh which varies with the local Peclet number Pe. For a first-order irreversible reaction, the standard film theory yields [22]

$$E = \text{Ha} \coth \text{Ha}, \tag{7}$$

where Ha is the Hatta number. For vanishing Ha, the hyperbolic tangent tends to zero, so that the above equation should be replaced by the series

$$E = 1 + \frac{1}{3}\text{Ha}^2 - \frac{1}{45}\text{Ha}^4 + \frac{2}{945}\text{Ha}^6 - \frac{1}{4725}\text{Ha}^8 + \dots \tag{8}$$

The enhancement factor for second-order reactions $A + \nu_B B \to \nu_P P$ can be derived analytically only in a few limiting cases, e.g. for an instantaneous irreversible reaction:

$$E_i = 1 + \frac{d_B c_B}{\nu_B d_A c_A^*}, \tag{9}$$

$d_A$ and $d_B$ being the molecular diffusion coefficients. For slow and intermediate reactions, equations of the film model have to be solved approximately. A widely used correlation of numerical results was introduced by Van Krevelen and Hoftijzer [31]:

$$E = M \coth M, \qquad M = \mathrm{Ha}\sqrt{\frac{E_i - E}{E_i - 1}}. \tag{10}$$

Note that this equation is nonlinear and has to be solved iteratively or approximated by a simpler formula due to Wellek *et al.* [32]

$$\frac{1}{(E-1)^{1.35}} = \frac{1}{(E_i-1)^{1.35}} + \frac{1}{(E_1-1)^{1.35}}, \qquad E_1 = \mathrm{Ha}\coth \mathrm{Ha}. \tag{11}$$

Recall that the mass transfer term (5) contains the interfacial area $a_S$ which still is to be defined. This key variable depends on the number and size of bubbles which may vary in space and time. If there is no coalescence and breakup of bubbles, their number density is conserved and satisfies the continuity equation

$$\frac{\partial n}{\partial t} + \nabla \cdot (n\mathbf{u}_G) = 0. \tag{12}$$

Let us assume that the gas-liquid mixture is locally monodisperse, i.e. the bubbles are spherical with an average radius $r$. As long as $n$ represents the number of bubbles per unit volume, multiplying it by the volume of a single bubble yields the local gas holdup $\varepsilon = \frac{4}{3}\pi r^3 n$. Hence, the specific interfacial area $a_S = 4\pi r^2 n$ is given by

$$a_S = (4\pi n)^{1/3}(3\varepsilon)^{2/3}, \qquad \text{where } \varepsilon = \frac{\tilde\rho_G}{\rho_G} = \tilde\rho_G \frac{RT}{p\eta}. \tag{13}$$

Here the physical density $\rho_G$ is evaluated using the ideal gas law. The effective density $\tilde\rho_G$ is computed from the continuity equation (1) for the gas phase.

### 2.2 Chemical reactions

The concentrations of all species dissolved in the liquid phase are described by a system of scalar transport equations with source and/or sink terms due to mass transfer and chemical reactions. For homogeneous reactions of second order it reads

$$\frac{\partial \tilde c_A}{\partial t} + \nabla \cdot (\tilde c_A \mathbf{u}_L - \tilde d_A \nabla c_A) = \frac{m_{\mathrm{int}}}{\eta} - \tilde k_2 c_A c_B, \tag{14}$$

$$\frac{\partial \tilde c_B}{\partial t} + \nabla \cdot (\tilde c_B \mathbf{u}_L - \tilde d_B \nabla c_B) = -\nu_B \tilde k_2 c_A c_B, \tag{15}$$

$$\frac{\partial \tilde c_P}{\partial t} + \nabla \cdot (\tilde c_P \mathbf{u}_L - \tilde d_P \nabla c_P) = \nu_P \tilde k_2 c_A c_B, \tag{16}$$

where the subscript $A$ denotes the dissolved gas, $B$ the component it reacts with and $P$ the resulting product. The effective diffusion coefficients $\tilde{d}_A$, $\tilde{d}_B$, $\tilde{d}_P$ may involve contributions of the turbulent dispersion in the liquid phase.

If the reactant $B$ is present in excess, its virtually constant concentration can be built into the reaction rate $k_1 = k_2 c_B$, and the reaction is said to be a pseudo-first-order one. In this case, the system of mass balances shrinks to

$$\frac{\partial \tilde{c}_A}{\partial t} + \nabla \cdot (\tilde{c}_A \mathbf{u}_L - \tilde{d}_A \nabla c_A) = \frac{m_{\text{int}}}{\eta} - \tilde{k}_1 c_A, \tag{17}$$

$$\frac{\partial \tilde{c}_P}{\partial t} + \nabla \cdot (\tilde{c}_P \mathbf{u}_L - \tilde{d}_P \nabla c_P) = \nu_P \tilde{k}_1 c_A, \tag{18}$$

and the enhancement factor is defined by equation (7) rather than (10) or (11). In any event, the concentration fields have little or no effect on the liquid velocity, so the above transport equations are essentially decoupled from the rest of the model.

### 2.3 Closure of the momentum equations

Both phases are considered to be Newtonian fluids sharing the same pressure field. Hence, the stress tensor for either phase can be represented as

$$\mathcal{S}(\mathbf{u}) = -p\mathcal{I} + \mu_{\text{eff}} \mathcal{D}(\mathbf{u}), \tag{19}$$

where the pressure $p$ and the deformation rate tensor $\mathcal{D}$ are defined by

$$p = \rho_G \frac{R}{\eta} T, \qquad \mathcal{D}(\mathbf{u}) = \frac{1}{2}[\nabla \mathbf{u} + \nabla \mathbf{u}^T].$$

For simplicity, the effective viscosity $\mu_{\text{eff}}$ is assumed to be constant. In general, it consists of the laminar viscosity $\mu$ depending solely on the physical properties of the fluid and the turbulent eddy viscosity $\mu_T$ modeling the effect of microscopic velocity fluctuations which are eliminated during the averaging process. The standard $k$-$\varepsilon$ turbulence model seems to perform reasonably well [2], [24] and can be extended to allow for the bubble-induced turbulence [9], [19]. Nevertheless, turbulence modeling for two-phase flows remains an open question and no reliable model is available to date.

The coupling term $\mathbf{f}_{\text{int}}$ in the balance equations (3) and (4) represents the interphase momentum transfer due to forces exerted by the liquid phase on the rising bubbles. In spite of the ongoing controversy in the literature, it is generally accepted that $\mathbf{f}_{\text{int}}$ involves contributions from the drag force, the virtual mass force and the lift force [5]:

$$\mathbf{f}_{\text{int}} = \mathbf{f}_D + \mathbf{f}_{\text{VM}} + \mathbf{f}_L. \tag{20}$$

By far the most important constituent is the drag force $\mathbf{f}_D$ experienced by a bubble as it moves steadily in the surrounding liquid. The ensuing interphase friction is given by

$$\mathbf{f}_D = -\varepsilon C_D \frac{3}{8} \frac{\rho_L}{r} |\mathbf{u}_G - \mathbf{u}_L| (\mathbf{u}_G - \mathbf{u}_L), \tag{21}$$

where $C_D$ is a dimensionless drag coefficient which can be assessed by measuring the terminal rise velocity of a single bubble. Numerous empirical correlations are available for $C_D$ as a function of the Reynolds number. Note that the drag force is proportional to the relative velocity so that the bubbles slow down whereas the liquid gains momentum.

The 'added mass' of liquid entrained in the wake of accelerating bubbles results in a virtual mass force of the form

$$\mathbf{f}_{\rm VM} = -\varepsilon C_{\rm VM} \rho_L \left( \frac{d\mathbf{u}_G}{dt} - \frac{d\mathbf{u}_L}{dt} \right). \tag{22}$$

Depending on the properties of the gas-liquid mixture, the virtual mass coefficient $C_{\rm VM}$ may deviate appreciably from the default value $C_{\rm VM} = 0.5$ valid for spherical particles under some idealistic assumptions.

Additional forces transverse to the direction of motion are typically represented by

$$\mathbf{f}_L = -\varepsilon C_L \rho_L (\mathbf{u}_G - \mathbf{u}_L) \times (\nabla \times \mathbf{u}_L), \tag{23}$$

where the lift coefficient $C_L$ equals 0.25 for dilute flows of spheres.

## 2.4 Algebraic slip relation

The implications of the three contributions to the interphase force were analyzed in detail by Sokolichin and Eigenberger [23]. They demonstrated that the influence of the virtual mass force $\mathbf{f}_{\rm VM}$ on the bubble motion can be neglected. At the same time, they found the lift force to be rather significant but questioned its existence in the form (23) which was originally derived for solid particles. In the literature, $\mathbf{f}_L$ is often misused for fitting the results to experimental data. In some publications even the sign of the lift coefficient is reversed. Sokolichin and Eigenberger condemned this practice and argued that the lift force should be omitted as long as its origins, form and magnitude are unclear.

The interphase momentum transfer is dominated by the drag force $\mathbf{f}_D$ and the above arguments indicate that it is the only one which needs to be taken into account. Moreover, the density of gas is much smaller than that of liquid, so that the inertia and gravity terms in the gas phase momentum balance (3) can be neglected [23]. Assuming the gas phase to be inviscid, we insert the stress tensor $\mathcal{S}_G = p\mathcal{I}$ and obtain the equilibrium relation

$$0 = -\varepsilon \nabla p + \mathbf{f}_D, \tag{24}$$

which can be used to compute the relative bubble velocity $\mathbf{u}_{\rm slip}$. To this end, it is expedient to linearize the drag force as proposed by Schwarz and Turner [21]:

$$\mathbf{f}_D = -\varepsilon C_D \frac{3}{8} \frac{\rho_L}{r} |\mathbf{u}_{\rm slip}| \mathbf{u}_{\rm slip} \approx -\varepsilon C_W \mathbf{u}_{\rm slip} \Rightarrow \mathbf{u}_{\rm slip} = -\frac{\nabla p}{C_W}. \tag{25}$$

Unlike the standard drag coefficient $C_D$, the linearization parameter $C_W$ is a dimensional quantity. It is commonly assigned the constant value $5 \cdot 10^4$ $\mathrm{kg\,m^{-3}\,s^{-1}}$ which corresponds to the bubble rise velocity of approximately 20 $\mathrm{cm\,s^{-1}}$.

The above model simplifications make it possible to replace the gas phase momentum balance by the algebraic slip relation

$$\mathbf{u}_G = \mathbf{u}_L + \mathbf{u}_{\mathrm{slip}} + \mathbf{u}_{\mathrm{drift}}, \tag{26}$$

where the drift velocity $\mathbf{u}_{\mathrm{drift}}$ was introduced to model the transport of bubbles by turbulent eddies. The net flux of gas due to turbulent dispersion is proportional to the gradient of the number density and directed opposite to it. The corresponding drift velocity reads

$$\mathbf{u}_{\mathrm{drift}} = -d_G \frac{\nabla n}{n}. \tag{27}$$

In general, the bubble path dispersion coefficient $d_G$ depends on the eddy viscosity $\mu_T$ and on the turbulent Schmidt number [23].

If the velocity $\mathbf{u}_G$ is computed from the slip relation (26) making use of (27), the gas phase continuity equations (1) and (12) can be rewritten the form

$$\frac{\partial \tilde{\rho}_G}{\partial t} + \nabla \cdot (\tilde{\rho}_G(\mathbf{u}_L + \mathbf{u}_{\mathrm{slip}}) - d_G \nabla \tilde{\rho}_G) = -m_{\mathrm{int}} \tag{28}$$

and

$$\frac{\partial n}{\partial t} + \nabla \cdot (n(\mathbf{u}_L + \mathbf{u}_{\mathrm{slip}}) - d_G \nabla n) = 0, \tag{29}$$

respectively. Strictly speaking, the diffusive term in (28) is valid only for a constant bubble mass but this form can be justified if one substitutes the drift velocity after decoupling the transport and mass transfer (see the description of operator splitting tools below).

## 2.5 Drift-flux model

The two-fluid model (1)–(4) is not the only way to describe dispersed two-phase flows using a fully continuous representation. Alternatively, the gas-liquid mixture can be considered as a single fluid whose effective density and momentum are defined by

$$\rho = \tilde{\rho}_L + \tilde{\rho}_G, \qquad \rho \mathbf{u} = \tilde{\rho}_L \mathbf{u}_L + \tilde{\rho}_G \mathbf{u}_G. \tag{30}$$

The weakly compressible Navier-Stokes equations for these quantities constitute the so-called drift-flux model. As soon as the mixture velocity $\mathbf{u}$ is available, the velocities of both phases can be recovered using the algebraic slip relation (26).

For bubbly flows, the two-fluid model and the drift-flux model are essentially equivalent [23]. Indeed, the sum of phasic continuity equations (1) and (2) yields

$$\frac{\partial \rho}{\partial t} + \nabla \cdot (\rho \mathbf{u}) = 0. \tag{31}$$

Furthermore, if the simplified gas phase momentum balance (24) is added to equation (4) with the stress tensor $\mathcal{S}_L$ given by (19), the interphase force term vanishes and we obtain

$$\tilde{\rho}_L \left[ \frac{\partial \mathbf{u}_L}{\partial t} + \mathbf{u}_L \cdot \nabla \mathbf{u}_L \right] = -\nabla p + \nabla \cdot (\tilde{\mu}_{\text{eff}} \mathcal{D}(\mathbf{u}_L)) + \tilde{\rho}_L \mathbf{g}. \tag{32}$$

Recall that the gas density is very small as compared to that of liquid. Therefore, its contribution to the mixture density and momentum can be neglected:

$$\rho \approx \tilde{\rho}_L, \qquad \mathbf{u} \approx \mathbf{u}_L. \tag{33}$$

Under these assumptions, equation (32) reduces to the momentum balance for the mixture.

## 2.6 Boussinesq approximation

The effective density $\tilde{\rho}_L = (1-\varepsilon)\rho_L$ depends on the local gas holdup, which leads to liquid circulation due to the buoyancy in aerated regions. Similar processes occur in single-phase flows in the presence of temperature gradients. Following Sokolichin and Eigenberger [23], we use an analog of the Boussinesq approximation for natural convection problems and retain the variable density only in the gravity term of equation (32) dropping the tilde elsewhere. After these modifications, we recover the incompressible Navier-Stokes equations with an extra buoyancy term proportional to the gas holdup:

$$\frac{\partial \mathbf{u}_L}{\partial t} + \mathbf{u}_L \cdot \nabla \mathbf{u}_L = -\nabla p_* + \nu_{\text{eff}} \Delta \mathbf{u}_L - \varepsilon \mathbf{g}, \tag{34}$$

$$\nabla \cdot \mathbf{u}_L = 0, \qquad p_* = \frac{p - p_{\text{atm}}}{\rho_L} + \mathbf{g} \cdot \mathbf{x} - gh.$$

At high gas throughputs it is advisable to relax the Boussinesq approximation in the momentum equation and solve it in its original form (32). As far as the continuity equation is concerned, the constant density assumption can be retained. This model simplification is particularly advantageous from the computational viewpoint, since it eliminates the need for dealing with the evolution of the free surface on top of the reactor [23].

### 2.7 Summary of equations

As a bottomline, let us collect the equations to be solved. Our simplified Euler-Euler model of drift-flux type consists of the following subproblems:

- Navier-Stokes/'Boussinesq' equations

$$\frac{\partial \mathbf{u}_L}{\partial t} + \mathbf{u}_L \cdot \nabla \mathbf{u}_L = -\nabla p_* + \nu_{\text{eff}} \Delta \mathbf{u} - \varepsilon \mathbf{g}, \tag{35}$$
$$\nabla \cdot \mathbf{u}_L = 0.$$

- Continuity equations for the gas phase

$$\frac{\partial \tilde{\rho}_G}{\partial t} + \nabla \cdot (\tilde{\rho}_G(\mathbf{u}_L + \mathbf{u}_{\text{slip}}) - d_G \nabla \tilde{\rho}_G) = -m_{\text{int}}, \tag{36}$$

$$\frac{\partial n}{\partial t} + \nabla \cdot (n(\mathbf{u}_L + \mathbf{u}_{\text{slip}}) - d_G \nabla n) = 0. \tag{37}$$

- Transport equations for the liquid phase

$$\frac{\partial \tilde{c}_A}{\partial t} + \nabla \cdot (\tilde{c}_A \mathbf{u}_L - \tilde{d}_A \nabla c_A) = \frac{m_{\text{int}}}{\eta} - \tilde{k}_2 c_A c_B, \tag{38}$$

$$\frac{\partial \tilde{c}_B}{\partial t} + \nabla \cdot (\tilde{c}_B \mathbf{u}_L - \tilde{d}_B \nabla c_B) = -\nu_B \tilde{k}_2 c_A c_B, \tag{39}$$

$$\frac{\partial \tilde{c}_P}{\partial t} + \nabla \cdot (\tilde{c}_P \mathbf{u}_L - \tilde{d}_P \nabla c_P) = \nu_P \tilde{k}_2 c_A c_B. \tag{40}$$

These partial differential equations are supplemented by the algebraic closure relations

$$\mathbf{u}_{\text{slip}} = -\frac{\nabla p}{C_W}, \qquad m_{\text{int}} = E k_L^0 a_S \left(\frac{p}{H} - c_A\right)\eta,$$
$$\varepsilon = \frac{\tilde{\rho}_G RT}{p\eta}, \quad a_S = (4\pi n)^{1/3}(3\varepsilon)^{2/3}, \quad \nu_{\text{eff}} = \text{const}.$$

The problem statement is to be completed by specifying appropriate initial and boundary conditions which depend on the particular application.

## 3 Numerical algorithm

The main features of the numerical algorithm are summarized below. The discretization in space is performed by an unstructured grid finite element method in order to provide an accurate treatment of non-Cartesian geometries with internal obstacles. The incompressible Navier-Stokes equations call for the use of an LBB-stable finite element pair. A suitable candidate is the nonconforming Rannacher-Turek element (rotated multilinear velocity, piecewise-constant pressure) [20]. Standard multilinear elements are employed for other variables. The manually generated coarse mesh is successively refined to produce hierarchical data structures for the multigrid solver [27], [28].

The coupled subproblems are solved one at a time within a block-iterative loop of Gauss-Seidel type. A two-way coupling is implemented only for the hydrodynamic part $(\mathbf{u}_L, p, \varepsilon)$. We solve the Navier-Stokes equations using the gas holdup from the last outer iteration, then update the gas holdup and repeat this procedure until the residual of the momentum equation or the relative changes of the gas holdup distribution are small enough. The resulting velocity field is substituted into the transport equations (37)–(40) and remains fixed until the end of the time step.

Operator splitting is employed to separate convection-diffusion and absorption-reaction processes. First, all scalar quantities are transported without taking the sources/sinks into account. The homogeneous equations (41)–(44) are decoupled and can be processed in parallel. The updated concentration fields serve as initial data for an ODE system which describes the accumulation and consumption of species. Equations (45)–(48) are nonlinear and strongly coupled but the nodal ODE systems are independent of one another.

*Step 1: convection-diffusion*

$$\frac{\partial \tilde{\rho}_G}{\partial t} + \nabla \cdot (\tilde{\rho}_G(\mathbf{u}_L + \mathbf{u}_{\text{slip}}) - d_G \nabla \tilde{\rho}_G) = 0, \tag{41}$$

$$\frac{\partial \tilde{c}_A}{\partial t} + \nabla \cdot (\tilde{c}_A \mathbf{u}_L - \tilde{d}_A \nabla c_A) = 0, \tag{42}$$

$$\frac{\partial \tilde{c}_B}{\partial t} + \nabla \cdot (\tilde{c}_B \mathbf{u}_L - \tilde{d}_B \nabla c_B) = 0, \tag{43}$$

$$\frac{\partial \tilde{c}_P}{\partial t} + \nabla \cdot (\tilde{c}_P \mathbf{u}_L - \tilde{d}_P \nabla c_P) = 0. \tag{44}$$

*Step 2: absorption-reaction*

$$\frac{\mathrm{d}\tilde{\rho}_G}{\mathrm{d}t} = -m_{\text{int}}, \tag{45}$$

$$\frac{\mathrm{d}\tilde{c}_A}{\mathrm{d}t} = \frac{m_{\text{int}}}{\eta} - \tilde{k}_2 c_A c_B, \tag{46}$$

$$\frac{\mathrm{d}\tilde{c}_B}{\mathrm{d}t} = -\nu_B \tilde{k}_2 c_A c_B, \tag{47}$$

$$\frac{\mathrm{d}\tilde{c}_P}{\mathrm{d}t} = \nu_P \tilde{k}_2 c_A c_B. \tag{48}$$

Note that operator splitting is applied locally in time, i.e. within each time step. The solution is integrated in time from $t_n$ to $t_{n+1} = t_n + \Delta t_n$ by the following algorithm:

1. Solve the Navier-Stokes equations (35) for $\mathbf{u}_L$ and $p$.
2. Recover the pressure gradient $\nabla p$ using an $L_2$-projection.
3. Calculate the slip velocity as $\mathbf{u}_{\text{slip}} = -\frac{\nabla p}{C_W}$.
4. Advance $\tilde{\rho}_G$ using (41) and recompute $\varepsilon = \frac{\tilde{\rho}_G RT}{p\eta}$.
5. Substitute $\varepsilon$ into (35) and evaluate the residual.
6. If converged then proceed to 7, otherwise go to 1.
7. Update the number density $n$ according to (37).
8. Solve the homogeneous transport equations (42)–(44).
9. Compute the interfacial area $a_S = (4\pi n)^{1/3}(3\varepsilon)^{2/3}$.
10. Solve the ODE system (45)–(48) for each node.

An implicit time discretization of Crank-Nicolson or backward Euler type is employed for all equations. The value of the implicitness parameter $\theta$ and of the local time step can be selected individually for each subproblem so as to maximize accuracy and/or stability. This provides an efficient treatment of physical processes occurring simultaneously but on entirely different time scales. The communication between the subproblem blocks takes place at the end of the common macro time step $\Delta t_n$ which is chosen adaptively so as to control the changes of the gas holdup distribution (see below). The linear systems are solved using preconditioned BiCGSTAB or multigrid methods. In what follows, we elucidate some constituents of the numerical algorithm in detail.

### 3.1 Treatment of incompressibility

Boussinesq system (35) is solved using the academic code FEATFLOW [27]. This open-source software package represents a collection of robust discretization tools and optimized multigrid solvers for incompressible flow problems. Below, we briefly outline the Pressure Schur Complement method adopted for gas-liquid flow simulations.

The fully discretized Navier-Stokes equations represent a saddle point problem, in which the pressure acts as a Lagrange multiplier for the incompressibility constraint:

$$\begin{bmatrix} S & \Delta t B \\ B^T & 0 \end{bmatrix} \begin{bmatrix} \mathbf{u} \\ p \end{bmatrix} = \begin{bmatrix} \mathbf{f} \\ 0 \end{bmatrix}. \tag{49}$$

The velocity can be formally written as $\mathbf{u} = S^{-1}(\mathbf{f} - \Delta t B p)$ and plugged into the discrete continuity equation $B^T\mathbf{u} = 0$ to derive the missing equation for the pressure:

$$B^T S^{-1} B p = \Delta t^{-1} B^T S^{-1} \mathbf{f}. \tag{50}$$

This discrete problem can be solved by the preconditioned Richardson iteration [26]

$$p^{(l+1)} = p^{(l)} + [B^T M_L^{-1} B]^{-1} \Delta t^{-1} B^T S^{-1} (\mathbf{f} - \Delta t B p^{(l)}), \qquad (51)$$

where $l$ is the outer iteration counter and $M_L$ is the lumped mass matrix which proves to be a reasonable approximation to the evolution operator $S$ at high Reynolds numbers.

Each Pressure Schur Complement iteration can be interpreted and implemented as a projection cycle which consists of the following algorithmic steps:

1. Compute the velocity $\tilde{\mathbf{u}}$ from the Burgers equation
$$S\,\tilde{\mathbf{u}} = \mathbf{f} - \Delta t B p^{(l)}$$
2. Solve the discrete 'Pressure-Poisson' problem
$$B^T M_L^{-1} B q = \Delta t^{-1} B^T \tilde{\mathbf{u}}$$
3. Correct the pressure and the intermediate velocity
$$p^{(l+1)} = p^{(l)} + q, \quad \mathbf{u}^{(l+1)} = \tilde{\mathbf{u}} - \Delta t M_L^{-1} B q$$

In a nutshell, the right-hand side of the momentum equation is assembled using the old pressure iterate and the resulting intermediate velocity $\tilde{\mathbf{u}}$ is projected onto the subspace of divergence-free functions so as to satisfy the condition $B^T \mathbf{u}^{(l+1)} = 0$. The Pressure Schur Complement approach constitutes a very general framework which unites coupled solution techniques and classical projection schemes, the degree of coupling being determined by the number of outer iterations. For details the interested reader is referred to [29].

### 3.2 Treatment of convection

Convection is notoriously difficult to treat numerically. The standard Galerkin method is a 'centered' scheme which gives rise to an unstable discretization of convective terms. A common remedy is to add streamline diffusion which provides the necessary stabilization without reducing the order of approximation. However, even stabilized high-order methods tend to produce nonphysical undershoots and overshoots in the vicinity of steep gradients. As a result, negative gas holdups or concentrations may arise, which is clearly unacceptable. It is possible to get rid of oscillations by adding adaptive artificial diffusion depending on the local solution behavior. The high-order scheme can be used in smooth regions but near discontinuities it should be replaced by a low-order scheme like 'upwind' which is diffusive enough to prevent the formation of wiggles.

The first discretization procedure to utilize the idea of adaptive switching between high- and low-order methods was the flux-corrected-transport algorithm introduced in the early 1970s by Boris and Book [3]. The state-of-the-art generalization proposed by Zalesak [33] has made it possible to incorporate FCT into unstructured grid methods. The foundations of flux correction for finite elements were laid by Löhner *et al.* [17]. In a series of recent publications [11], [13], [14] we refined the FEM-FCT methodology and extended it to implicit time stepping. Unfortunately, Zalesak's limiter leads to some loss of accuracy for large time steps, which compromises the advantages of implicit schemes.

A class of high-resolution methods with artificial diffusion independent of the time step was developed by Harten [7]. His total variation diminishing schemes and extensions thereof rest on a firm mathematical basis and are widely used in CFD computations. At the same time, it has been largely unclear how to apply them in the finite element context. In the scarce publications on that subject, TVD-like artificial viscosities were designed using some reconstruction of a local 1D stencil [18]. The use of $P_1$ elements was essential to the derivation of the underlying edge-based data structure. Below we propose an alternative approach which is applicable to arbitrary Galerkin discretizations. The (anti-)diffusive terms are represented as an array of internodal fluxes. A complete transition to an edge-based data structure is feasible [14] but not mandatory.

### 3.2.1 Galerkin discretization

Consider a time-dependent conservation law for a scalar quantity $u$:

$$\frac{\partial u}{\partial t} + \nabla \cdot (\mathbf{f} - d\nabla u) = 0 \quad \text{in } \Omega, \tag{52}$$

where $\mathbf{f} = \mathbf{v}u$ stands for the convective flux and $d$ denotes the diffusion coefficient. All of our homogeneous transport equations (37), (41)–(44) can be written in this form.

The concomitant boundary conditions are specified in terms of normal fluxes:

$$\begin{aligned} -\mathbf{n} \cdot (\mathbf{f} - d\nabla u) &= g \quad \text{on } \Gamma_1, \\ \mathbf{n} \cdot (\mathbf{f} - d\nabla u) &= 0 \quad \text{on } \Gamma_2, \\ \mathbf{n} \cdot \nabla u &= 0 \quad \text{on } \Gamma_3, \end{aligned}$$

where $\mathbf{n}$ denotes the outward unit normal. On the feed boundary $\Gamma_1$ the incoming flux is determined by the operating conditions. In particular, the total gas throughput divided by the aerated area yields the value of $g$ for the continuity equation (41). Obviously, there is no flux through the solid wall $\Gamma_2$. On the outflow boundary $\Gamma_3$ the diffusive flux is cancelled while the convective one is left free as required by the hyperbolic limit of pure convection ($d = 0$). It is implied that $\mathbf{v} \cdot \mathbf{n} < 0$ on $\Gamma_1$ while $\mathbf{v} \cdot \mathbf{n} > 0$ on $\Gamma_3$.

Using the divergence theorem to integrate the spatial derivatives in the weak form of equation (52) by parts and inserting the natural boundary conditions, we obtain

$$\int_\Omega w \frac{\partial u}{\partial t} \,\mathrm{d}\mathbf{x} - \int_\Omega \nabla w \cdot (\mathbf{f} - d\nabla u) \,\mathrm{d}\mathbf{x} + \int_{\Gamma_3} w \mathbf{f} \cdot \mathbf{n} \,\mathrm{d}s = \int_{\Gamma_1} wg \,\mathrm{d}s. \tag{53}$$

A common practice in finite element methods for conservation laws is to interpolate the fluxes in the same way as the approximate solution:

$$u_h = \sum_j u_j \varphi_j, \qquad \mathbf{f}_h = \sum_j \mathbf{f}_j \varphi_j = \sum_j (\mathbf{v}_j u_j) \varphi_j, \tag{54}$$

where $\varphi_i$ are the basis functions spanning the finite-dimensional subspace. The resulting Galerkin discretization of equation (52) reads

$$\sum_j \left[ \int_\Omega \varphi_i \varphi_j \,\mathrm{d}\mathbf{x} \right] \frac{\mathrm{d}u_j}{\mathrm{d}t} - \sum_j \left[ \int_\Omega \nabla \varphi_i \cdot (\mathbf{v}_j \varphi_j - \mathrm{d}\nabla \varphi_j) \,\mathrm{d}\mathbf{x} \right] u_j \tag{55}$$
$$+ \sum_j \left[ \int_{\Gamma_3} \varphi_i \varphi_j \mathbf{v}_j \cdot \mathbf{n} \,\mathrm{d}s \right] u_j = \int_{\Gamma_1} \varphi_i g \,\mathrm{d}s.$$

This ODE system can be written in compact matrix form as follows

$$M_C \frac{\mathrm{d}u}{\mathrm{d}t} = (K - B)u + q, \tag{56}$$

where $M_C$ is the consistent mass matrix, $K$ is the discrete transport operator, $B$ is the contribution of the surface integral over the Neumann boundary and $q$ is the source term due to incoming fluxes. The matrix entries are given by

$$m_{ij} = \int_\Omega \varphi_i \varphi_j \,\mathrm{d}\mathbf{x}, \quad k_{ij} = \mathbf{v}_j \cdot \mathbf{c}_{ij} - \mathrm{d}s_{ij}, \tag{57}$$

where $\mathbf{c}_{ij}$ and $s_{ij}$ stand for the constant coefficients

$$\mathbf{c}_{ij} = \int_\Omega \nabla \varphi_i \varphi_j \,\mathrm{d}\mathbf{x}, \quad s_{ij} = \int_\Omega \nabla \varphi_i \cdot \nabla \varphi_j \,\mathrm{d}\mathbf{x}. \tag{58}$$

corresponding to the first- and second-order derivatives, respectively. For customary finite elements, the sum of basis functions is identically one so that the sum of their derivatives vanishes. Hence, the above coefficient matrices have zero column sums and so does the discrete transport operator: $\sum_i k_{ij} = \mathbf{v}_j \cdot \sum_i \mathbf{c}_{ij} - d\sum_i s_{ij} = 0$.

Note that the differential operators defined by equation (58) do not change and need to be assembled just once. As long as they are available, the matrix $K$ can be updated in an efficient way by computing the off-diagonal entries $k_{ij}$ from the formula (57) without resorting to costly numerical integration. Due to the zero column sum property, the diagonal matrix entries can be recovered as $k_{jj} = -\sum_{i\neq j} k_{ij}$.

The boundary terms are assembled separately. It is expedient to calculate the surface integrals approximately using a one-point quadrature rule. For each boundary edge/face $\partial E_k$ we evaluate the normal flux at the midpoint, multiply it by the respective area $|\partial E_k|$ and distribute between the nodes. This leads to a diagonal matrix $B$ with entries

$$b_i = \frac{1}{m} \sum_{k \in \mathcal{J}_1^i} \mathbf{v}_i \cdot \mathbf{n}_k |\partial E_k|, \quad \mathcal{J}_1^i = \{j \colon \mathbf{x}_i \in \partial E_j \cap \Gamma_1\}, \tag{59}$$

where $m$ denotes the number of local degrees of freedom per edge/face and $\mathbf{n}_k$ is the (unique) outward normal to $\partial E_k$. The source term is assembled in a similar way:

$$q_i = \frac{1}{m} \sum_{k \in \mathcal{J}_3^i} g_k |\partial E_k|, \quad \mathcal{J}_3^i = \{j \colon \mathbf{x}_i \in \partial E_j \cap \Gamma_3\}. \tag{60}$$

Clearly, $q$ does not have to be recomputed as long as the boundary conditions are fixed.

An important advantage of using piecewise-constant normal fluxes in the assembly process is that the disjoint boundary components $\Gamma_1, \Gamma_2, \Gamma_3$ are unambiguously defined and the boundary conditions are allowed to have a jump at the interface.

### 3.2.2 Construction of a linear LED scheme

Let us perform mass lumping and represent the semi-discrete problem in the form

$$M_L \frac{\mathrm{d}u}{\mathrm{d}t} = (K - B)u + q \Leftrightarrow m_i \frac{\mathrm{d}u_i}{\mathrm{d}t} = \sum_{j \neq i} k_{ij}(u_j - u_i) + r_i u_i + q_i, \tag{61}$$

where $m_i = \sum_j m_{ij}$ and $r_i = \sum_j k_{ij} - b_i$. The term $r_i u_i$ is a discrete counterpart of $u\nabla \cdot \mathbf{v}$ which vanishes for incompressible flows. It is instructive to consider the special case $r_i = q_i = 0$ in which the discrete transport operator $K$ has zero row sum. Such a discretization would be local extremum diminishing [8] if all off-diagonal coefficients $k_{ij}$ were nonnegative. Indeed, if $u_i$ is a maximum, then $k_{ij}(u_j - u_i) \leq 0$, so that $\frac{\mathrm{d}u_i}{\mathrm{d}t} \leq 0$. Hence, a maximum cannot increase, and similarly a minimum cannot decrease. In one dimension, a LED scheme is total variation diminishing and monotonicity preserving.

The Galerkin transport operator $K$ can be rendered LED by adding a tensor of artificial diffusion $D$ designed so as to eliminate negative off-diagonal entries [13], [14]

$$L = K + D, \quad d_{ii} = -\sum_{k\neq i} d_{ik}, \quad d_{ij} = d_{ji} = \max\{0, -k_{ij}, -k_{ji}\}. \tag{62}$$

The diffusion coefficients $d_{ij}$ are associated with the edges of the graph representing the connectivity of the matrix. For each pair of neighboring nodes $i$ and $j$, a negative off-diagonal coefficient (say, $k_{ij}$) is set to zero, while its mirror image $k_{ji}$ is incremented by $-k_{ij}$ and the diagonal entries $k_{ii}$, $k_{jj}$ are updated so as to maintain zero column sums. This yields the least diffusive linear LED scheme which reduces to the standard upwind approximation for scalar convection problems in one dimension [13]. The row number of the nullified off-diagonal entry determines which of the two nodes is located 'upwind'. Without loss of generality, we will assume that $i$ is the upwind node for the numerical edge $\vec{ij}$, so that $l_{ij} = 0$ whereas $l_{ji} = |k_{ji} - k_{ij}|$. The diffusive flux $d_{ij}(u_j - u_i)$ reduces the difference between the nodal values so as to enforce the positivity constraint.

Note that the artificial diffusion coefficients $d_{ij}$ as defined in (62) are independent of the diagonal part $r_i$ which allows for the physical growth and decay of extrema due to compressibility. The boundary terms $q_i$ are assumed to be nonnegative and may only increase the value of $u$ in the interior. Physical diffusion (if any) built into the coefficients $k_{ij}$ is automatically detected and the amount of artificial diffusion is reduced accordingly. Alternatively, discrete upwinding can be performed only for the convective part of $K$.

### 3.2.3 Defect correction

According to the well-known Godunov theorem, linear monotonicity preserving schemes can be at most first order accurate. To obtain a nonlinear high-resolution scheme, we combine the Galerkin discretization and the associated low-order method within an iterative defect correction loop for the fully discretized problem:

$$u^{(l+1)} = u^{(l)} + \omega^{(l)}[C(u^{(l)})]^{-1}R(u^{(l)}), \quad l = 0, 1, 2, \ldots \tag{63}$$

In a practical implementation one solves a linear subproblem for the solution increment and applies the weighted correction to the last iterate:

$$C(u^{(l)})\Delta u^{(l)} = R(u^{(l)}), \quad u^{(l+1)} = u^{(l)} + \omega^{(l)}\Delta u^{(l)}, \quad u^{(0)} = u^n. \tag{64}$$

Specifically, we take the 'preconditioner' to be the low-order evolution operator

$$C(u^{(l)}) = M_L - \theta\Delta t L(u^{(l)}), \quad 0 \leq \theta \leq 1,$$

where $\theta$ is the implicitness parameter. By construction, $C$ is an M-matrix which makes it amenable to iterative solution. The residual vector is given by

$$R(u^{(l)}) = M_L(u^n - u^{(l)}) + \theta\Delta t[L(u^{(l)}) - A(u^{(l)})]u^{(l)} \\ + (1-\theta)\Delta t[L(u^n) - A(u^n)]u^n + q^{n+\theta},$$

where $A$ is a nonlinear antidiffusion operator to be defined below.

The difference between $K^* = L - A = K + D - A$ and the original transport operator $K$ is a matrix with zero row and column sums. Hence, the diffusive-antidiffusive terms can be decomposed into skew-symmetric internodal fluxes:

$$(K^*u)_i = \sum_j k_{ij}u_j + \sum_{j\neq i}(f_{ij}^d + f_{ij}^a), \quad \text{where } f_{ij}^d = d_{ij}(u_j - u_i) = -f_{ji}^d. \tag{65}$$

The antidiffusive flux $f_{ij}^a$ from node $j$ into node $i$ is designed so as to comply with the LED principle for the discrete scheme. The objective is to add as much antidiffusion as possible without generating wiggles. This can be accomplished by applying flux limiters of TVD and FCT type presented in the next two subsections.

### 3.2.4 TVD flux limiter

For the discrete scheme to be LED, the antidiffusive flux $f_{ij}^a$ must be interpretable as a diffusive flux from some other node(s). To provide this property, we may consider

$$f_{ij}^a = a_{ij}\mathcal{L}(1,\theta_i)(u_i - u_j) = a_{ij}\mathcal{L}(u_i - u_j, \Delta u_{ij}), \quad f_{ji}^a = -f_{ij}^a. \tag{66}$$

Extra antidiffusion due to the consistent mass matrix can be included and limited as in the FEM-FCT framework (see below). Let us leave the antidiffusion coefficient $a_{ij}$ and the difference $\Delta u_{ij} = \theta_i(u_i - u_j)$ unspecified for the time being. The flux limiter $\mathcal{L}$ represents a limited average operator satisfying certain properties which guarantee the positivity of coefficients [8], [18]. In particular, $\mathcal{L}(a,b) = 0$ if $a$ and $b$ have opposite sign. Another useful property is symmetry: $\mathcal{L}(a,b) = \mathcal{L}(b,a)$. Some popular TVD limiters are:

- $\mathcal{L}(a,b) = \mathcal{S}(a,b) \cdot \min\{|a|,|b|\}$ — minmod
- $\mathcal{L}(a,b) = \mathcal{S}(a,b) \cdot \dfrac{2ab}{|a+b|}$ — Van Leer
- $\mathcal{L}(a,b) = \mathcal{S}(a,b) \cdot \min\left\{\dfrac{|a+b|}{2}, 2|a|, 2|b|\right\}$ — 2-mean
- $\mathcal{L}(a,b) = \mathcal{S}(a,b) \cdot \max\{\min\{2|a|,|b|\}, \min\{|a|,2|b|\}\}$ — superbee

where $\mathcal{S}(a,b) = (\text{sign}(a) + \text{sign}(b))/2$.

Recall that the numerical edge $\vec{ij}$ connects an upwind node $i$ and a downwind node $j$. Its contributions to the modified convective term can be written as

$$\begin{aligned}\text{node } i: \quad & k_{ij}^*(u_j - u_i) = a_{ij}\mathcal{L}(1,\theta_i)(u_i - u_j) = a_{ij}\mathcal{L}(1,1/\theta_i)\Delta u_{ij},\\ \text{node } j: \quad & k_{ji}^*(u_i - u_j) = (|k_{ji} - k_{ij}| - a_{ij}\mathcal{L}(1,\theta_i))(u_i - u_j).\end{aligned}$$

The increment to node $j$ exhibits a LED structure provided that the coefficient $k_{ji}^*$ is nonnegative. All of the flux limiters presented above satisfy the inequality $0 \leq \mathcal{L}(1,\theta) \leq 2$. Hence, the antidiffusion coefficient is taken to be

$$a_{ij} = \min\{d_{ij}, |k_{ji} - k_{ij}|/2\}. \tag{67}$$

The increment to node $i$ will also be local extremum diminishing as long as the upwind difference $\Delta u_{ij}$ can be cast in the form $\Delta u_{ij} = \sum_{k \neq i} c_{ik}(u_k - u_i)$ with $c_{ik} \geq 0$.

In one-dimensional TVD schemes, $\theta_i$ represents the slope ratio at the upwind node, so that $\Delta u_{ij} = u_k - u_i$, where $k \neq j$ refers to the second neighbor of node $i$. However, the choice of $\Delta u_{ij}$ for finite element discretizations on unstructured meshes is nontrivial. A geometric approach commonly employed in the literature is to reconstruct a local one-dimensional stencil by introducing two dummy nodes on the continuation of the edge $\vec{ij}$ [8], [18]. The difference $\Delta u_{ij}$ is defined as in the 1D case using the interpolated or extrapolated solution value at the dummy node $k$ adjacent to the upwind node.

A detailed comparison of various techniques for the recovery of $u_k$ can be found in [18]. It was shown that the numerical results strongly depend on the employed procedure. In particular, those obtained by means of the gradient reconstruction

$$\Delta u_{ij} = (\mathbf{x}_i - \mathbf{x}_j) \cdot \nabla_h u_i, \quad \text{where } \nabla_h u_i = \frac{1}{m_i} \sum_{k \neq i} \mathbf{c}_{ki}(u_k - u_i) \tag{68}$$

were found to be quite poor. It is quite clear why this choice of the upwind difference may fail to produce non-oscillatory results. The $L_2$-projection of the discrete gradient using Galerkin differential operators is not monotone. Some of the coefficients in the formula for $\Delta u_{ij}$ may be negative, so that it will not possess the desired form. To rectify this, we employ a low-order projection operator designed by resorting to discrete upwinding as proposed above. Note that $\mathbf{c}_{ki} = -\mathbf{c}_{ik}$ (for internal nodes), so that the elimination of negative off-diagonal coefficients leads to the following LED-type expression:

$$\Delta u_{ij} = \frac{2}{m_i} \sum_{k \neq i} \max\{0, \mathbf{c}_{ki} \cdot (\mathbf{x}_i - \mathbf{x}_j)\}(u_k - u_i).$$

For uniform meshes in one dimension, this kind of extrapolation corresponds to using the upwind gradient and yields the standard value $\Delta u_{ij} = u_k - u_i$.

### 3.2.5 FCT flux limiter

As an alternative, we outline another flux limiter based on the generalized FEM-FCT formulation [11], [13]. In this case, the raw antidiffusion is given by

$$f_{ij} = a_{ij}^{(l)}(u_i^{(l)} - u_j^{(l)}) + a_{ij}^n(u_i^n - u_j^n), \quad f_{ji} = -f_{ij}, \tag{69}$$

where

$$a_{ij}^{(l)} = m_{ij} + \theta \Delta t d_{ij}^{(l)}, \quad a_{ij}^n = -m_{ij} + (1-\theta)\Delta t d_{ij}^n. \tag{70}$$

Solution-dependent correction factors $\alpha_{ij} \in [0,1]$ are applied to $f_{ij}$ in order to preclude the arising of nonphysical oscillations. The admissible antidiffusive fluxes $f_{ij}^a = \alpha_{ij} f_{ij}$ are inserted into the defect vector $R(u^{(l)})$ to reduce the errors induced by mass lumping and discrete upwinding. The unified limiting strategy proposed in [11], [13] is as follows.

At the beginning of each time step, the explicit subproblem

$$M_L \tilde{u} = [M_L + (1-\theta)\Delta t L(u^n)]u^n \tag{71}$$

is solved for $\tilde{u}$ to determine the local extrema $\tilde{u}_i^{\max}$ and $\tilde{u}_i^{\min}$ . Note that the auxiliary solution $\tilde{u}$ is associated with the time instant $t^{n+1-\theta}$ and reduces to $u^n$ for the backward Euler method. Following Zalesak [33], we introduce

$$P_i^{\pm} = \frac{1}{m_i} \sum_{j \neq i} \begin{matrix}\max\\ \min\end{matrix}\{0, f_{ij}\}, \quad Q_i^{\pm} = \tilde{u}_i^{\substack{\max\\ \min}} - \tilde{u}_i. \tag{72}$$

For the scheme to be positivity-preserving, the flux into node $i$ should be multiplied by

$$R_i^{\pm} = \begin{cases} \min\{1, Q_i^{\pm}/P_i^{\pm}\}, & \text{if } P_i^{\pm} \neq 0, \\ 0, & \text{if } P_i^{\pm} = 0. \end{cases} \tag{73}$$

Due to the fact that $f_{ji} = -f_{ij}$ the 'optimal' correction factors are given by

$$\alpha_{ij} = \begin{cases} \min\{R_i^+, R_j^-\}, & \text{if } f_{ij} \geq 0, \\ \min\{R_j^+, R_i^-\}, & \text{if } f_{ij} < 0. \end{cases} \tag{74}$$

The ins and outs of the FEM-FCT methodology are elucidated in [13], [14], [16].

### 3.3 Solver and time step

Finally, let us briefly discuss the choice of the iterative solver and of the time discretization. In general, explicit schemes are rather inefficient due to severe stability limitations which require taking impractically small time steps. For this reason, we restrict ourselves to the implicit Crank-Nicolson and backward Euler methods which are unconditionally stable and permit large time steps at the expense of solving nonsymmetric linear systems.

In our experience, BiCGSTAB and geometric multigrid constitute excellent solvers as long as the parameters are properly tuned and the underlying smoothers/preconditioners are consistent with the size of the time step. If $\Delta t$ is rather small, standard components like Jacobi, Gauss-Seidel and SOR schemes will suffice. For large time steps, the condition number of the matrix deteriorates and convergence may fail. This can be rectified by resorting to an ILU factorization in conjunction with an appropriate renumbering scheme.

In order to capture the dynamics of the two-phase flow in a computationally efficient way, one needs a smart strategy for the time step control. So far we have used a fairly simple PID controller due to Valli *et al.* [30]. In our case, the algorithm reads

1. Monitor the relative changes of the gas holdup distribution

$$e_n = \frac{\|\varepsilon^{n+1} - \varepsilon^n\|}{\|\varepsilon^{n+1}\|}$$

2. If $e_n > \delta$ reject the solution and repeat the time step using

$$\Delta t_* = \frac{\delta}{e_n} \Delta t_n$$

3. Adjust the time step smoothly so as to approach the prescribed tolerance for the relative changes

$$\Delta t_{n+1} = \left(\frac{e_{n-1}}{e_n}\right)^{k_P} \left(\frac{\mathrm{TOL}}{e_n}\right)^{k_I} \left(\frac{e_{n-1}^2}{e_n e_{n-2}}\right)^{k_D} \Delta t_n$$

4. Limit the growth and reduction of the time step so that

$$\Delta t_{\min} \le \Delta t_{n+1} \le \Delta t_{\max}, \quad m \le \frac{\Delta t_{n+1}}{\Delta t_n} \le M$$

Note that the local gas holdup $\varepsilon$ is a perfect indicator variable because the bubble-induced buoyancy represents a major driving force behind the flows in gas-liquid reactors.

## 4 Numerical examples

The developed finite element software builds on the FEATFLOW package [27] and on the underlying FEAT libraries [28]. The multidimensional TVD limiter was incorporated into the matrix assembly routine for the momentum equation and has proved its worth for the nonconforming $\tilde{Q}_1$ finite elements. The subroutines for the solution of scalar transport equations were equipped with both FCT and TVD limiters. The performance of these high-resolution schemes was found to be similar. Let us substantiate the mathematical model and the numerical algorithm by preliminary simulation results.

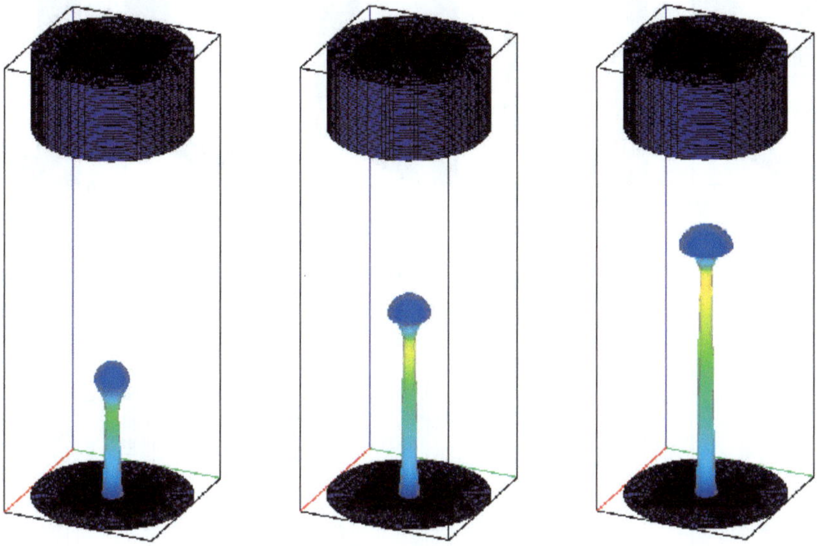

Figure 1. Startup of a locally aerated cylindrical bubble column.

The first example deals with the startup of a locally aerated cylindrical bubble column. The gas is injected at the center into the initially quiescent liquid. The buoyancy induced by the ascending bubbles results in a circulation of liquid with upflow in the middle. The evolution of the gas holdup distribution during the first seconds after the onset of aeration is shown in Fig. 1 along with fragments of the employed computational mesh. Similar mushroom-like shapes were observed previously in 2D simulations [10], [15], [25].

Next, we present our three-dimensional results for the airlift loop reactor which was investigated both experimentally and numerically by Becker *at al.* [1].The aeration takes place at the bottom of the riser section where both phases flow cocurrently in the upward direction. At the upper surface, the bubbles escape while the liquid is diverted into the gas-free downcomer forming a closed loop. The two-phase flow reaches a steady state within

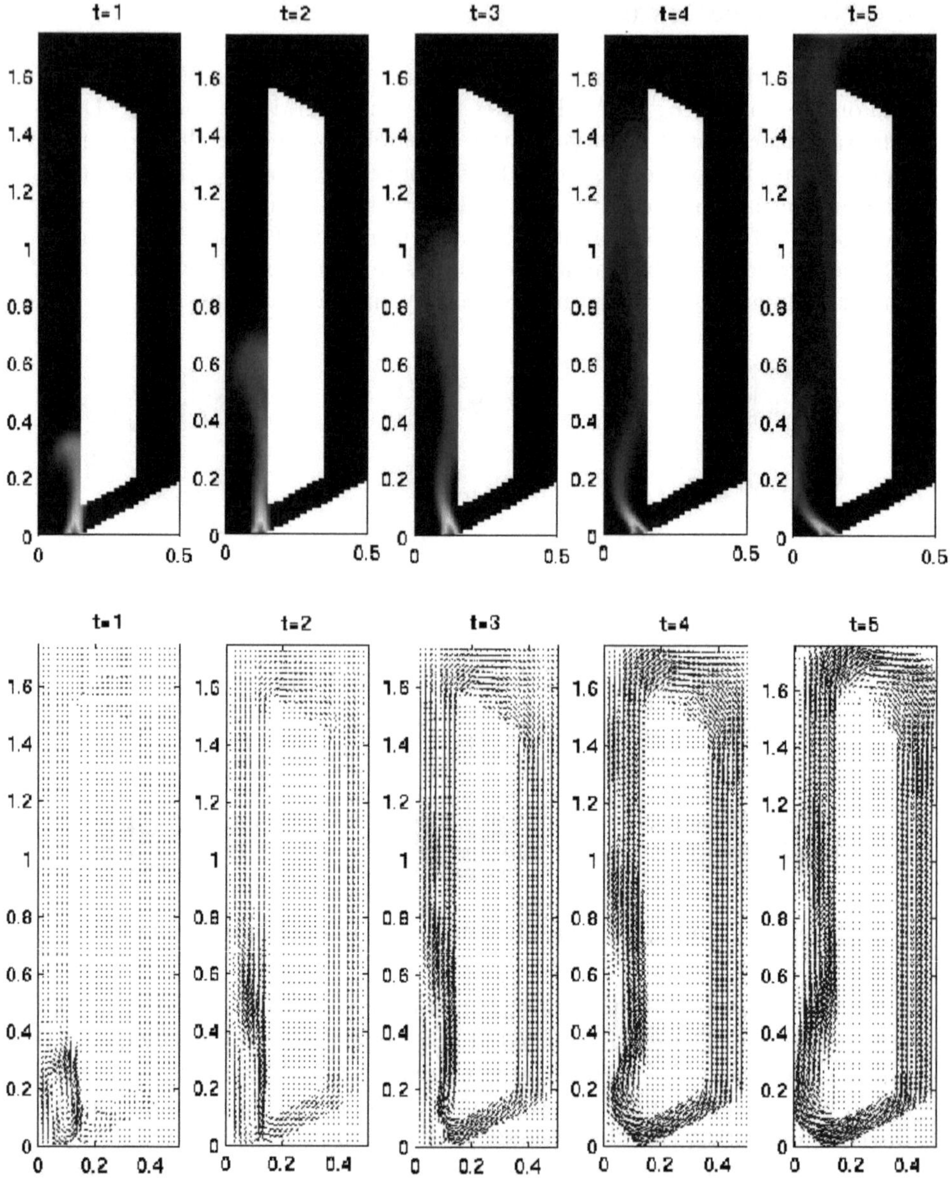


Figure 2. Airlift loop reactor: startup (left) and steady state (right).

a few seconds after the startup of the reactor. Fig. 2 (left) displays snapshots of the developing gas holdup distribution (top) and instantaneous velocity fields (bottom) in the middle cross section. The ragged boundary of the internal part is due to the Cartesian mesh used for visualization. The three-

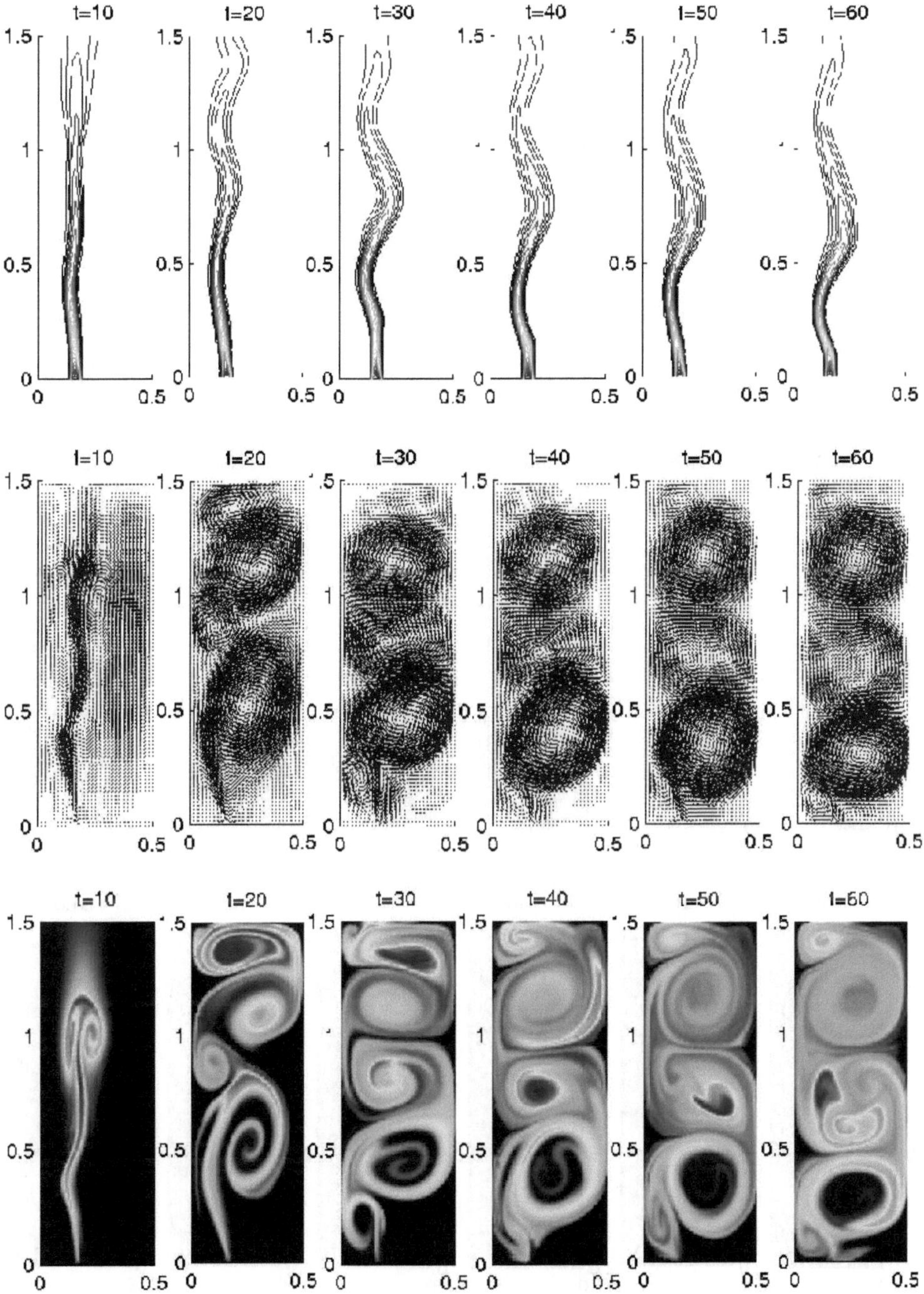


Figure 3. Physical absorption of $CO_2$ (2D simulation).

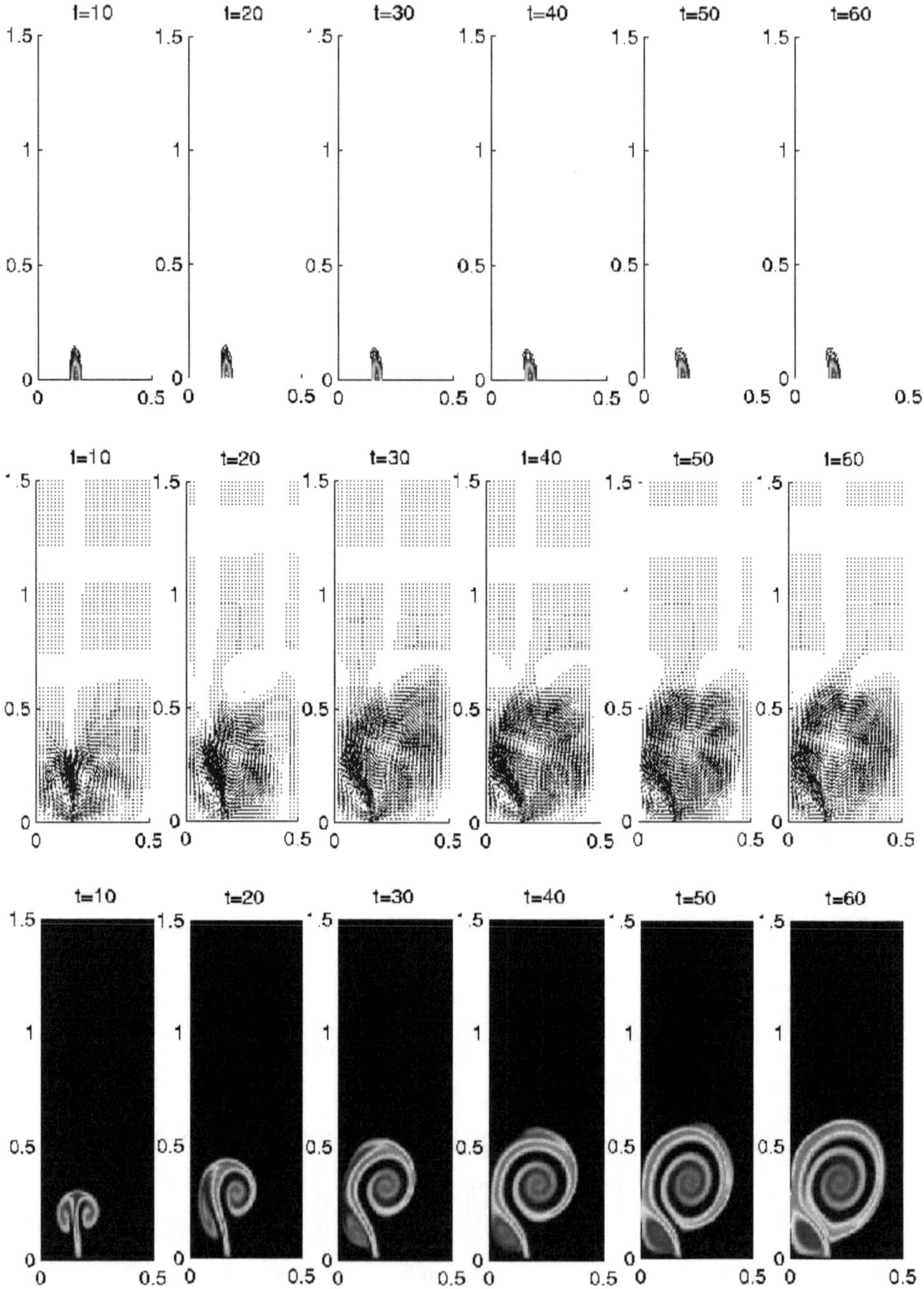


Figure 4. Reaction-enhanced absorption of $CO_2$ (2D simulation).

Figure 5. Reaction-enhanced absorption of $CO_2$ (3D simulation).

dimensional stationary solution superimposed on the actual block-structured computational mesh is shown on the right.

To give an insight into the complex interplay of physical and chemical phenomena, let us consider the absorption of $CO_2$ in a locally aerated bubble column filled with water. The construction parameters are those defined by

Becker *et al.* [1] and adopted in our first 2D simulations for this test case [10], [12]. In the absence of mass transfer, the rising bubbles would expand due to the fall of hydrostatic pressure. However, carbon dioxide is highly soluble in water so that the growth is diminished and the bubbles may even shrink.

These effects are nicely reproduced by our model. In the case of physical absorption, the gas dissolves rather slowly and neither the instantaneous gas holdup distribution (Fig. 3, top) nor the velocity fields (Fig. 3, middle) are significantly affected by the absorption process [10]. The flow structure consisting of three large vortices is reflected by the concentration fields for the dissolved carbon dioxide (Fig. 3, bottom).

The flow pattern is totally different if the mass transfer is accelerated by chemical reactions in the liquid phase. Let us demonstrate this by taking an aqueous solution of sodium hydroxide instead of water. The reaction $CO_2+2NaOH \rightarrow Na_2CO_3+H_2O$ is so fast that the bubbles are completely dissolved within 20 cm from the inlet (Fig. 4, top). There is no liquid circulation in the upper part of the reactor since the gas does not reach it. A large vortex develops in the vicinity of the gas sparger (Fig. 4, middle) twisting the concentration of the produced sodium carbonate into a spiral shape (Fig. 4, bottom). A 3D simulation predicts essentially the same flow behavior depicted in Fig. 5.

The results are in good agreement with experimental data reported by Fleischer *et al.* [6].

## 5 Conclusions

The drift-flux model for buoyancy-driven bubbly flows was coupled with scalar transport equations describing the absorption of gas followed (and enhanced) by chemical reactions in the liquid phase. An unstructured grid finite element method was proposed for the numerical solution. The discretization of the troublesome convective terms was performed by a novel high-resolution scheme. Nested iterations were used to provide the coupling of model equations, get rid of nonlinearities and solve the linear systems. The subtleties of the algorithm were discussed in detail. The strong interdependence of the hydrodynamics, mass transfer and chemical reaction processes was illustrated by numerical examples. Turbulence effects were not considered and will be addressed in a forthcoming paper.

## References

[1] S. Becker, A. Sokolichin, and G. Eigenberger, *Gas-liquid flow in bubble columns and loop reactors: Part II. Comparison of detailed experiments and flow simulations*, Chem. Eng. Sci. **49** (1994), 5747–5762.

[2] O. Borchers, C. Busch, A. Sokolichin, and G. Eigenberger, *Applicability of the standard turbulence model to the dynamic simulation of*

*bubble columns: Part II. Comparison of detailed experiments and flow simulations*, Chem. Eng. Sci. **54** (1999), 5927–5935.

[3] J. P. Boris, D. L. Book, *Flux-corrected transport. I. SHASTA, A fluid transport algorithm that works*, J. Comput. Phys. **11** (1973), 38–69.

[4] D. A. Drew, *Mathematical modeling of two-phase flow*, Ann. Rev. Fluid Mech. **15** (1983), 261–291.

[5] D. A. Drew, S. L. Passman, *Theory of Multicomponent Fluids*, Springer, 1999.

[6] C. Fleischer, S. Becker, and G. Eigenberger, *Detailed modeling of the chemisorption of $CO_2$ into NaOH in a bubble column*, Chem. Eng. Sci. **51** (1996), 1715–1724.

[7] A. Harten, *High resolution schemes for hyperbolic conservation laws*, J. Comput. Phys. **49** (1983), 357–393.

[8] A. Jameson, *Positive schemes and shock modelling for compressible flows*, Int. J. Numer. Methods Fluids **20** (1995), 743–776.

[9] I. Kataoka, A. Serizawa, *Basic equations of turbulence in gas-liquid two-phase flow*, Int. J. Multiphase Flow (1989), 843–855.

[10] D. Kuzmin, *Numerical Simulation of Reactive Bubbly Flows. PhD. Thesis, Jyväskylä Studies in Computing 2*, University of Jyväskylä, 1999.

[11] D. Kuzmin, *Positive finite element schemes based on the flux-corrected transport procedure*, in: Computational Fluid and Solid Mechanics, Elsevier, 2001, pp. 887–888.

[12] D. Kuzmin, S. Turek, *Efficient numerical techniques for flow simulation in bubble column reactors*, Preprints of the 5th German-Japanese Symposium on Bubble Columns, VDI/GVC, 2000, pp. 99–104.

[13] D. Kuzmin, S. Turek, *Flux correction tools for finite elements*, J. Comput. Phys. **175** (2002), 525–558.

[14] D. Kuzmin, M. Möller, and S. Turek, *Multidimensional FEM-FCT schemes for arbitrary time-stepping. Technical report 215: University of Dortmund, 2002*, Int. J. Numer Methods Fluids, submitted.

[15] A. Lapin, A. Lübbert, *Numerical simulation of the dynamics of two-phase gas-liquid flows in bubble columns*, Chem. Eng. Sci. **49** (1994), 3661–3674.

[16] R. Löhner, *Adaptive CFD Techniques*, Wiley, 2001.

[17] R. Löhner, K. Morgan, J. Peraire, and M. Vahdati, *Finite element flux-corrected transport (FEM-FCT) for the Euler and Navier-Stokes equations*, Int. J. Numer. Methods Fluids **7** (1987), 1093–1109.

[18] P. R. M. Lyra, *Unstructured Grid Adaptive Algorithms for Fluid Dynamics and Heat Conduction*, PhD. Thesis, University of Wales, Swansea, 1994.

[19] D. Pfleger, S. Gomes, N. Gilbert, and H.-G. Wagner, *Hydrodynamic simulations of laboratory scale bubble columns – Fundamental studies of the Eulerian-Eulerian modelling approach*, Chem. Eng. Sci. **54** (1999), 5091–5098.

[20] R. Rannacher, S. Turek, *A simple nonconforming quadrilateral Stokes element*, Numer. Methods Partial Differ. Equations **8** (1992), 97–111.

[21] M. P. Schwarz, W. J. Turner, *Applicability of the standard $k$-$\varepsilon$ turbulence model to gas-stirred baths*, Appl. Math. Modelling **12** (1988), 273–279.

[22] T. K. Sherwood, R. L. Pigford, and C. R. Wilke, *Mass Transfer*, Mc Graw-Hill, New York, 1975.

[23] A. Sokolichin, G. Eigenberger, *Modellierung und effiziente numerische Simulation von Gas-Flüssigkeits-Reaktoren mit Blasenströmungen nach dem Euler-Euler-Konzept. DFG technical report*, University of Stuttgart, 1995–1997; available at *http://pcvt12.verfahrenstechnik.uni-stuttgart.de/alex/icvt2.html.*

[24] A. Sokolichin, G. Eigenberger, *Applicability of the standard turbulence model to the dynamic simulation of bubble columns: Part I. Detailed numerical simulations*, Chem. Eng. Sci. **54** (1999), 2273–2284.

[25] A. Sokolichin, G. Eigenberger, A. Lapin, and A. Lübbert, *Dynamic numerical simulation of gas-liquid two-phase flows: Euler-Euler versus Euler-Lagrange*, Chem. Eng. Sci. **52** (1997), 611–626.

[26] S. Turek, *On discrete projection methods for the incompressible Navier-Stokes equations: An algorithmical approach*, Comput. Methods Appl. Mech. Eng. **143** (1997), 271–288.

[27] S. Turek *et al.*, *FEATFLOW: finite element software for the incompressible Navier-Stokes equations*, User manual, University of Dortmund, 2000, available at *http://www.featflow.de.*

[28] S. Turek *et al.*, *FEAT2D/FEAT3D: Finite Element Analysis Tools. User manual*, University of Heidelberg, 1995, available at *http://www.featflow.de.*

[29] S. Turek, *Efficient Solvers for Incompressible Flow Problems: An Algorithmic and Computational Approach. LNCSE* 6, Springer, 1999.

[30] A. M. P. Valli, G. F. Carey, and A. L. G. A. Coutinho, *Control strategies for timestep selection in simulation of coupled viscous flow and heat transfer*, Commun. Numer. Methods Eng. **18** (2002), 131–139.

[31] D. W. Van Krevelen, P. J. Hoftijzer, *Kinetics of gas-liquid reactions. Part I: General Theory*, Rec. Trav. Chim. **67** (1948), 563–586.

[32] R. M. Wellek, R. J. Brunson, and F. H. Law, Can. J. Chem. Eng. **56** (1978), 181.

[33] S. T. Zalesak, *Fully multidimensional flux-corrected transport algorithms for fluids*, J. Comput. Phys. **31** (1979), 335–362.

# Implicit Flux-Corrected Transport Algorithm for Finite Element Simulation of the Compressible Euler Equations

Matthias Möller,[1] Dmitri Kuzmin,[2] Stefan Turek[3]

[1]Institute of Applied Mathematics (LS III), University of Dortmund, Vogelpothsweg 87, D–44227 Dortmund, Germany
matthias.moeller@math.uni-dortmund.de
[2]Institute of Applied Mathematics (LS III), University of Dortmund, Vogelpothsweg 87, D–44227 Dortmund, Germany
kuzmin@math.uni-dortmund.de
[3]Institute of Applied Mathematics (LS III), University of Dortmund, Vogelpothsweg 87, D–44227 Dortmund, Germany
ture@featflow.de

## 1 Introduction

Even today, the accurate treatment of convection-dominated transport problems remains a challenging task in numerical simulation of both compressible and incompressible flows. The discrepancy arises between high accuracy and good resolution of singularities on the one hand and preventing the growth and birth of nonphysical oscillations on the other hand. In 1959 it was proven [6], that linear methods are restricted to be at most first order if they are to preserve monotonicity. Thus, the use of nonlinear methods is indispensable to overcome smearing by numerical diffusion without sacrificing important properties of the exact solution such as positivity and monotonicity. The advent of the promising methodology of *flux-corrected transport* (FCT) can be traced back to the pioneering work of Boris and Book [3]. Even though their original FCT algorithm named SHASTA was a rather specialized one-dimensional finite difference scheme, the cornerstone for a variety of high-resolution schemes was laid. Strictly speaking, the authors recommended using a high-order discretization in regions of smooth solutions and switching to a low-order method in the vicinity of steep gradients. This idea of adaptive toggling between methods of high and low order was dramatically improved by Zalesak [3] who proposed a multi-dimensional generalization applicable to arbitrary combinations of high- and low-order discretizations but still remaining in the realm of finite differences. This barrier was first crossed by Parrott and Christie [21] who settled the idea of flux-correction in the framework of finite elements. Finally, FEM-FCT reached maturity by the considerable contributions of Löhner and his coworkers [16], [17]. Beside the classical formulation of Zalesak's limiter in terms of element contributions, an alternative approach is available limiting the fluxes edge-by-edge [26], [27].

Pursuing this idea, some modern compressible flow solvers completely abandon the conventional finite element data structure in favor of an edge-based version. The foundations were laid by Peraire *et al.* [23] but their conservative decomposition of Galerkin integrals into fluxes assigned to the edges is applicable only to linear finite elements on triangular and tetrahedral meshes.

In [12] we presented a coherent methodology for the decomposition of both convective and diffusive terms into internodal fluxes independently of the underlying discretization. In the same paper we proposed a multi-dimensional FCT algorithm which is applicable to explicit and implicit schemes. The construction of the positivity-preserving low-order scheme based on the elimination of negative off-diagonal entries was addressed. In this paper we will advance the new approach to implicit schemes applied to systems of equations. The property of having diagonalizable Jacobians inheriting the hyperbolicity will prove to be of great use for the design of the low-order method. We will review its construction for systems in general and elucidate in particular how to obtain a positive low-order method by adding artificial diffusion proportional to the spectral radius of the Roe matrix for hyperbolic systems. Following the concept of edge-based data structures, we propose an efficient algorithm for the matrix assembly but without the need of storing them completely. The flux-correction step will be polished by a synchronization between the distinct variables. Furthermore, we will address the treatment of the fully discretized Euler equations within a defect-correction loop which benefits from the use of a decoupling block-diagonal preconditioner. Last but not least, we elaborate on various implementation aspects and present several numerical examples to stress the potential of the implicit formulation generalized to systems of equations.

## 2 Governing equations

Consider the conservative form of the Euler equations without source terms in the three-dimensional space

$$\frac{\partial U}{\partial t} + \nabla \cdot \mathbf{F} = 0 \tag{1}$$

with the conservative variables and fluxes given by

$$U = \begin{bmatrix} \rho \\ \rho v_1 \\ \rho v_2 \\ \rho v_3 \\ E \end{bmatrix}, \quad F^{x_1} = \begin{bmatrix} \rho v_1 \\ \rho v_1^2 + p \\ \rho v_1 v_2 \\ \rho v_1 v_3 \\ v_1(E+p) \end{bmatrix}, \tag{2}$$

$$F^{x_2} = \begin{bmatrix} \rho v_2 \\ \rho v_1 v_2 \\ \rho v_2^2 + p \\ \rho v_2 v_3 \\ v_2(E+p) \end{bmatrix}, \quad F^{x_3} = \begin{bmatrix} \rho v_3 \\ \rho v_1 v_3 \\ \rho v_2 v_3 \\ \rho v_3^2 + p \\ v_3(E+p) \end{bmatrix},$$

where $\rho$, $v_d$, $p$ and $E$ represent the density, velocity, pressure and total energy of the medium, respectively. Furthermore, we make the assumption of a polytropic gas, thus the equation of state reads

$$E = \frac{p}{\gamma - 1} + \frac{\rho}{2}|\mathbf{v}|^2, \tag{3}$$

in which $\mathbf{v}$ stands for the velocity vector and $\gamma$ denotes the ratio of specific heats ($\gamma = 1.4$ for air). Beside the standard divergence form, the Euler equations can be rewritten in an equivalent nonconservative formulation by introducing the Jacobian matrices for the fluxes in each coordinate direction $A^{x_d} = \partial F^{x_d}/\partial U$ which can be found for instance in [8]. As a result, the quasi-linear Euler equations can be reformulated as

$$\frac{\partial U}{\partial t} + \sum_d A^{x_d} \frac{\partial U}{\partial x_d} = 0. \tag{4}$$

Interestingly enough, the flux vectors $F^{x_d}$ for compressible inviscid flow are homogeneous functions of the conservative variables $U$ independently of the spatial dimension, so that the following useful identity holds [18]

$$F^{x_d} = \frac{\partial F^{x_d}}{\partial U} U = A^{x_d} U. \tag{5}$$

In contrast to the one-dimensional Euler equations which are strictly hyperbolic, the situation in multi-dimensions is more complicated. It can be shown [8] that for any linear combination of the Jacobian matrices

$$A(\kappa) = \sum_d A^{x_d} \kappa^{x_d} \tag{6}$$

there exists a nonsingular matrix $R(\kappa)$ such that

$$A(\kappa) = R(\kappa)\Lambda(\kappa)R(\kappa)^{-1} \tag{7}$$

is diagonalizable, where $\Lambda$ is the diagonal matrix of the *real* eigenvalues which are only distinct in one space dimension. Nevertheless, it *is not* possible to diagonalize each Jacobian simultaneously with the same matrix $R$. For the one-dimensional Euler equations the diagonalization procedure simply reduces to

$$A = R\Lambda R^{-1}, \text{ where } \Lambda = \mathrm{diag}(v - c, v, v + c) \tag{8}$$

is the diagonal matrix of eigenvalues, and

$$R = \begin{bmatrix} 1 & 1 & 1 \\ v-c & v & v+c \\ h-vc & v^2/2 & h+vc \end{bmatrix} \tag{9}$$

is the matrix of the corresponding right eigenvectors. Here $c = \sqrt{\gamma p/\rho}$ stands for the local speed of sound and $h$ denotes the total enthalpy which is given by $h = (E+p)/\rho$.

## 3 Galerkin flux decomposition

For deducing the basic theory, let us go back to a generic time-dependent conservation law for a scalar quantity $u$

$$\frac{\partial u}{\partial t} + \nabla \cdot \mathbf{f} = q \quad \text{in } \Omega, \tag{10}$$

where $q$ is a source term, and $\mathbf{f}$ is a flux function. The weak form of this equation reads

$$\int_\Omega w\left[\frac{\partial u}{\partial t} + \nabla \cdot \mathbf{f} - q\right] \mathrm{d}\mathbf{x} = 0, \quad \forall w. \tag{11}$$

It is well known from the theory of finite difference methods that a numerical scheme is conservative if it admits the decomposition into a sum of fluxes from one node into another. Indeed, as long as the internodal fluxes are equal in magnitude and opposite in direction, the total mass of the system may only change due to boundary fluxes. Hence, it is highly desirable to represent the numerical method in conservative form whenever possible. At the same time, it has been largely unclear how to accomplish this in the context of finite elements on unstructured meshes.

Many authors have concentrated on this topic promoting edge-based data structures [15], [18], [20] and [23]. Unfortunately, the underlying flux decomposition is feasible only for simplex elements with linear basis functions which have a constant gradient. In [12] we propose an alternative flux decomposition technique which is applicable to general finite element approximations on arbitrary meshes including quadrilateral and hexahedral ones. Hence, let us just point out the main ideas in this paper.

Integration by parts of the weak formulation (11) yields

$$\int_\Omega w\frac{\partial u}{\partial t}\,\mathrm{d}\mathbf{x} - \int_\Omega \nabla w \cdot \mathbf{f}\mathrm{d}\mathbf{x} + \int_{\partial\Omega} w\mathbf{f}\cdot\mathbf{n}\,\mathrm{d}s - \int_\Omega wq\,\mathrm{d}\mathbf{x} = 0, \quad \forall w. \tag{12}$$

A common practice in finite element computations of compressible flow is to approximate the fluxes in the same way as the desired solution which is referred to as the *group finite element formulation* [5].

Let the solution, the fluxes and the source terms be represented in the form

$$u = \sum_j u_j \varphi_j, \quad \mathbf{f} = \sum_j \mathbf{f}_j \varphi_j, \quad q = \sum_j q_j \varphi_j. \tag{13}$$

In fact, it is not compelling to use the same approximations for $u$ and $\mathbf{f}$. For instance, one can consider using nonconforming Crouzeix-Raviart or Rannacher-Turek elements for the fluxes.

After the substitution of expressions (13) and the weighting functions $w = \varphi_i$ into the variational formulation (12), we obtain

$$\sum_j \left[\int_\Omega \varphi_i \varphi_j \,\mathrm{d}\mathbf{x}\right](\dot{u}_j - q_j) - \sum_j \left[\int_\Omega \nabla\varphi_i \varphi_j \,\mathrm{d}\mathbf{x} - \int_{\partial\Omega} \varphi_i \varphi_j \mathbf{n} \,\mathrm{d}s\right] \cdot \mathbf{f}_j = 0, \tag{14}$$

which can be written in compact matrix form as

$$M_C(\dot{u} - q) = \sum_{d=1}^{3} K^{x_d} f^{x_d} \tag{15}$$

in the three-dimensional case. The utility of the group formulation is illustrated by the fact that the three matrices $K^{x_d}$ engendered by the corresponding first-order derivatives can be assembled once and for all at the beginning of the simulation, as long as the mesh does not change. This is in contrast to the standard finite element approach, whereby the discrete operators for the linearized convective terms have to be updated in each time step.

By construction, the discretized flux term consists of an interior part and a boundary part. The former is given by the integral

$$\sum_j \left[\int_\Omega \nabla\varphi_i \varphi_j \,\mathrm{d}\mathbf{x}\right] \cdot \mathbf{f}_j = \sum_j \mathbf{c}_{ij} \cdot \mathbf{f}_j, \quad \mathbf{c}_{ij} = \int_\Omega \nabla\varphi_i \varphi_j \,\mathrm{d}\mathbf{x}, \tag{16}$$

where the coefficient matrices $c_{ij}^{x_d}$ possess the zero column sum property, since it is assumed that the sum of basis functions equals unity. Therefore, it is possible to express the diagonal coefficients in terms of off-diagonal ones

$$\sum_i \mathbf{c}_{ij} = 0 \Rightarrow \mathbf{c}_{ii} = -\sum_{j \neq i} \mathbf{c}_{ji}. \tag{17}$$

It follows that the interior flux term (16) can be rewritten as

$$\sum_j \mathbf{c}_{ij} \cdot \mathbf{f}_j = \sum_{j \neq i} g_{ij}, \quad \text{where} \quad g_{ij} := \mathbf{c}_{ij} \cdot \mathbf{f}_j - \mathbf{c}_{ji} \cdot \mathbf{f}_i. \tag{18}$$

The newly introduced quantity $g_{ij}$ represents the *Galerkin flux* from node $j$ into node $i$. It is obvious that $g_{ji} = -g_{ij}$, so that node $j$ receives the same contribution with the opposite sign. Importantly, flux decomposition is also possible for *generalized diffusion operators* [11] which are defined as symmetric matrices having zero row and column sums. The purely diffusive Galerkin flux assumes a remarkably simple form

$$\sum_i d_{ij} = \sum_j d_{ij} = 0, \quad d_{ij} = d_{ji} \Rightarrow g_{ij} = d_{ij}(u_j - u_i). \tag{19}$$

Note that generalized diffusion operators are not required to have continuous counterparts. Some typical examples are the discrete Laplacian or the matrix $M_C - M_L$ sometimes referred to as 'mass diffusion'. As we will see shortly, the properties of discrete diffusion operators render them a valuable tool for the design of nonoscillatory low-order methods to be combined with high-order ones within the flux-corrected-transport algorithm.

Another promising approach to the derivation of high-resolution finite element schemes involves the replacement of the original Galerkin flux by another *consistent numerical flux*. Its potential is demonstrated by numerous publications [18], [19], [20] in which one-dimensional limiters are successfully applied on unstructured meshes in conjunction with the edge-based data structure of Peraire *et al.* [23].

As a matter of fact, it is not necessary to implement the Galerkin flux decomposition en bloc. On the contrary, it is sufficient to restrict this procedure to the diffusive terms. Thus, our methodology can be integrated easily into existing element-based codes. However, this decomposition constitutes an excellent tool for the derivation of new flux-manipulating schemes.

## 4 Discrete positivity criteria

Let us recall that most real-world applications are described by inherently nonlinear systems of partial differential equations modeling physical phenomena. It is therefore quite natural to impose some constraints on the numerical method which are dictated by the continuous model such as positivity-preservation of physical quantities. We will see that the concept of an M-matrix provides a handy criterion to guarantee positivity. The second fundamental issue concerning numerical methods is related to convergence. Convergence proofs for nonlinear systems take their pivotal ideas from the nonlinear stability analysis, the functional analytic framework of the theory of compactness. Combining both issues leads to the class of *monotone* methods, as a subset of the positive schemes, which guarantee that a converged solution of any conservation law does satisfy the entropy inequality ensuring that the numerical solution converges to the physically relevant weak solution.

### 4.1 Positivity

**Definition.** A nonsingular discrete operator $A \in \mathbb{R}^{n\times n}$ is called an M-matrix if $a_{ij} \leq 0$ for $i \neq j$ and all the entries of $A^{-1}$ are nonnegative.

If $A$ is strictly diagonally dominant and $a_{ii} > 0$, while $a_{ij} \leq 0$ for $i \neq j$, then $A$ is an M-matrix. Note that for $M$-matrices $Ax \geq 0$ implies that $x \geq 0$. This property leads to the following fundamental lemma:

**Lemma.** *Let the numerical scheme be presented in abstract matrix operator form as*

$$Lu^{n+1} = Ru^n. \tag{20}$$

*A sufficient condition for such a scheme to preserve positivity is that L be an M-matrix and all entries of R be nonnegative ($R \geq 0$).*

The proof of this lemma is almost trivial and can be found in [11].

To introduce another useful concept, consider a semi-discrete problem of the form

$$\frac{\mathrm{d}u_i}{\mathrm{d}t} = \sum_j c_{ij}u_j, \quad \sum_j c_{ij} = 0, \tag{21}$$

where $u_i$ are the nodal values, and $c_{ij}$ are some coefficients depending on the procedure employed for the spatial discretization. Actually, for incompressible flows the lumped-mass Galerkin discretization of the transport equation admits such a representation.

Since the coefficient matrix has zero row sum, the scheme can be rewritten as

$$\frac{\mathrm{d}u_i}{\mathrm{d}t} = \sum_{j\neq i} c_{ij}(u_j - u_i). \tag{22}$$

Furthermore, by supposing all coefficients to be nonnegative ($c_{ij} \geq 0,\ j \neq i$) we can show stability in the $L_\infty$-norm for this scheme.

Consider $u_i$ to be a maximum. It follows that for all $j$ $u_j - u_i \leq 0$, so that $\mathrm{d}u_i/\mathrm{d}t \leq 0$. Hence, a maximum cannot increase, and similarly a minimum cannot decrease. Recall that coefficient matrices are sparse in such a way that $c_{ij} \neq 0$ only if $i$ and $j$ are adjacent nodes. Restricting ourselves to a local neighborhood of $u_i$ the same consideration as above reveals that the method is *local extremum diminishing*. The LED criterion was introduced by Jameson [9], [10] as a convenient tool for the design of high-resolution schemes on unstructured meshes. It implies positivity, since if the initial solution is positive everywhere, then so is the global minimum which cannot decrease by definition. Hence, the LED property provides an effective mechanism for preventing the birth and growth of nonphysical oscillations.

### 4.2 Monotonicity

In one dimension the LED property ensures that the *total variation* of the solution which is defined as

$$TV(u) = \int_{-\infty}^{+\infty} \left| \frac{\partial u}{\partial x} \right| \mathrm{d}x \tag{23}$$

does not increase. As a matter of fact, equation (23) remains valid even for discontinuous $u(x)$ in the sense of distribution theory. For the sake of simplicity, consider homogeneous Dirichlet boundary conditions at both endpoints. Then the discrete total variation is given by

$$TV(u) = 2\Big(\sum \max u \; - \sum \min u\Big). \tag{24}$$

Thus, a one-dimensional LED scheme is necessarily *total variation diminishing*. This is a highly advantageous property, which has led to the development of a whole class of nonoscillatory TVD schemes and which is sufficient for *total variation stability*. Thus, existing convergence theorems for consistent TV-stable methods can easily be generalized [7], [13]. Let us remark that the TVD feature can be consulted for the construction of high-order schemes in contrast to just 'mimicking' other properties of the true solution which renders the accuracy to be at most first order [13].

Recall that equations (21) and (22) correspond to the problem discretized in space only. Let us now investigate the conditions under which a LED scheme will remain positive after the time discretization. If the standard one-step $\theta$-scheme is employed, the fully discretized equation reads

$$\begin{aligned} \frac{u_i^{n+1} - u_i^n}{\Delta t} &= \theta \sum_{j \neq i} c_{ij} (u_j^{n+1} - u_i^{n+1}) \\ &\quad + (1-\theta) \sum_{j \neq i} c_{ij} (u_j^n - u_i^n), \quad 0 \leq \theta \leq 1. \end{aligned} \tag{25}$$

The choice of the parameter $\theta$ specifies the type of time-stepping. The extreme cases $\theta = 0$ and $\theta = 1$ define the well-known forward and backward Euler methods. Both of them are first-order accurate with respect to the time step $\Delta t$. The method corresponding to $\theta = 0.5$ is known as the Crank-Nicolson scheme, which is second-order accurate.

The application of our lemma to equation (25) yields the following [11]

**Positivity Theorem.** *Any LED scheme discretized in time by the backward Euler method is unconditionally positive. Other time-stepping schemes* ($0 \leq \theta < 1$) *preserve positivity under the CFL-like condition*

$$1 + \Delta t (1-\theta) \min_i c_{ii} \geq 0. \tag{26}$$

An important message delivered by this theorem is that the positivity criterion at our disposal makes it possible to obtain rigorous estimates of the largest admissible time step for explicit and semi-implicit schemes. Remarkably, the derivation of the upper bound does not require any knowledge of the underlying partial differential equation and of the employed spatial mesh. It is sufficient to examine the diagonal coefficients $c_{ii}$ of the semi-discrete scheme. Upper bounds for non-LED schemes can be readily derived in much the same way.

## 5 Low-order discretization

To a large extent, the performance of the flux-corrected-transport procedure depends on the quality of the underlying low-order method which is supposed to preserve positivity and refrain from forming numerical wiggles. In the context of finite difference and finite volume discretizations, a perfect candidate for this job certainly is the upwind scheme. At the same time, it has been largely unclear how to perform upwinding in the finite element framework. Most upwind-like finite element methods encountered in the literature resort to a finite volume discretization for the convective terms [2], [29]. An alternative derivation of the least diffusive positivity-preserving scheme can be carried out by adding discrete diffusion depending solely on the magnitude and position of negative entries in the finite element matrix [11].

### 5.1 Discrete diffusion

Let the scalar conservation law (10), (11) be discretized in space by the Galerkin method

$$M_C \frac{\mathrm{d}u}{\mathrm{d}t} = K^H u + M_C q. \tag{27}$$

On the way of rendering this semi-discrete scheme positive, any implicit antidiffusion disguised in the consistent mass matrix must be removed by performing mass lumping. In the next step, the discrete transport operator $K^H$ has to be modified by applying a proper amount of artificial diffusion. To this end, we define a discrete dissipation tensor $D$ as a symmetric matrix with zero row and column sum which is designed so as to eliminate all negative off-diagonal entries of the high-order operator

$$d_{ii} = -\sum_{k \neq i} d_{ik}, \quad d_{ij} = d_{ji} = \max\{0, -k_{ij}^H, -k_{ji}^H\}, \quad \forall i < j. \tag{28}$$

In essence, this corresponds to applying one-dimensional diffusion operators associated with the (fictitious) edges connecting the adjacent nodes. As a result, the diffusive term can be decomposed into a sum of internodal fluxes, such that each flux is proportional to the difference of corresponding nodal values

$$(Du)_i = \sum_j d_{ij} u_j = \sum_{j \neq i} d_{ij}(u_j - u_i) = \sum_{j \neq i} f_{ij}. \tag{29}$$

Obviously, the two participating nodes receive the same flux but with opposite sign. Thus, the artificial diffusion operator ensures the conservation of mass. By adding the tensorial dissipation to the high-order transport operator its low-order counterpart $K^L = K^H + D$ is obtained.

It is worth mentioning, that if physical diffusion is strong enough the coefficients are nonnegative from the outset, and thus no artificial diffusion is introduced. Hence, for diffusion-dominated problems $K^H$ and $K^L$ coincide. Note that this approach can be carried over to the Euler equations by defining the modulation parameter $d_{ij}$ to be proportional to the spectral radius of the Roe matrix (see below).

After the discretization in time the modified scheme reads

$$[M_L - \theta \Delta t K^L] u^{n+1} = [M_L + (1-\theta)\Delta t K^L] u^n + \Delta t M_C q^{n+\theta}. \tag{30}$$

Note that the presence of (negative) source terms, which may originate from both chemical reactions and the transition to spherical or cylindrical coordinates, threatens to violate the positivity constraint. In principle, this can be overcome by following the linearizing splitting approach proposed by Patankar [22]

$$q = q_C + q_P u, \quad \text{where} \quad q_C \geq 0, \; q_P \leq 0. \tag{31}$$

Equation (30) can be transformed into the convenient form (20) with

$$L = M_L - \theta \Delta t K^L + \Delta t S^-$$
$$R = M_L + (1-\theta)\Delta t K^L + \Delta t S^+,$$

where $S^+$ and $S^-$ are diagonal matrices engendered by the source term. A splitting which ensures that $L$ and $R$ possess the properties required by the lemma can readily be derived [12].

By construction all off-diagonal entries of $L$ are nonpositive, while those of $R$ are nonnegative. It remains to indemnify the positivity of the diagonal coefficients. Since the elements of $M_L$ and the contributions of the source terms $q_C$ (if any) are positive this can always be achieved by choosing the time step small enough. Dropping the source terms for simplicity, the time step for the low-order discretization of the incompressible convection-diffusion equation is bounded by

$$\Delta t \leq \frac{1}{1-\theta} \min_i \{-m_i / k_{ii}^L \mid k_{ii}^L < 0\}, \tag{32}$$

where $m_i$ denotes the diagonal entries of the lumped mass matrix and $q_i^+$ those residing in the diagonal matrix $S^+$ [12]. From this CFL-like condition we get a sharp estimate of the largest admissible time step that can be used to steer adaptive time-stepping for (semi-)explicit schemes. In essence, the upper bound is determined by the ratio $m_i/k_{ii}^L$. It follows, that excessive artificial diffusion not only degrades the accuracy of the method but also requires taking impractically small time steps.

### 5.2 1D example: Pure convection

Let us illustrate the construction of the low-order operator by considering the one-dimensional scalar equation of pure convection with constant positive velocity

$$\frac{\partial u}{\partial t} + v\frac{\partial u}{\partial x} = 0. \tag{33}$$

Employing a discretization with linear elements on a uniform mesh yields the following element matrices

$$\hat{M}_L = \frac{\Delta x}{2}\begin{bmatrix} 1 & 0 \\ 0 & 1 \end{bmatrix}, \quad \hat{K}^H = \frac{v}{2}\begin{bmatrix} 1 & -1 \\ 1 & -1 \end{bmatrix}. \tag{34}$$

After the global assembly, the central difference approximation of the convective terms is recovered at the interior nodes

$$\frac{\mathrm{d}u_i}{\mathrm{d}t} = -v\,\frac{u_{i+1} - u_{i-1}}{2\Delta x}. \tag{35}$$

In order to eliminate the negative entries of $K^H$, the artificial dissipation has to be taken proportional to $\hat{d}_{12} = v/2$ to produce the least diffusive results but to ensure positivity. Consequently, the corresponding diffusion operator restricted to one element is given by

$$\hat{D} = \frac{v}{2}\begin{bmatrix} -1 & 1 \\ 1 & -1 \end{bmatrix} \Rightarrow \hat{K}^L = v\begin{bmatrix} 0 & 0 \\ 1 & -1 \end{bmatrix}. \tag{36}$$

Putting it all together, the low-order scheme coincides with the upwind finite difference method

$$\frac{\mathrm{d}u_i}{\mathrm{d}t} = -v\,\frac{u_i - u_{i-1}}{\Delta x} \tag{37}$$

which preserves positivity provided that the time step satisfies this CFL-like condition

$$v\frac{\Delta t}{\Delta x} \leq \frac{1}{1-\theta}. \tag{38}$$

In a nutshell, this example demonstrates that our low-order discretization reduces to standard upwinding for pure convection in one spatial dimension. At the same time, its derivation based on the postprocessing of the discrete transport operator remains valid for arbitrary meshes and multi-dimensional problems. Moreover, physical diffusion (if any) is automatically detected, and the amount of artificial diffusion is reduced accordingly.

## 6 Flux-based FEM-FCT formulation

Another cornerstone of the FEM-FCT algorithm is the linear high-order method. A variety of finite element schemes employing streamline diffusion to stabilize the troublesome convective terms were proposed in the literature, e.g. [4]. For instance, Taylor-Galerkin methods attribute this stabilization to high-order time derivatives in the Taylor series expansion. This leads to improved time-stepping schemes which are combined with the standard Galerkin spatial discretization. The most popular representative of such stabilized methods is the well-known Lax-Wendroff scheme. An investigation of the modified equation for its finite element counterpart reveals that the introduced dissipation just counterbalances the intrinsic negative diffusion which renders the explicit Euler/Galerkin scheme unstable for pure convection problems. For an in-depth study of the Lax-Wendroff and higher order Taylor-Galerkin methods the reader is referred to [4].

In this paper we will concentrate on implicit finite element schemes based on the Crank-Nicolson and backward Euler time-stepping which are unconditionally stable. We just mention that a stabilization of convective terms is mandatory for the fully explicit time discretization. Linear Galerkin schemes are of little use, because they are prone to nonphysical oscillations. The incorporation of a flux limiter makes it possible to get rid of oscillations in the framework of nonlinear Crank-Nicolson/FCT and backward Euler/FCT methods.

The high-order transport operator can be transformed into a low-order one as explained in the previous section. For simplicity, let us omit the (linearized) source terms. The resulting methods of high and low order discretized in time by the standard one-step $\theta$-scheme are related by the following formula

$$[M_L - \theta\Delta t K^L]u^H = [M_L + (1-\theta)\Delta t K^L]u^n + F(u^H, u^n), \qquad (39)$$

where the diffusion responsible for high spatial accuracy is given by

$$\begin{aligned} F(u^H, u^n) = &- [(M_C - M_L) + \theta\Delta t(K^L - K^H)]u^H \\ &+ [(M_C - M_L) - (1-\theta)\Delta t(K^L - K^H)]u^n. \end{aligned} \qquad (40)$$

Here the superscript $H$ refers to the high-order solution. If the antidiffusive term $F(u^H, u^n)$ is omitted, then the positive low-order scheme will be recovered, whereas retaining it yields the original high-order method.

It can readily be seen that all the matrices in (40) represent discrete (anti-)diffusion operators featuring zero row and column sums. Hence, they admit the decomposition into a sum of internodal fluxes

$$\begin{aligned} f_{ij} &= -(m_{ij} + \theta\Delta t d_{ij})(u_j^H - u_i^H) + (m_{ij} - (1-\theta)\Delta t d_{ij})(u_j^n - u_i^n) \\ f_{ji} &= -f_{ij}, \quad i \neq j. \end{aligned} \qquad (41)$$

These raw antidiffusive fluxes counterbalance the errors induced by mass lumping and 'upwinding'. The coefficients $m_{ij}$ and $d_{ij}$ denote the entries of the consistent mass matrix and the artificial diffusion, respectively.

The crucial step of the FCT procedure consists in adding as much antidiffusion as possible without generating nonphysical undershoots and overshoots. Incorporating the limiter, the implicit FEM-FCT algorithm for equation (39) reads

$$[M_L - \theta \Delta t K^L]u^{n+1} = M_L \tilde{u} + F^*(u^H, u^n), \quad F^*(u^H, u^n) = \sum_{j \neq i} \alpha_{ij} f_{ij} \qquad (42)$$

where $F^*(u^H, u^n)$ denotes the limited antidiffusion ($\alpha_{ij}$ will be defined later), while $\tilde{u}$ represents the positivity-preserving solution to the explicit subproblem

$$M_L \tilde{u} = [M_L + (1-\theta)\Delta t K^L]u^n. \qquad (43)$$

In essence, $\tilde{u}$ corresponds to an intermediate solution computed at the time instant $t^{n+1-\theta}$ by the explicit low-order scheme which reduces to the old solution $u^n$ for the backward Euler method.

It is obvious that the success of the FCT algorithm depends on the positivity of the provisional solution $\tilde{u}$ and on the choice of the correction factors $\alpha_{ij}$. For $\tilde{u}$ to be positive, the time step must satisfy the CFL-like condition (32) unless the scheme is fully implicit. As long as the left-hand side operator is an M-matrix, our positivity criterion ensures that scheme (41) can be rendered positive by tuning the correction factors.

This new family of FEM-FCT schemes distinguishes itself in that it is applicable to explicit and implicit time discretizations alike. Note that implicit schemes require solving *two* nonsymmetric linear systems per time step: one for the high-order solution (which is needed to compute the antidiffusive fluxes) and one for the final solution. However, implicit methods are typically more efficient than explicit ones for they allow to use larger time steps.

The majority of practical applications are described by *nonlinear* conservation laws. In this case, the matrices $K^H$ and $K^L$ depend on the unknown solution, so that additional outer iterations are necessary for implicit schemes. The simplest iterative treatment of nonlinearities is afforded by a fixed point defect correction method. If we consider an abstract nonlinear system of the form

$$S(u)u = g, \qquad (44)$$

then the basic nonlinear iteration can be formulated as

$$u^{(m+1)} = u^{(m)} + [C(u^{(m)})]^{-1}(g - S(u^{(m)})u^{(m)}), \qquad (45)$$

where $m$ is the outer iteration counter, and $C$ is a suitably chosen 'preconditioner' (an approximate Fréchet derivative) which should be easy to invert. The iteration process is terminated when the relative solution changes are small enough or $m$ exceeds a given limit. As a rule, the 'inversion' of $C$ is also performed by some iterative procedure. Hence, a certain number of inner iterations per cycle is required. It is worth mentioning that the problem does not have to be solved very accurately at each outer iteration. A moderate improvement of the residual (1–2 digits) is sufficient to obtain a good overall accuracy.

For a nonlinear problem of the form (39), it is reasonable to use the low-order operator as preconditioner

$$C(u^{(m)}) = M_L - \theta\Delta t K^L(u^{(m)}). \tag{46}$$

This yields an iterative FEM-FCT algorithm, whereby the approximate solution and the transport operator are successively updated as follows:

$$[M_L - \theta\Delta t K^L(u^{(m)})]u^{(m+1)} = [M_L + (1-\theta)\Delta t K^L(u^n)]u^n \\ + F^*(u^{(m)}, u^n). \tag{47}$$

Flux correction can be performed after each outer iteration or just once after the high-order solution has converged. In either case, positivity of the numerical solution is guaranteed.

## 7 Limiting strategy

The flux limiter is a key element of the FEM-FCT paradigm. By varying the correction factors $\alpha_{ij}$ between zero and unity, it it possible to obtain the diffusive low-order solution, the oscillatory high-order one or anything in-between. Obviously, it is desirable to utilize the antidiffusive terms to the largest extent possible without generating wiggles and violating the positivity constraint. Kuzmin and Turek [11] demonstrated that Zalesak's multi-dimensional limiter can be generalized to implicit time discretizations.

### 7.1 Zalesak's limiter

The multi-dimensional version of the original limiter [3] was proposed by Zalesak [30] remaining in the realm of finite differences. Parrott and Christie [21] were the first who utilized flux-correction in the context of finite elements. FEM-FCT reached maturity by the efforts done by Löhner [16]. A quite comprehensive description of Zalesak' s limiter containing a geometrical interpretation can be found in [12]. However, the geometrical interpretation fails if we consider implicit time discretizations. Consequently, we have to apply the mathematical theory of positivity-preserving schemes to prove the positivity of Zaselak's limiter for arbitrary time-stepping. Let us just give a brief outline of the theoretical aspects.

Considering a local maximum or minimum $\tilde{u}_i^{\max/\min}$, any incoming flux $f_{ij}$ which would accentuate the extremum has to be canceled out completely.

Afterwards, we can represent the right-hand side of our FEM-FCT scheme in the following form

$$RHS = m_i\tilde{u}_i + \sum_{j\neq i} \alpha_{ij} f_{ij} = m_i\tilde{u}_i + c_i Q_i, \quad c_i = \frac{\sum_{j\neq i} \alpha_{ij} f_{ij}}{Q_i}, \tag{48}$$

where the intermediate solution $\tilde{u} = u^L(t^{n+1-\theta})$ is subject to the concrete time-stepping employed and the multiplier $Q_i$ is chosen as follows

$$Q_i = \begin{cases} Q_i^+ = \tilde{u}_i^{\max} - \tilde{u}_i, & \text{if } \sum_{j\neq i} \alpha_{ij} f_{ij} > 0, \\ Q_i^- = \tilde{u}_i^{\min} - \tilde{u}_i, & \text{if } \sum_{j\neq i} \alpha_{ij} f_{ij} < 0, \\ 1, & \text{if } \sum_{j\neq i} \alpha_{ij} f_{ij} = 0. \end{cases} \tag{49}$$

Keep in mind that we took the necessary precautions that all $Q_i \neq 0$ by canceling out the local extremum accentuating fluxes completely.

Let us consider the local extremum $\tilde{u}_i^{\max/\min}$ to be attained at some node $k$ adjacent to node $i$. The LED property of the antidiffusive term makes it possible to rewrite the equation at hand as

$$RHS = m_i\tilde{u}_i + c_i(\tilde{u}_k - \tilde{u}_i) = (m_i - c_i)\tilde{u}_i + c_i\tilde{u}_k, \quad c_i \geq 0 \tag{50}$$

which states that positivity is preserved as long as $m_i \geq c_i$. Obviously, Zalesak's limiter picks out the proper correction factors $\alpha_{ij}$ for this job as can be seen from

$$m_i Q_i^- \leq m_i R_i^- P_i^- \leq \sum_{j\neq i} \alpha_{ij} f_{ij} \leq m_i R_i^+ P_i^+ \leq m_i Q_i^+, \tag{51}$$

where the sum of contributions to node $i$ is denoted by

$$P_i^\pm = \frac{1}{m_i} \sum_{j\neq i} \max_{\min}\{0, f_{ij}\}. \tag{52}$$

Finally, the auxiliary quantities $R_i^\pm$ which determine the maximum percentage of allowable flux into node $i$ are defined as

$$R_i^\pm = \begin{cases} \min\{1, Q_i^\pm / P_i^\pm\}, & \text{if } P_i^\pm \neq 0, \\ 0, & \text{if } P_i^\pm = 0. \end{cases} \tag{53}$$

As a result, the proper correction factors have to be chosen according to

$$\alpha_{ij} = \begin{cases} \min\{R_i^+, R_j^-\}, & \text{if } f_{ij} \geq 0, \\ \min\{R_j^+, R_i^-\}, & \text{if } f_{ij} < 0. \end{cases} \tag{54}$$

### 7.2 Limiting for systems of equations

Despite the remarkable progress in the development of FCT schemes for scalar equations, it remained largely unclear how to apply flux correction to systems of hyperbolic conservation laws. Neglecting the special character of coupled systems, a conventional operator-splitting approach is feasible by treating each scalar subproblem in a block-iterative loop. However, limiting each variable independently turned out to be of no good. Consequently, the FCT community has devised a common limiter for the entire system by merging individual limiters for different variables.

Flux correction for the system of Euler equations was addressed by Löhner [15], [16]. He singled out the following approaches to the design of a synchronized limiter:

- Apply correction factors corresponding to a single 'indicator variable'.
- Apply the minimum of correction factors for a group of variables.

Some clues for a suitable choice of variables will be given in the subsequent section.

## 8 Application to the Euler equations

The new methodology for scalar transport equations can be generalized to hyperbolic systems of conservation laws. Nevertheless, a rigorous positivity proof is still outstanding and meanwhile some heuristic considerations have to be employed. In what follows, we build on the concept of a Roe matrix to devise an efficient algorithm for the edge-by-edge matrix assembly and discrete upwinding. Another option is to apply the Galerkin flux decomposition and modify the numerical flux as explained above. The resulting large nonlinear systems call for the use of a smart iteration technique to minimize the computational costs. As in the scalar case, we resort to defect correction with a block-diagonal low-order preconditioner.

### 8.1 Matrix assembly

Let us start with the conservative formulation of the Euler equations. If we apply the Galerkin discretization to the weak formulation of equation (1) without integrating by parts and recall that customary basis functions sum to unity, the semi-discrete problem can be written in the following form

$$\begin{aligned}\left[M\frac{\mathrm{d}U}{\mathrm{d}t}\right]_i &= \sum_{j\neq i}\mathbf{k}_{ij}\cdot(\mathbf{F}_j-\mathbf{F}_i) \qquad (55)\\ &= \sum_{j\neq i}\left[\frac{\mathbf{k}_{ij}-\mathbf{k}_{ji}}{2}+\frac{\mathbf{k}_{ij}+\mathbf{k}_{ji}}{2}\right]\cdot\hat{\mathbf{A}}_{ij}(U_j-U_i),\end{aligned}$$

$$\left[M\frac{\mathrm{d}U}{\mathrm{d}t}\right]_j = \sum_{j\neq i} \mathbf{k}_{ji}\cdot(\mathbf{F}_i - \mathbf{F}_j)$$
$$= \sum_{j\neq i}\left[\frac{\mathbf{k}_{ji}-\mathbf{k}_{ij}}{2} + \frac{\mathbf{k}_{ij}+\mathbf{k}_{ji}}{2}\right]\cdot \hat{\mathbf{A}}_{ij}(U_i - U_j)$$

for node $i$ and $j$, respectively. Here $\mathbf{k}_{ij}$ denotes the vector of coefficients coming from the Galerkin discretization, i.e. $-\mathbf{c}_{ji}$ in the notation of Section 3. In particular, $k_{ij} = \pm 1/2$ in one dimension. Integration by parts yields the following relation (see [15], p. 195):

$$\mathbf{k}_{ij} + \mathbf{k}_{ji} = -\int_\Gamma \mathbf{n}\varphi_i\varphi_j \,\mathrm{d}s, \tag{56}$$

where $\mathbf{n}$ is the outward unit normal. In equation (55), $\mathbf{k}_{ij}$ was decomposed into

$$\mathbf{k}^I_{ij} := \frac{\mathbf{k}_{ij}-\mathbf{k}_{ji}}{2} = \mathbf{k}_{ij} + \frac{1}{2}\int_\Gamma \mathbf{n}\varphi_i\varphi_j\,\mathrm{d}s, \tag{57}$$
$$\mathbf{k}^B_{ij} := \frac{\mathbf{k}_{ij}+\mathbf{k}_{ji}}{2} = -\frac{1}{2}\int_\Gamma \mathbf{n}\varphi_i\varphi_j\,\mathrm{d}s.$$

Basis functions corresponding to internal nodes vanish on the boundary so that $\mathbf{k}^I_{ij} = \mathbf{k}_{ij}$ while $\mathbf{k}^B_{ij} = 0$ in this case. For linear and multilinear finite elements we have $\mathbf{k}^B_{ij} = -\mathbf{n}s_{ij}/2$ which is nonzero iff both nodes are on the boundary. Here $s_{ij} = \int_\Gamma \varphi_i\varphi_j\,\mathrm{d}s$ denotes an entry of the mass matrix for the surface triangulation. For instance, $s_{ij} = |\Gamma_{ij}|/6$ in 2D. Shortly we will see why the splitting of coefficients into a sum of internal and boundary contributions is of use.

The vector $\mathbf{F}$ in equation (55) denotes the nodal values of fluxes. Furthermore, the matrix $\hat{\mathbf{A}}_{ij}$ is the so-called Roe matrix which is obtained by evaluating the Jacobians at the intermediate state [24]

$$\hat{\rho}_{ij} = \sqrt{\rho_i\rho_j}, \quad \hat{\mathbf{v}}_{ij} = \frac{\sqrt{\rho_i}\mathbf{v}_i + \sqrt{\rho_j}\mathbf{v}_j}{\sqrt{\rho_i}+\sqrt{\rho_j}}, \quad \hat{h}_{ij} = \frac{\sqrt{\rho_i}h_i + \sqrt{\rho_j}h_j}{\sqrt{\rho_i}+\sqrt{\rho_j}}. \tag{58}$$

The resulting density-averaged quantities $\hat{\rho}_{ij}$, $\hat{\mathbf{v}}_{ij}$ and $\hat{h}_{ij}$ are called the Roe mean values.

While in the one-dimensional case there is only one Jacobian matrix to be assembled for each pair of nodes, in multi-dimensions it is replaced by a suitable linear combination of the averaged Jacobians $\hat{A}^{x_d}_{ij}$ for each coordinate direction $d = 1, 2, 3$. To this end, we introduce the so-called cumulative Roe matrix

$$\mathrm{A}_{ij} := \mathbf{k}_{ij}\cdot\hat{\mathbf{A}}_{ij} = k^{x_1}_{ij}\hat{A}^{x_1}_{ij} + k^{x_2}_{ij}\hat{A}^{x_2}_{ij} + k^{x_3}_{ij}\hat{A}^{x_3}_{ij}$$

which consists of the interior and boundary contributions:

$$\mathrm{A}_{ij} = \mathrm{A}^I_{ij} + \mathrm{A}^B_{ij}, \quad \text{where} \quad \mathrm{A}^I_{ij} := \mathbf{k}^I_{ij}\cdot\hat{\mathbf{A}}_{ij}, \quad \mathrm{A}^B_{ij} := \mathbf{k}^B_{ij}\cdot\hat{\mathbf{A}}_{ij}. \tag{59}$$

Let us remark that $\mathrm{A}_{ij}$ can be interpreted as a projection of the multidimensional Jacobian tensor onto the corresponding 'numerical edge' connecting the nodes $i$ and $j$. Each edge corresponds to a pair of basis functions with overlapping supports. Hence, the list of edges is determined by the sparsity pattern of the global finite element matrix.

Similarly, the Roe matrix $\mathrm{A}_{ji}$ for the coefficient vector $\mathbf{k}_{ji}$ is given by

$$\mathrm{A}_{ji} := \mathbf{k}_{ji} \cdot \hat{\mathbf{A}}_{ij} = -\mathrm{A}_{ij}^{I} + \mathrm{A}_{ij}^{B}.$$

For internal nodes $\mathrm{A}_{ji} = -\mathrm{A}_{ij}$ since the boundary contribution $\mathrm{A}_{ij}^{B}$ vanishes. Hence, just one cumulative Jacobian per edge needs to be evaluated in this case. Otherwise, the local Roe matrices associated with the numerical edge will consist of a skew-symmetric internal part and a symmetric boundary part.

The transition from the fluxes to the nodal values of the conservative variables in (55) enables us to assemble finite element matrices explicitly. The connectivity of the global matrix depends on the underlying mesh and on the type of finite element approximation. For systems of equations, the array of edges remains the same as in the scalar case. However, there are interactions not only between basis functions for different nodes but also between basis functions for different variables. Hence, each coefficient of the discrete operator turns into a matrix of size equal to the squared number of variables.

For the three-dimensional Euler equations, the contribution of a numerical edge to the global Jacobian matrix $K$ is represented by the following $5 \times 5$ blocks:

$$\begin{aligned} \mathrm{K}_{ii} &= -\mathrm{A}_{ij}, \quad & \mathrm{K}_{ij} &= \mathrm{A}_{ij}, \\ \mathrm{K}_{ji} &= \mathrm{A}_{ji}, \quad & \mathrm{K}_{jj} &= -\mathrm{A}_{ji}. \end{aligned} \tag{60}$$

These local Jacobians are evaluated edge-by-edge, and their entries $\mathrm{K}_{ij}^{kl}$ $(k, l = 1, \ldots, 5)$ are scattered to the corresponding positions in the blocks $K_{kl} \in \mathbb{R}^{N \times N}$, where $N$ stands for the total number of nodes. The fully discretized high-order scheme can be written in the compact form $S^H U^H = G^H$. The constituent blocks of the overall stiffness matrix $S^H$ and of the load vector $G^H$ are defined as

$$S_{kl}^{H} = M_C \delta_{kl} - \theta \Delta t K_{kl}, \quad g_k^H = M_C u_k^n + (1-\theta)\Delta t \sum_l K_{kl} u_l^n. \tag{61}$$

Due to the nonlinearity of the Euler equations, the discrete system has to be solved iteratively. Hence, it is not necessary to assemble and store the global stiffness matrix. The contributions of edges can be inserted directly into the defect vector. If a block-diagonal preconditioner is used (see below), the storage requirement will reduce from 25 to 5 blocks. A distinct advantage of the edge-by-edge assembly algorithm proposed in this section is that the projection coefficients $\mathbf{k}_{ij}$ are typically fixed, so that the matrices can be assembled efficiently without resorting to numerical integration.

## 8.2 Discrete diffusion

A usable low-order method can be constructed by adding tensorial artificial diffusion to the local Roe matrices. In the framework of scalar theory for positivity-preserving schemes, the artificial dissipation was tailored so as to eliminate all negative off-diagonal entries. For systems of equations, the off-diagonal entries are no longer scalar quantities but matrices themselves. Adapting the procedure which underlies the scalar theory, we have to design the dissipation tensor so as to render all off-diagonal matrix blocks positive definite. To this end, we perform mass lumping and apply artificial diffusion to the edge contributions before inserting them into the global matrix and/or the defect vector:

$$\begin{aligned} \mathrm{K}_{ii} &= \mathrm{K}_{ii} - \mathrm{D}_{ij}, \quad \mathrm{K}_{ij} = \mathrm{K}_{ij} + \mathrm{D}_{ij}, \\ \mathrm{K}_{ji} &= \mathrm{K}_{ji} + \mathrm{D}_{ij}, \quad \mathrm{K}_{jj} = \mathrm{K}_{jj} - \mathrm{D}_{ij}. \end{aligned} \tag{62}$$

As a result, the low-order scheme can be represented as $S^L U^L = G^L$ with the corresponding block entries

$$\begin{aligned} S_{kl}^L &= M_L \delta_{kl} - \theta \Delta t (K_{kl} + D_{kl}), \\ g_k^L &= M_L u_k^n + (1-\theta)\Delta t \sum_l (K_{kl} + D_{kl}) u_l^n. \end{aligned} \tag{63}$$

In the rest of this section, we will concentrate on the specification of the quantity $\mathrm{D}_{ij}$.

The hyperbolicity of the Euler equations implies that any linear combination of the Jacobian matrices is diagonalizable with real, not necessarily distinct eigenvalues (see above). Thus, for an arbitrary coefficient vector $\mathbf{e}_{ij}$, there exists a regular matrix $\hat{R}(\mathbf{e}_{ij})$ of right eigenvectors which diagonalizes the cumulative Roe matrix

$$\mathbf{e}_{ij} \cdot \hat{\mathbf{A}}_{ij} = \hat{R}(\mathbf{e}_{ij}) \hat{\Lambda}(\mathbf{e}_{ij}) \hat{R}(\mathbf{e}_{ij})^{-1}, \tag{64}$$

where the diagonal matrix of eigenvalues is given by

$$\hat{\Lambda}(\mathbf{e}_{ij}) = |\mathbf{e}_{ij}| \operatorname{diag}(\hat{v}_{ij} - \hat{c}_{ij}, \hat{v}_{ij}, \hat{v}_{ij}, \hat{v}_{ij}, \hat{v}_{ij} + \hat{c}_{ij}). \tag{65}$$

The projected mean velocity $\hat{v}_{ij}$ and the characteristic speed of sound $\hat{c}_{ij}$ read

$$\hat{v}_{ij} = \frac{\mathbf{e}_{ij} \cdot \hat{\mathbf{v}}_{ij}}{|\mathbf{e}_{ij}|}, \quad \hat{c}_{ij} = \sqrt{(\gamma - 1)(\hat{h}_{ij} - \hat{v}_{ij}^2/2)}. \tag{66}$$

Recall that the Galerkin discretization with matrix assembly carried out in an edge-by-edge fashion leads to off-diagonal Jacobian blocks which consist of skew-symmetric internal contributions and symmetric boundary contributions:

$$\mathrm{K}_{ij} = \mathrm{A}_{ij}^I + \mathrm{A}_{ij}^B, \quad \mathrm{K}_{ji} = -\mathrm{A}_{ij}^I + \mathrm{A}_{ij}^B. \tag{67}$$

To enforce the positive definiteness of all components, we construct the tensor of artificial diffusion in an additive way so as to eliminate the negative eigenvalues. Making use of the characteristic decomposition procedure (64), we arrive at

$$\mathrm{D}_{ij} = |\mathrm{A}_{ij}^I| - \langle \mathrm{A}_{ij}^B \rangle^-, \tag{68}$$

i.e. the interior and boundary contributions (if any) are modified by

$$|\mathrm{A}_{ij}^I| = \hat{R}(\mathbf{k}_{ij}^I)|\hat{\Lambda}(\mathbf{k}_{ij}^I)|\hat{R}(\mathbf{k}_{ij}^I)^{-1} \tag{69}$$

and

$$\langle \mathrm{A}_{ij}^B \rangle^- = \hat{R}(\mathbf{k}_{ij}^B)\langle \hat{\Lambda}(\mathbf{k}_{ij}^B)\rangle^- \hat{R}(\mathbf{k}_{ij}^B)^{-1}, \tag{70}$$

respectively, to achieve the 'upwinding' effect. Here $\langle \hat{\Lambda}(\mathbf{k}_{ij}^B)\rangle^-$ denotes the diagonal matrix of negative eigenvalues which is obtained by setting all positive entries of $\hat{\Lambda}(\mathbf{k}_{ij}^B)$ equal to zero. Alternatively, we can assemble the boundary integral corresponding to $\mathrm{A}_{ij}^B$ separately using a lumped mass matrix with entries $s_i = \int_\Gamma \varphi_i \, ds$ for the surface triangulation.

For the purpose of illustration, let us consider the one-dimensional Euler equations whereby no neighboring nodes are situated on the boundary simultaneously. In this case, the proposed technique reduces to Roe's approximate Riemann solver [24], which represents one of the most popular upwind-biased methods for the numerical solution of the Euler equations. It can be interpreted as the replacement of the centered Galerkin flux $G_{ij} = (F_i + F_j)/2$ in decomposition (18) by the consistent numerical flux

$$G_{ij}^* = \frac{F_i + F_j}{2} + \frac{1}{2}|\hat{A}_{ij}|(U_j - U_i), \tag{71}$$

where

$$|\hat{A}_{ij}| = \hat{R}_{ij}|\hat{\Lambda}_{ij}|\hat{R}_{ij}^{-1} \quad \text{and} \quad |\hat{\Lambda}_{ij}| = \mathrm{diag}(|\hat{v}_{ij} - \hat{c}_{ij}|, |\hat{v}_{ij}|, |\hat{v}_{ij} + \hat{c}_{ij}|). \tag{72}$$

Roe's approximate Riemann solver and its various extensions [25] constitute fairly good low-order methods *per se* but they are not to be recommended for the use in the FEM-FCT environment due to considerable overhead costs. A much cheaper alternative is to add scalar dissipation. The local tensor of artificial diffusion is taken to be a diagonal matrix whose internal part is proportional to the spectral radius of the Roe matrix:

$$\mathrm{D}_{ij} = d_{ij} I_5, \quad \text{where} \quad d_{ij} = |\hat{\lambda}(\mathbf{k}_{ij}^I)| - \langle \hat{\lambda}(\mathbf{k}_{ij}^B)\rangle^-. \tag{73}$$

Here $I_5$ stands for the $5 \times 5$ identity matrix, while $|\hat{\lambda}(\mathbf{k}_{ij}^I)|$ and $\langle\hat{\lambda}(\mathbf{k}_{ij}^B)\rangle^-$ denote the largest in magnitude diagonal entries of $|\hat{\Lambda}(\mathbf{k}_{ij}^I)|$ and $\langle\hat{\Lambda}(\mathbf{k}_{ij}^B)\rangle^-$, respectively. According to (65), $|\hat{\lambda}(\mathbf{k}_{ij}^I)| = (|\hat{v}_{ij}| + \hat{c}_{ij})|\mathbf{k}_{ij}^I|$. In one space dimension the diffusion coefficient is simply $d_{ij} = (|\hat{v}_{ij}| + \hat{c}_{ij})/2$, while in the scalar case the proposed technique reduces to the 'discrete upwinding' approach introduced in Section 5 and illustrated by the 1D example.

Note that during the global assembly (62) the artificial diffusion operator has to be applied only to the 5 diagonal blocks of the global matrix. Moreover, it is the same for all components which results in considerable savings in terms of both computational costs and storage requirements. As long as excessive artificial diffusion is removed in the flux correction step, a slightly better accuracy of a consistent Riemann solver does not pay off. Hence, it is expedient to treat hyperbolic systems by adding scalar dissipation.

### 8.3 FEM-FCT algorithm

After the discretization in time by the standard $\theta$-scheme, the methods of high and low order can be combined similarly to relation (39). Let us consider the resulting nonlinear system to be given in the generic form $SU = G$ which can be decomposed into

$$\begin{bmatrix} S_{11} & S_{12} & S_{13} & S_{14} & S_{15} \\ S_{21} & S_{22} & S_{23} & S_{24} & S_{25} \\ S_{31} & S_{32} & S_{33} & S_{34} & S_{35} \\ S_{41} & S_{42} & S_{43} & S_{44} & S_{45} \\ S_{51} & S_{52} & S_{53} & S_{54} & S_{55} \end{bmatrix} \begin{bmatrix} u_1 \\ u_2 \\ u_3 \\ u_4 \\ u_5 \end{bmatrix} = \begin{bmatrix} g_1 \\ g_2 \\ g_3 \\ g_4 \\ g_5 \end{bmatrix}. \tag{74}$$

Note that equation (74) is just an abstract representation and nothing to be implemented in a real code. As pointed out earlier for the scalar case the iterative treatment of nonlinearities can be afforded by a fixed point defect correction method. As a consequence, the update can be performed in much the same way as in equation (45):

$$U^{(m+1)} = U^{(m)} + [C^{(m)}]^{-1}(G - S^{(m)}U^{(m)}). \tag{75}$$

The choice of a suitable 'preconditioner' $C$ becomes much more important for systems than in the scalar case. Considering $S$ as a global matrix without remembering where it came from would result in solving *one* huge nonlinear and extremely ill-conditioned system. However, recall that the abstract vector $U$ denotes the conservative variables which are governed by a set of strongly coupled but *scalar* conservation laws. The choice of a block-diagonal 'preconditioner' turns out to be of great use. Let the 5 diagonal blocks of $C$ be given by the appropriate low-order operators

$$C_k^{(m)} = M_L - \theta\Delta t[K_{kk}^{(m)} + D_{kk}^{(m)}], \quad k = 1, \dots, 5, \tag{76}$$

which constitutes a generalization of equation (46) to coupled systems. This kind of a preconditioner leads to 'well-behaved' linear systems for the solution increments.

Applying the sum of blocks $S_{kl}$ multiplied by $u_l$ to the defect vector in equation (75) yields a sequence of scalar subproblems for the individual variables

$$\begin{aligned} C_k^{(m)} \delta u_k^{(m)} = {} & M_L u_k^n + (1-\theta)\Delta t \sum_l [K_{kl}^n + D_{kl}^n] u_l^n \\ & - M_L u_k^{(m)} + \theta \Delta t \sum_l [K_{kl}^{(m)} + D_{kl}^{(m)}] u_l^{(m)} \\ & + F^*(u_k^{(m)}, u_k^n). \end{aligned} \tag{77}$$

Here the auxiliary quantity $\delta u_k^{(m)}$ denotes the correction to be applied to the current iterate $u_k^{(m)}$. The right-hand side of the above system is represented by the defect vector which consists of a low-order contribution and compensating antidiffusion.

The corresponding raw antidiffusive terms are determined in much the same way as in the scalar case (40), namely,

$$\begin{aligned} F(u_k^{(m)}, u_k^n) = & - [(M_C - M_L) + \theta \Delta t D_{kk}^{(m)}] u_k^{(m)} \\ & + [(M_C - M_L) - (1-\theta)\Delta t D_{kk}^n] u_k^n. \end{aligned} \tag{78}$$

At this point recall that we use scalar dissipation, so that the local diffusion operator $\mathrm{D}_{ij}$ used in the edge-by-edge assembly of the global blocks $D_{kl}$ is defined by a single diffusion coefficient valid for all variables. Hence, just the diagonal blocks $D_{kk}$ need to be assembled. An approximate Riemann solver based on tensorial artificial diffusion would require the generation of all $5 \times 5$ blocks. To clarify the notation, we remark that the subscripts $k$ and $l$ refer to the numbers of the conservative variables, while $i$ and $j$ are reserved for the numbering of mesh nodes.

Let us briefly address some implementation details. Generally speaking, the assembly of fluxes is carried out following the scalar algorithm. Each nonzero entry of the upper/lower triangular part of the global finite element matrix can be attributed to a numerical edge. Using the sparsity pattern to determine the list of edges, we compute the local Roe matrices and apply artificial diffusion. The resulting matrices $\mathrm{K}_{ii}$, $\mathrm{K}_{ij}$, $\mathrm{K}_{ji}$, $\mathrm{K}_{jj}$ represent the contribution of the edge to the blocks $K_{kl}$ of the global Jacobian. In a practical implementation, we abstain from manipulations with large matrices and insert edge contributions directly into the five preconditioner blocks $C_k$ and into the defect vector.

The array of antidiffusive fluxes is generated on the fly in the same loop over the edges but extra efforts have to be invested into the limiting process. If the correction factors $\alpha_{ij}^k$ were determined separately for each variable $k$, the lack of synchronization would give rise to oscillations and violate the principle of mass conservation. To overcome this drawback, a suitable combination of the limiting coefficients $\alpha_{ij} = \mathcal{L}(\alpha_{ij}^1, \dots, \alpha_{ij}^5)$ has to be taken [15]. For highly unsteady flows taking the minimum of those for the density and energy is recommended. By transition to the set of primitive variables the minimum of the correction factors for the density and pressure provides a reasonable choice especially for stationary problems.

As a matter of fact, this kind of transition can be generalized [15] to an arbitrary set of variables $U'$. For this purpose we have to pursue the following algorithm. First, convert the intermediate solution values $\tilde{U}_i$ into their nonconservative counterparts $\tilde{U}_i' := T(\tilde{U}_i)\tilde{U}_i$. Next, compute the associated upper and lower bounds. By considering the transformed fluxes $F_{ij}' := T(\tilde{U}_i)F_{ij}$ for each edge, we can estimate the correction factors $\beta_{ij}^k$ and assemble the corrected fluxes following

$$F_i^*(u_k^{(m)}, u_k^n) = T(\tilde{U}_i)^{-1} \sum_{j\neq i} \mathcal{L}(\beta_{ij}^1, \dots, \beta_{ij}^5) F_{ij}'. \tag{79}$$

However, the skew-symmetry of the fluxes may be lost for the nonconservative variables. The use of a synchronization procedure implies that the correction factors reduce to a scalar quantity and thus (79) simplifies dramatically. Specifically, the coefficients $\beta_{ij}^k$ are determined for the set of control variables $\tilde{U}'$ but applied to the conservative fluxes $F_{ij}$.

As the final step of the defect correction loop, the update (75) is performed by adding the increments to the solution from the previous iteration. A suitable underrelaxation strategy can be employed to aid convergence

$$u_k^{(m+1)} = u_k^{(m)} + \omega_k^{(m)} \delta u_k^{(m)}, \quad u_k^{(0)} = u_k^n. \tag{80}$$

## 9 Summary of the algorithm

As we have seen, the generalized FEM-FCT formulation in combination with Zalesak's limiter can be applied to a wide range of problems described by (systems of) conservation laws of the form (10). The proposed high-resolution finite element scheme can be implemented on arbitrary unstructured grids using either the conventional or the edge-based data structure. Let us summarize the main algorithm by describing one cycle of the nonlinear defect correction loop:

1. Do an edge-by-edge assembly of the diagonal preconditioner blocks $C_k^{(m)}$ by considering the low-order operator.
2. Apply the low-order contribution to the defect vector in an edge-based manner.
3. Assemble the raw antidiffusive terms $F(u_k^{(m)}, u_k^n)$ for each node.

4. Compute the positivity-preserving auxiliary solution $\tilde{u}_k = u_k^L(t^{n+1-\theta})$ but only in the first iteration step.
5. Apply Zalesak's limiter and synchronize the correction factors $\alpha_{ij}^k$.
6. Add the limited antidiffusive fluxes $F^*(u_k^{(m)}, u_k^n)$ to the defect vector.
7. Solve the sequence of linear systems for the resulting scalar subproblems.
8. Use a suitable underrelaxation strategy, apply the increments $\delta u_k^{(m)}$ to the variables $u_k$ and proceed to the next nonlinear iteration / time step.

A short remark is in order concerning the iterative solution of linear systems. While for relatively small time steps the nonsymmetric linear systems engendered by implicit schemes can be solved by a simple Jacobi-like iteration with the lumped mass matrix as preconditioner, the matrices become increasingly ill-conditioned for larger time steps. Hence, it is advisable to utilize BiCGSTAB or multigrid methods with basic components like Jacobi, Gauss-Seidel or SOR smoothers. Compromising the benefits of the unconditionally positive backward Euler/FCT methods, standard smoothers and preconditioners fail for very large time steps and aspect ratios. In our experience, an ILU-decomposition with an appropriate renumbering scheme like Cuthill-Mckee turns out to be of great use in this situation.

## 10 Numerical examples

In the examples which follow, we study the behavior of the implicit Crank-Nicolson (CN/FCT) and the backward Euler (BE/FCT) schemes. Throughout all computations we used quadrilateral elements $Q_1$. However, our FEM-FCT algorithm is applicable to triangular elements $P_1$ as well as nonconforming elements $\tilde{Q}_1$. Many other test problems were considered in [11], [12] which should also be consulted especially for scalar problems.

### 10.1 Shock tube problem

One of the standard benchmarks for compressible flow solvers is the shock tube problem proposed by Sod [28]. Its physical prototype is a closed tube initially filled with a quiescent gas separated by a membrane into two regions. A higher pressure is maintained on the left of the tube than on the right. In addition, the initial data for this two dimensional Riemann problem are given as

$$\begin{bmatrix} \rho_L \\ \mathbf{v}_L \\ p_L \end{bmatrix} = \begin{bmatrix} 1.0 \\ 0.0 \\ 1.0 \end{bmatrix} \quad \text{for } x \in [0, 0.5], \qquad \begin{bmatrix} \rho_R \\ \mathbf{v}_R \\ p_R \end{bmatrix} = \begin{bmatrix} 0.125 \\ 0.0 \\ 0.1 \end{bmatrix} \quad \text{for } x \in (0.5, 1].$$

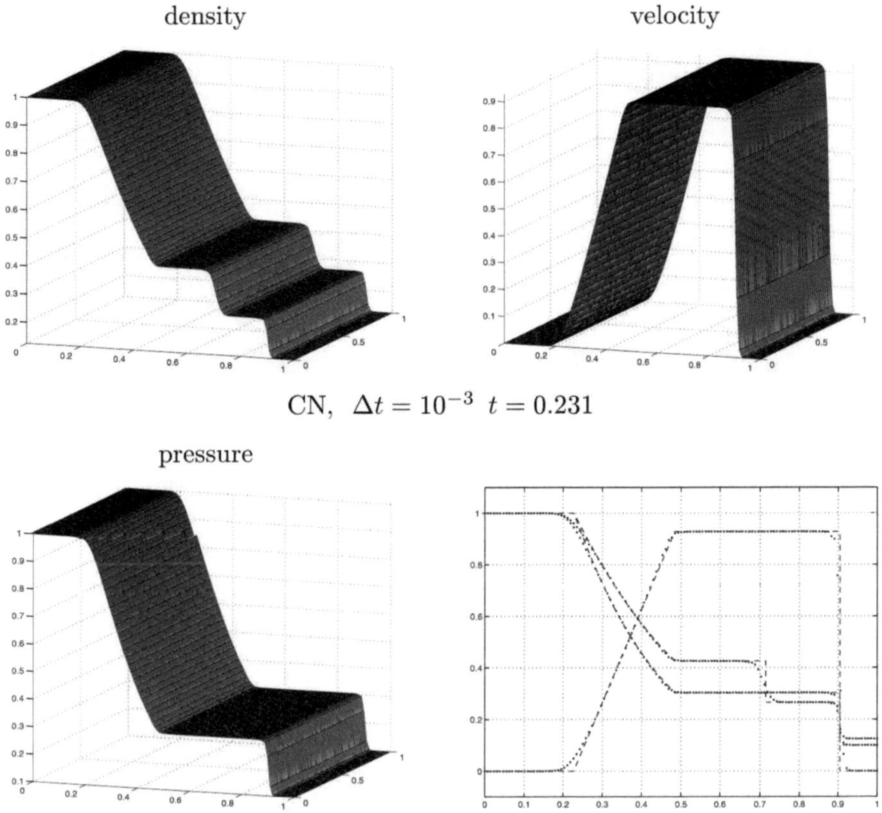


Figure 1. Shock tube problem. Numerical solution at $t = 0.231$.

The abrupt removal of the membrane gives rise to a motion of gas into the direction of lower pressure initiating three distinct waves, namely a normal shock wave, a contact discontinuity wave and a rarefaction wave. While some variables are discontinuous across the first two waves, all variables remain continuous across the rarefaction fan. In fact, this flow structure prevails only until the waves impinge on either the left or the right wall.

We computed the numerical solution displayed in Fig. 1 on a uniform mesh of $129 \times 129$ grid points. The time step was fixed to $\Delta t = 10^{-3}$. All snapshots show the solution obtained with the second order Crank-Nicolson scheme taken at the time $t = 0.231$. The limiter was synchronized by taking the minimum of the correction factors for the density and energy.

The lower right plot shows a cutline for $y = 0.5$ through the solution. Here, the dashed line indicates the exact solution while the dotted lines correspond to the numerical one for all variables. For a more detailed description of this benchmark problem including a comparison between the low-order solutions obtained from scalar artificial diffusion and from Roe's approximate Riemann solver the interested reader is referred to [12].

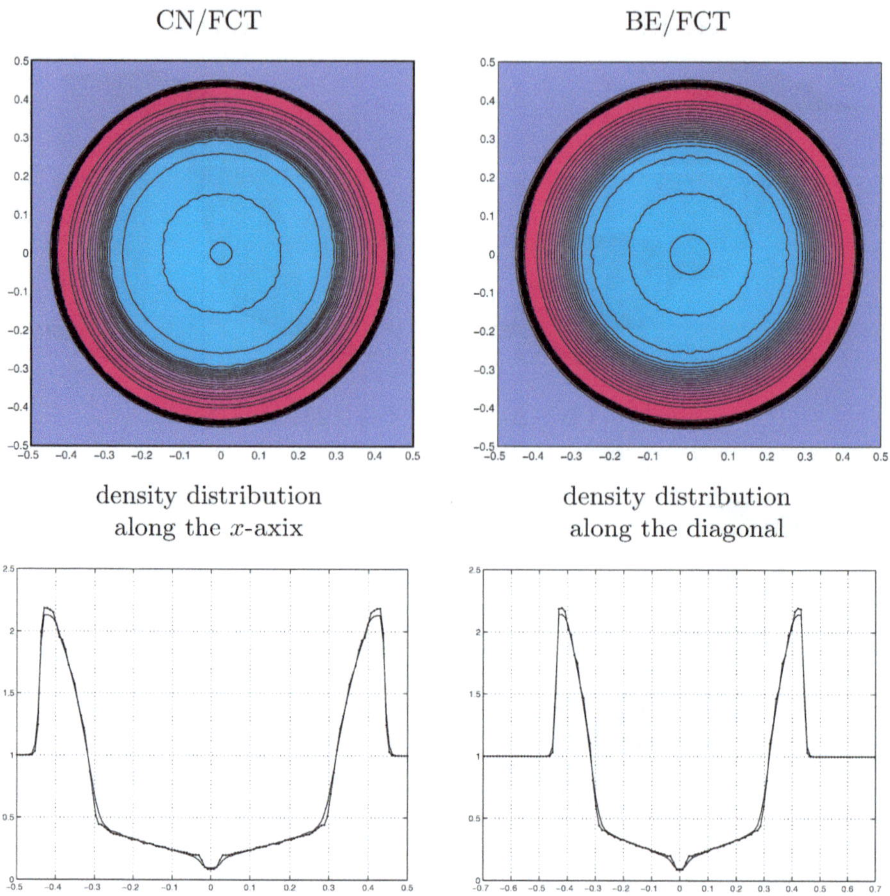


Figure 2. Radially symmetric Riemann problem, $t = 0.13$.

Let us remark that we investigated the same benchmark with the computational domain rotated 45° counterclockwise, whereby the waves propagate along the line $x = y$. In contrast to adhoc application of one-dimensional schemes by using directional splitting our method proved to be truly multidimensional producing the same solution as those shown above.

## 10.2 Radially symmetric Riemann problem

The next benchmark was proposed by LeVeque [14] to assess the ability of methods to conserve radial symmetry. At the beginning, a circular region of higher density and pressure is embedded into a region of unit pressure and density. Initially, the medium is at rest in both regions. In our simulation the initial data was taken to be

$$\begin{bmatrix} \rho_{\text{int}} \\ \mathbf{v}_{\text{int}} \\ p_{\text{int}} \end{bmatrix} = \begin{bmatrix} 2.0 \\ 0.0 \\ 15.0 \end{bmatrix} \quad \text{for } r < 0.13, \qquad \begin{bmatrix} \rho_{\text{ext}} \\ \mathbf{v}_{\text{ext}} \\ p_{\text{ext}} \end{bmatrix} = \begin{bmatrix} 1.0 \\ 0.0 \\ 1.0 \end{bmatrix} \quad \text{otherwise.}$$

The pressure difference induces a radially expanding shock wave whose discontinuities have to be captured accurately. Let us remark that the 'exact' solution can be derived by reducing this benchmark to a one dimensional Riemann problem with geometric source terms [14].

Our results were computed on the same uniform mesh as in the preceding example. The time step was fixed to $\Delta t = 10^{-3}$. The results displayed in Fig. 2 were taken at the instant $t = 0.13$.

The contour plot (top) exhibits radial symmetry for both Crank-Nicolson and backward Euler. However, a closer look reveals that the first order scheme is slightly more diffusive with the result that it fails to capture the critical discontinuity near the origin with high accuracy.

This can best be recognized when considering the density distribution along the x-axis and the diagonal presented in Fig. 2 (bottom). Moreover, the second order accurate Crank-Nicolson scheme resolves the peaks slightly better. Nevertheless, the comparison between the solution structure along the different 'axes' confirms the fact that both methods produce perfectly symmetric results.

### 10.3 Cones and wedges

Let us investigate our algorithm when applied to stationary problems. We enter this group of benchmarks considering a supersonic flow at Mach 2.5 impinging on a corner with an angle of 15°. This kind of flow is understood very well and the solution to this setup can be derived analytically from oblique shock theory [1]. For this purpose one has to consider the so-called *$\theta - \beta - M$ relation*

$$\tan\theta = 2\cot\beta\frac{M^2\sin^2\beta - 1}{M^2(\gamma + \cos^2 2\beta) + 2}, \tag{81}$$

where $\theta$, $\beta$ and $M$ specify the deflection angle, the shock wave angle and the Mach number in front of the shock, respectively. Provided that the deflection angle remains below some maximum $\theta_{\max}$ (which is Mach number dependent) the shock keeps attached to the wedge. Increasing either the Mach number or the deflection angle would start detaching the shock from the obstacle, rendering its behavior more complicated. However, the exact solution to our benchmark problem is a weak shock with $\beta = 36.94°$ and Mach 1.87 behind the shock.

The solution depicted in Fig. 3 was computed on a structured mesh of $129 \times 129$ quadrilaterals without any adaptive refinement. It is worth mentioning, that we performed the same computation on an adaptive mesh using 2.500 elements with the same accuracy. Flux correction was performed by employing the synchronized correction factors for the density and energy. The transition to the set of primitive variables taking the density and pressure as indicator variables did not show any improvements.

BE/FCT Low-order method

Figure 3. Oblique shock at Mach 2.5 with $\theta = 15°$.

## 11 Conclusions

A new family of high-resolution finite element schemes based on the idea of flux-corrected transport was presented. The scalar theory for positive methods was generalized to systems of equations. A new procedure for the construction of the low-order method was proposed. By adding artificial diffusion proportional to the spectral radius of the Roe matrix, the scalar strategy of eliminating all negative off-diagonal entries from the high-order operator was mimicked by rendering all off-diagonal blocks positive definite.

We proposed a universal approach to the flux decomposition of convective and diffusive terms resulting from the Galerkin discretization. This enabled us to represent the antidiffusive terms as a sum of internodal fluxes which could be limited in an essentially one-dimensional manner. In doing this, the skew-symmetry of the conservative fluxes guaranteed strict mass conservation. Furthermore, we suggested an efficient algorithm for an edge-by-edge matrix assembly.

A fixed point defect correction was employed for an iterative treatment of nonlinearities. By utilizing the low-order operator as block-diagonal preconditioner, the nonlinear system was split into a sequence of scalar subproblems. Furthermore, this relieved us from storing huge matrices except of the 5 diagonal blocks of the sparse low-order operator.

Attention was paid to the need of synchronizing the correction factors of the flux-limiter. As a consequence, the limiting coefficients reduced to a scalar quantity even for systems. The idea of mixing two sets of variables in the flux-limiter amounted to estimating the correction factors for a set of arbitrary variables and applying them to the conservative fluxes.

Generally speaking, the unified FEM-FCT formulation encompassing both explicit and implicit schemes has proved to be applicable to multi-dimensional hyperbolic systems with considerable success. Nevertheless, the time step dependency of Zalesak's limiter is alarming. The unconditional positivity of the backward Euler cannot be duly utilized without loss of accuracy. Further investigations will be necessary to develop an alternative limiter.

## References

[1] J. D. Anderson, Jr., *Modern Compressible Flow*, McGraw-Hill, 1990.

[2] P. Arminjon, A. Dervieux, *Construction of TVD-like artificial viscosities on 2-dimensional arbitrary FEM grids*, INRIA Research Report **1111** (1989).

[3] J. P. Boris, D. L. Book, *Flux-corrected transport. I. SHASTA, A fluid transport algorithm that works*, J. Comput. Phys. **11** (1973), 38–69.

[4] J. Donea, L. Quartapelle and V. Selmin, *An analysis of time discretization in the finite element solution of hyperbolic problems*, J. Comput. Phys. **70** (1987), 463–499.

[5] C. A. J. Fletcher, *The group finite element formulation*, Comput. Methods Appl. Mech. Engrg. **37** (1983), 225–243.

[6] S. K. Godunov, *Finite difference method for numerical computation of discontinuous solutions of the equations of fluid dynamics*, Mat. Sbornik **47** (1959), 271–306.

[7] A. Harten, *High resolution schemes for hyperbolic conservation laws. Discretization in the finite element solution of hyperbolic problems*, J. Comput. Phys. **49** (1983), 357–393.

[8] C. Hirsch, *Numerical Computation of Internal and External Flows, Vol.2*, Wiley, 1984.

[9] A. Jameson, *Computational algorithms for aerodynamic analysis and design*, Appl. Numer. Math. **13** (1993), 383–422.

[10] A. Jameson, *Positive schemes and shock modelling for compressible flows*, Int. J. Numer. Meth. Fluids **20** (1995), 743–776.

[11] D. Kuzmin, S. Turek, *Flux correction tools for finite elements*, J. Comput. Phys. **175** (2002), 525–558.

[12] D. Kuzmin, M. Möller, and S. Turek, *Multidimensional FEM-FCT schemes for arbitrary time-stepping, Technical report No. 215, University of Dortmund, 2002*, Int. J. Numer. Meth. Fluids, to appear.

[13] R. J. LeVeque, *Numerical Methods for Conservation Laws*, Birkhäuser, 1992.

[14] R. J. LeVeque, *Simplified multi-dimensional flux limiting methods*, Numerical Methods for Fluid Dynamics **IV** (1993), 175–190.

[15] R. Löhner, *Adaptive CFD Techniques*, Wiley, 2001.

[16] R. Löhner, K. Morgan, J. Peraire, and M. Vahdati, *Finite element flux-corrected transport (FEM-FCT) for the Euler and Navier-Stokes equations*, Int. J. Numer. Meth. Fluids **7** (1987), 1093–1109.

[17] R. Löhner, K. Morgan, M. Vahdati, J. P. Boris, and D. L. Book, *FEM-FCT: combining unstructured grids with high resolution*, Commun. Appl. Numer. Methods **4** (1988), 717–729.

[18] P. R. M. Lyra, *Unstructured Grid Adaptive Algorithms for Fluid Dynamics and Heat Conduction. PhD thesis*, University of Wales, Swansea, 1994.

[19] P. R. M. Lyra, K. Morgan, J. Peraire, and J. Peiro, *TVD algorithms for the solution of the compressible Euler equations on unstructured meshes*, Int. J. Numer. Meth. Fluids **19** (1994), 827–847.

[20] K. Morgan, J. Peraire, *Unstructured grid finite element methods for fluid mechanics*, Reports on Progress in Physics **61** (1998), 569–638.

[21] A. K. Parrott, M. A. Christie, *FCT applied to the 2-D finite element solution of tracer transport by single phase flow in a porous medium*, Proc. ICFD Conf. on Numerical Methods in Fluid Dynamics, Oxford University Press (1986), 609–619.

[22] S. V. Patankar, *Numerical Heat Transfer and Fluid Flow*, McGraw-Hill, New York, 1980.

[23] J. Peraire, M. Vahdati, J. Peiro, and K. Morgan, *The construction and behaviour of some unstructured grid algorithms for compressible flows. Numerical Methods for Fluid Dynamics Vol. IV*, Oxford University Press (1993), 221–239.

[24] P. L. Roe, *Approximate Riemann solvers, parameter vectors and difference schemes*, J. Comput. Phys. **43** (1981), 357–372.

[25] P. L. Roe, J. Pike, *Efficient construction and utilisation of Approximate Riemann solutions*, Comput. Methods Appl. Sci. Engrg. (1984), 499–518.

[26] V. Selmin, *Finite element solution of hyperbolic equations. I. One-dimensional case*, INRIA Research Report **655** (1987).

[27] V. Selmin, *Finite element solution of hyperbolic equations. II. Two-dimensional case*, INRIA Research Report **708** (1987).

[28] G. Sod, *A survey of several finite difference methods for systems of nonlinear hyperbolic conservation laws*, J. Comput. Phys. **27** (1978), 1–31.

[29] S. Turek, *Efficient Solvers for Incompressible Flow Problems: An Algorithmic and Computational Approach. LNCSE* 6, Springer, 1999.

[30] S. Zalesak, *Fully multi-dimensional flux-corrected transport algorithms for fluids*, J. Comput. Phys. **31** (1979), 335–362.

# Application of the PCG Method in Solution of a Nuclear Reactor Criticality Problem

Eva Neumanová

Department of Technical Mathematics, Faculty of Mechanical Engineering, Czech Technical University, Karlovo náměstí 13, 121 35 Praha 2, CZ–Czech Republic

neumanov@marian.fsik.cvut.cz

## 1 The model problem

Let $\Omega \subset E_2$ be a bounded domain with Lipschitz continuous boundary $\Gamma$. The $n$-group nuclear diffusion problem is to determine the largest (in modulus) eigenvalue $k_{\text{eff}}$ and the corresponding eigenvector $(\Phi_1(x,y), \dots, \Phi_n(x,y))$, satisfying the following system of partial differential equations

$$\begin{aligned} &-\frac{\partial}{\partial x}\Big(\mathcal{D}_k(x,y)\frac{\partial \Phi_k(x,y)}{\partial x}\Big) - \frac{\partial}{\partial y}\Big(\mathcal{D}_k(x,y)\frac{\partial \Phi_k(x,y)}{\partial y}\Big) \\ &+ \sigma_k^a(x,y)\Phi_k(x,y) - \sum_{l=1}^{k-1} \sigma_l^k(x,y)\Phi_l(x,y) \\ &= \frac{1}{k_{\text{eff}}}\sum_{l=1}^{n} \sigma_{kl}^f(x,y)\Phi_l(x,y), \quad (x,y) \in \Omega, \end{aligned} \tag{1}$$

and the homogeneous boundary conditions

$$\Phi_k(x,y) = 0, \quad (x,y) \in \Gamma, \tag{2}$$

for $k = 1, \dots, n$. Properties of all the coefficients in (1) are described bellow in Section 1.2.

### 1.1 Geometric assumptions

(a) The domain $\Omega$ is a circle with radius $R$ and the center at the origin of the Cartesian coordinate system, or $\Omega$ is a polygon with radius of circumscribed circle $R$, and with the center at the origin.

(b) The domain $\Omega$ is composed of a finite number of simply connected subdomains $\Omega_j$, $j = 1, \dots, P$, with Lipschitz boundaries such that

$$\bar{\Omega} = \bigcup_{j=1}^{P} \bar{\Omega}_j, \quad \Omega_j \cap \Omega_k = \emptyset \quad \text{for } j \neq k,$$

where $\bar{\Omega} = \Omega \cup \Gamma$, $\bar{\Omega}_j = \Omega_j \cup \Gamma_j$.

(c) The subdomain layout corresponds to the material structure of a nuclear reactor. Thus, the subdomains in the central part of $\Omega$ are regular hexagons. The numbering in these hexagonal subdomains corresponds to the physical properties of the active zone of the reactor. In the remaining part of the reactor (the reflector) we assume homogenous environment, i.e., constant coefficients (see an example in Fig. 1). Moreover, let the structure and the shape of the domain $\Omega$ be symmetric with respect to the lines $x = 0$, and $y = x\,\mathrm{tg}(j\pi/6)$, $j = 0, 1, 2, 4, 5$.

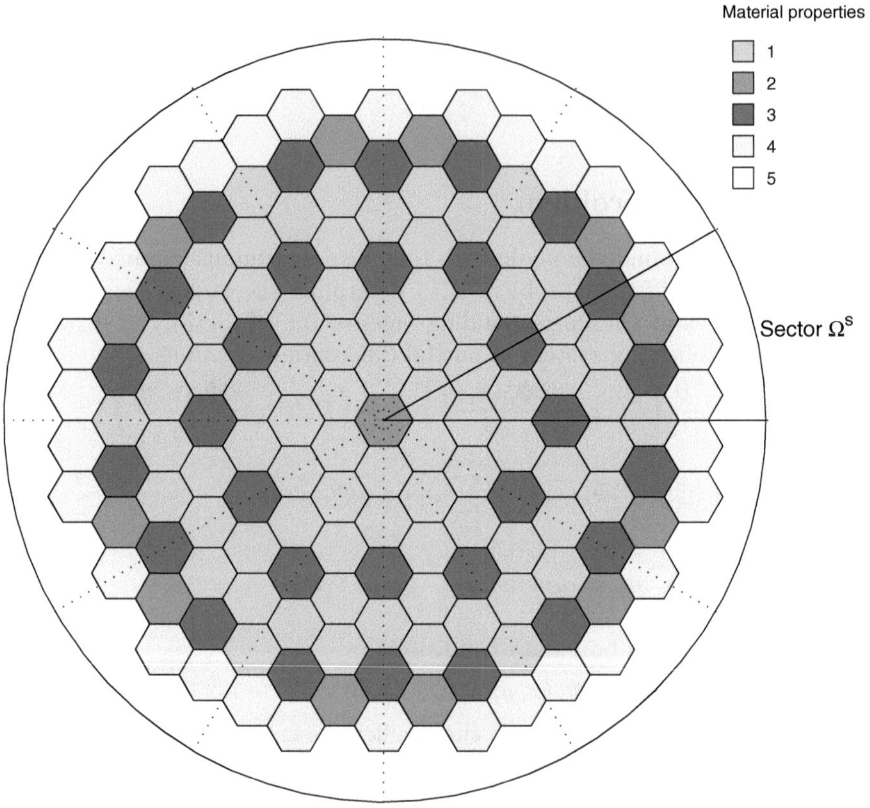


Figure 1. An example of a nuclear reactor structure with the lines of symmetry.

### 1.2 Physical assumptions

(d) $\mathcal{D}_k(x,y)$, $\sigma_k^a(x,y)$, $\sigma_l^k(x,y)$, and $\sigma_{kl}^f(x,y)$ are constant in each subdomain $\Omega_j$ and represent some physical properties (diffusion, absorption, scattering, and fission).

(e) $\mathcal{D}_k(x,y)$ and $\sigma_k^a(x,y)$ are positive, $\sigma_l^k(x,y)$ and $\sigma_{kl}^f(x,y)$ are non-negative in $\Omega$.

(f) $\Phi_k(x,y)$ and $\mathcal{D}_k(x,y)\nabla\Phi_k(x,y)\vec{\nu}$ are continuous in $\Omega$, $\vec{\nu}$ denotes the unit normal vector to a subdomain interface, and $\nabla$ stands for the gradient.

**Remark 1.1.** Let us note that the $k$th component $\Phi_k(x,y)$ of the eigenvector represents neutron flux in the $k$th energy group and the eigenvalue $k_{\text{eff}}$ is a criticality coefficient of a nuclear reactor.

### 1.3 The model problem on a subdomain

Considering the symmetries described in (c), model problem (1)–(2) may be formulated in one of the subdomains, i.e., in the sectors determined by the lines of symmetry.

Let us consider the sector $\Omega^s = \{(x,y) \in \Omega \mid x > 0,\ 0 < y < x\,\mathrm{tg}(\pi/6)\}$, and let $\Gamma^s = \Gamma_0 \cup \Gamma_1 \cup \Gamma_2$ denote the boundary of $\Omega^s$, where $\Gamma_0 = \Gamma^s \cap \Gamma$, $\Gamma_1 = \{(x,0) \in \Omega \mid x \geq 0\}$, $\Gamma_2 = \{(x, x\,\mathrm{tg}(\pi/6)) \in \Omega \mid x > 0\}$.

We formulate the model problem on the subdomain $\Omega^s$:
the system of equations (1) for $(x,y) \in \Omega^s$,
the homogenous boundary conditions (2) for $(x,y) \in \Gamma_0$,
and the Neumann type boundary conditions

$$\frac{\partial \Phi_k(x,y)}{\partial \nu_j} = 0 \quad \text{for } (x,y) \in \Gamma_j,\ j = 1,2,$$

where $\nu_j$ denotes an outer normal to the boundary $\Gamma_j$, and $k = 1,\ldots,n$.

*Unless otherwise stated, by the model problem we will understand the problem defined on the subdomain $\Omega^s$. Using the symmetry properties of the solution on the subdomain can be easily extended to the whole domain.*

The model problem in operator notation reads

$$(\mathbf{D} + \mathbf{A} - \mathbf{\Sigma})\mathbf{\Phi} = \frac{1}{k_{\text{eff}}}\mathbf{F}\mathbf{\Phi} \quad \text{in } \Omega^s, \tag{3}$$

$$\mathbf{\Phi} = \mathbf{0} \quad \text{on } \Gamma_0, \tag{4}$$

$$\frac{\partial \Phi_k(x,y)}{\partial \nu_j} = 0 \quad \text{on } \Gamma_j,\quad j = 1,2,\ k = 1,\ldots,n, \tag{5}$$

where

$$\mathbf{D} + \mathbf{A} = \begin{pmatrix} D_1 + A_1 & \ldots & 0 \\ \vdots & & \vdots \\ 0 & \ldots & D_n + A_n \end{pmatrix},$$

$$(D_k + A_k)\Phi_k = -\nabla(\mathcal{D}_k \nabla \Phi_k) + \sigma_k^a \Phi_k,$$

$$\mathbf{\Sigma} = \begin{pmatrix} 0 & \dots & \dots & 0 \\ C_{2\,1} & 0 & & \vdots \\ \vdots & \ddots & \ddots & \vdots \\ C_{n\,1} & \dots & C_{n\,n-1} & 0 \end{pmatrix}, \quad C_{j\,k}\Phi_k = \sigma_k^j \Phi_k,$$

$$\mathbf{F} = \begin{pmatrix} F_{1\,1} & \dots & F_{1\,n} \\ \vdots & \ddots & \vdots \\ F_{n\,1} & \dots & F_{n\,n} \end{pmatrix}, \quad F_{j\,k}\Phi_k = \sigma_{jk}^f \Phi_k.$$

Considering the assumptions (d)–(f), the operators $\mathbf{A}$, $\mathbf{\Sigma}$, and $\mathbf{F}$ are non-negative, bounded and linear, and $\mathbf{D}$ is a differential operator of elliptic type. The operators $\mathbf{D}$, $\mathbf{A}$, $\mathbf{\Sigma}$, $\mathbf{F}$ represent diffusion, absorption, scattering, and fission, respectively. The composed operator $\mathbf{D}+\mathbf{A}-\mathbf{\Sigma}$ is due to the term $\mathbf{\Sigma}$ non-selfadjoint. Habetler and Martino in [4] have shown that the maximum (in modulus) eigenvalue of the model problem is unique and all components of the corresponding eigenvector are positive almost everywhere.

## 2 The discrete problem

Let $H$ be the size of an edge of a hexagonal subdomain $\Omega_j$. We consider the mesh size $h = H/N$, where $N > 1$ is an integer. In the domain $\Omega^s$ we define a hexagonal mesh $\Omega^h$ so that the mesh points lie exactly on the interfaces of the subdomains $\Omega_j$ or their parts that belong to the domain $\Omega^s$, on the boundary parts $\Gamma_1$ and $\Gamma_2$, and on a piece-wise linear approximation of the boundary part $\Gamma_0$ (see Fig. 2).

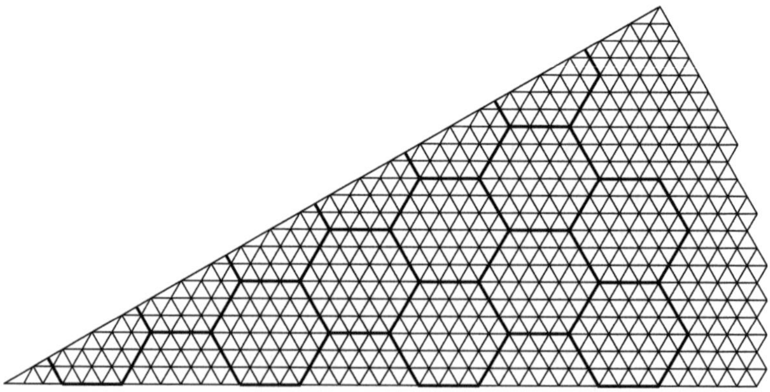

Figure 2. An example of a mesh $\Omega^h$ defined in $\Omega^s$.

To discretize the partial differential equations (1) the finite difference method has been used. A thorough description and derivation of the difference equations on hexagonal mesh (a seven point difference scheme) can be found in Neumanová [9]. Considering $m$ – the number of the mesh points, and $n$ – the number of the differential equations, we obtain $n \times$ m difference equations. The resulting system of the equations can be written in the following form

$$\mathbf{B}_h \mathbf{X}_h = \frac{1}{\lambda_h} \mathbf{F}_h \mathbf{X}_h, \tag{6}$$

where $\mathbf{B}_h$ is a lower block triangular matrix, and $\mathbf{F}_h$ is a non-negative block matrix. With the notation $\mathbf{X}_h = (X_1, X_2, \dots, X_n)^T$ the above equation then reads

$$\begin{pmatrix} \mathcal{D}_1 + \mathcal{A}_1 & 0 & \cdots & & 0 \\ -\mathcal{C}_{21} & \mathcal{D}_2 + \mathcal{A}_2 & & & \vdots \\ \vdots & & \ddots & & \vdots \\ \vdots & & & \ddots & \vdots \\ \vdots & & & \ddots & 0 \\ -\mathcal{C}_{n1} & & \cdots & -\mathcal{C}_{n\,n-1} & \mathcal{D}_n + \mathcal{A}_n \end{pmatrix} \cdot \begin{pmatrix} X_1 \\ X_2 \\ \vdots \\ \vdots \\ \vdots \\ X_n \end{pmatrix} \tag{7}$$

$$= \frac{1}{\lambda_h} \begin{pmatrix} \mathcal{F}_{11} & \cdots & \cdots & \mathcal{F}_{1n} \\ \vdots & & & \vdots \\ \vdots & & & \vdots \\ \vdots & & & \vdots \\ \vdots & & & \vdots \\ \mathcal{F}_{n1} & \cdots & \cdots & \mathcal{F}_{nn} \end{pmatrix} \cdot \begin{pmatrix} X_1 \\ X_2 \\ \vdots \\ \vdots \\ \vdots \\ X_n \end{pmatrix} .$$

The discrete problem is to determine the largest (in modulus) eigenvalue $\lambda_h$ and the corresponding eigenvector $\mathbf{X}_h$ satisfying the equation (6). Birkhoff and Varga [2], and also Marek [7] have shown that the largest (in modulus) eigenvalue $\lambda_h$ is unique and positive, and the corresponding eigenvector is positive. By refining the mesh, i.e., letting $h \to 0$, eigenvalues $\lambda_h$ and eigenvectors $\mathbf{X}_h$ converge to the solution of the model problem (3)–(5), i.e., to $k_{\text{eff}}$ and $\mathbf{\Phi}$, respectively.

## 3 Numerical solution of the discrete problem

Solving the discrete problem (6) requires two iteration loops, the outer loop to determine the maximum eigenvalue, and the inner loop to calculate the corresponding eigenvector.

### 3.1 Outer loop algorithm

To determine the eigenvalue $\lambda_h$ we have used the classical power method. Let $\lambda_i$ and $\mathbf{X}^i = (X_1^i, \dots, X_n^i)^T$ denote the $i$th iterate of the eigenvalue $\lambda_h$ and

corresponding eigenvector $\mathbf{X}_h$, respectively, $i = 0, 1, \ldots$. Since the largest (in modulus) eigenvalue $\lambda_h$ of problem (6) is unique, the sequence of the iterates $\{\lambda_i\}$ converges to $\lambda_h$, and the vectors $\{\mathbf{X}^i\}$ converge to the corresponding eigenvector $\mathbf{X}_h$, see Marek [7], [8].
**Given**: EPSO, OUTIT, $\lambda_0$, $\mathbf{X}^0$ – outer loop accuracy, maximum number of the outer loop iterations, initial guess of the eigenvalue, and initial approximation of the eigenvector, respectively.
So, the algorithm reads:

$i = 0$
$\lambda_{i+1} = \lambda_i + 1$
$\mathbf{S}^i = \mathbf{F}_h\mathbf{X}^i$
*while* $|\lambda_{i+1} - \lambda_i| > \text{EPSO}$ & $i \le \text{OUTIT}$
  **Solve** $\mathbf{B}_h\mathbf{X}^{i+1} = \mathbf{S}^i$ $\longleftrightarrow$ **INNER LOOP**
  $\mathbf{S}^{i+1} = \mathbf{F}_h\mathbf{X}^{i+1}$
  $\lambda^{i+1} = (\mathbf{S}^{i+1}, \mathbf{S}^{i+1})/(\mathbf{S}^{i+1}, \mathbf{S}^i)$
  $i = i + 1$
  $\mathbf{S}^i = \mathbf{S}^i/\lambda^i$
*end*
*if* $i >$ OUTIT
  'ERROR – required accuracy not achieved!'
*else*
  'END – algorithm successfully finished!'
*end*

**Remark 3.1.** The algorithm for computing the extreme eigenvalue can be improved, using an accelerating technique (see Hageman and Young [6], Chapter 11, Case studies).

### 3.2. Inner loop computational strategy

To design the method for computing the $(i+1)$st iterate of the vector $\mathbf{X}_h$ we rewrite the equation $\mathbf{B}_h\mathbf{X}^{i+1} = \mathbf{S}^i$ in the form

$$(\mathcal{D}_1 + \mathcal{A}_1)X_1^{i+1} = \frac{1}{\lambda_i}\sum_{k=1}^{n}\mathcal{F}_{1\,k}X_k^i, \tag{8}$$

$$(\mathcal{D}_j + \mathcal{A}_j)X_j^{i+1} = \frac{1}{\lambda_i}\sum_{k=1}^{n}\mathcal{F}_{j\,k}X_k^i + \sum_{k=1}^{j-1}\mathcal{C}_{j\,k}X_k^{i+1}, \quad j = 2, \ldots, n.$$

As we can see in (8), the computation of the vector $\mathbf{X}^{i+1}$ may be split into successive solving of $n$ subsystems. As a result of solving the $j$th subsystem we obtain $j$th component of the $(i+1)$st iteration vector, $j = 1, \ldots, n$. It is important to note that the blocks $\mathcal{D}_j + \mathcal{A}_j$ are positive definite M-matrices. They are strictly diagonally dominant (due to positivity of the terms $\sigma_k^a(x, y)$

in the equations (1)) and they have the same sparsity pattern. The blocks $\mathcal{C}_{ij}$ and $\mathcal{F}_{ij}$ are diagonal matrices with non-negative entries. Our goal is to find an efficient and reliable method for solving the subsystems of linear equations within the inner loops.

*Since the same method will be applied for solving all the subsystems, let the equation* $\mathbf{Ax} = \mathbf{b}$ *denote any subsystem to be solved within the inner loops.*

### PCG method for solving the subsystems

Let us consider the problem $\mathbf{Ax} = \mathbf{b}$ with a sparse, symmetric and positive definite M-matrix $\mathbf{A}$. If the problem is of moderate size, the direct method, e.g., the Cholesky factorization, in combination with sparse methods can be used. Wise reordering of the matrix $\mathbf{A}$, i.e., definition of a permutation matrix $\mathbf{P}$ in order to minimize fill-in in the factor $\mathbf{L}$, where $\mathbf{LL}^{\mathrm{T}} = \mathbf{PAP}^{\mathrm{T}}$, may be, and frequently is, an efficient and reliable tool for finding the solution of the problem. But, with increasing number of equations, i.e., as soon as the mesh size $h \to 0$, the matrix $\mathbf{A}$ becomes ill-conditioned and may affect numerical stability of the Cholesky method. Thus, such an inevitable consequence of the mesh refinement has been taken into account and the CG method in combination with algebraic type of preconditioner has been designed and numerically tested.

First of all, we introduce a few notations, definitions, and theorems. Then we briefly describe the technique of defining preconditioners, summarize spectral properties of the preconditioners, and finally we formulate the PCG algorithm.

### Notation

$\mathcal{E}^m$ denotes a space of real square $m \times m$ matrices.

$\mathbf{P} \in \mathcal{E}^m$ means a *permutation matrix*, i.e., the matrix which in each row and each column has one entry equal to unity and all the other entries zero.

$\sigma(\mathbf{A}) = \{\lambda_1, \dots, \lambda_m\}$ is the *spectrum* of the matrix $\mathbf{A} \in \mathcal{E}^m$.

$\kappa(\mathbf{A}) = \lambda_{\max}/\lambda_{\min}$ denotes a *spectral condition number* of a non-singular matrix $\mathbf{A} \in \mathcal{E}^m$, where $\lambda_{\max} = \max\{|\lambda_1|, \dots, |\lambda_m|\}$ and $\lambda_{\min} = \min\{|\lambda_1|, \dots, |\lambda_m|\}$.

**Definition 3.1** (Varga [10]). Let $\mathbf{A} \in \mathcal{E}^m$. The matrix $\mathbf{A}$ with $a_{ij} \leq 0$ for all $i \neq j$ is an *M-matrix* if $\mathbf{A}$ is nonsingular and $\mathbf{A}^{-1} \geq 0$.

**Definition 3.2** (Varga [10]). Let $\mathbf{A}$, $\mathbf{K}$, and $\mathbf{R} \in \mathcal{E}^m$. Then $\mathbf{A} = \mathbf{K} - \mathbf{R}$ is a *regular splitting of the matrix* $\mathbf{A}$ if $\mathbf{K}$ is nonsingular with $\mathbf{K}^{-1} \geq 0$, and $\mathbf{R} \geq 0$.

**Definition 3.3** (George, Liu [3]). Let $\mathbf{A} \in \mathcal{E}^m$ be symmetric. The *graph of the matrix* $\mathbf{A}$ denoted by $\mathbf{G(A)} = (\mathbf{X},\mathbf{E})$ consists of a set of $m$ nodes

$\mathbf{X} = \{x_1, \ldots, x_m\}$, and a set of edges $\mathbf{E}$, where an edge $\{x_j, x_k\} \in \mathbf{E}$ if and only if $a_{jk} = a_{kj} \neq 0$.

**Theorem 3.1** (Varga [10]). *Let $\mathbf{A} \in \mathcal{E}^m$ be an M-matrix and let $\mathbf{C}$ be any matrix obtained from $\mathbf{A}$ by setting certain off-diagonal entries of the matrix $\mathbf{A}$ to zero. The $\mathbf{C}$ is also an M-matrix.*

## Definition of a class of preconditioners

A graph of the matrix $\mathbf{A}$ coincides with the mesh generated for the finite difference method (see Fig. 2). Zero-nonzero pattern of the matrix $\mathbf{A}$ depends on the numbering of the mesh points, i.e., the nodes of the graph. A simple row-wise numbering of the nodes yields a block tridiagonal matrix that may be written in the following form

$$\mathbf{A} = \begin{pmatrix} K_1 & R_1^T & \cdots & \cdots & \cdots & 0 \\ R_1 & K_2 & R_2^T & & & \vdots \\ \vdots & \ddots & \ddots & \ddots & & \vdots \\ \vdots & & \ddots & \ddots & \ddots & \vdots \\ \vdots & & & R_{s-2} & K_{s-1} & R_{s-1}^T \\ 0 & \cdots & \cdots & \cdots & R_{s-1} & K_s \end{pmatrix}, \tag{9}$$

where $s$ is a number of rows of the mesh $\Omega_h$, in other words, of the graph $\mathbf{G(A)}$, and where the diagonal blocks $K_j$ are tridiagonal matrices (see an example in Fig. 3).

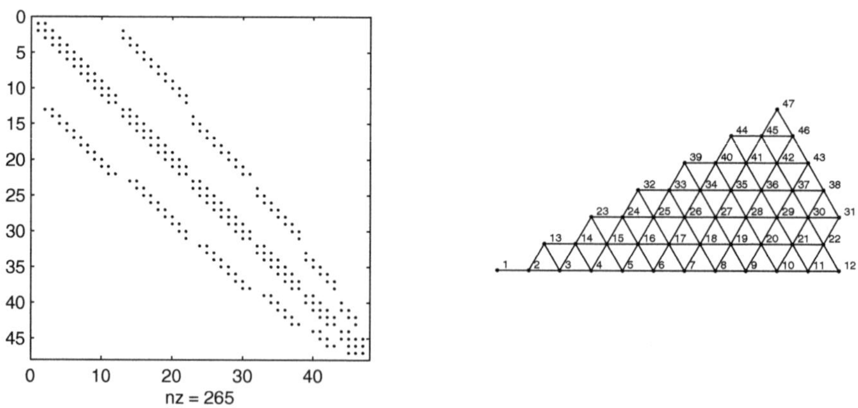


Figure 3. A matrix structure corresponding to a row-wise graph numbering.

Different numbering of the nodes yields different ordering (nonzero pattern) of the matrix. We seek a suitable ordering, i.e., a permutation matrix $\mathbf{P}$, so that most of the nonzero entries will be located within the diagonal blocks. Moreover, we intend to keep block tridiagonal structure of the matrix (but, of course, with different partitioning dependent on particular choice of the ordering). In Neumanová [9] we have proposed the way of ordering that meets the above stated requirements.

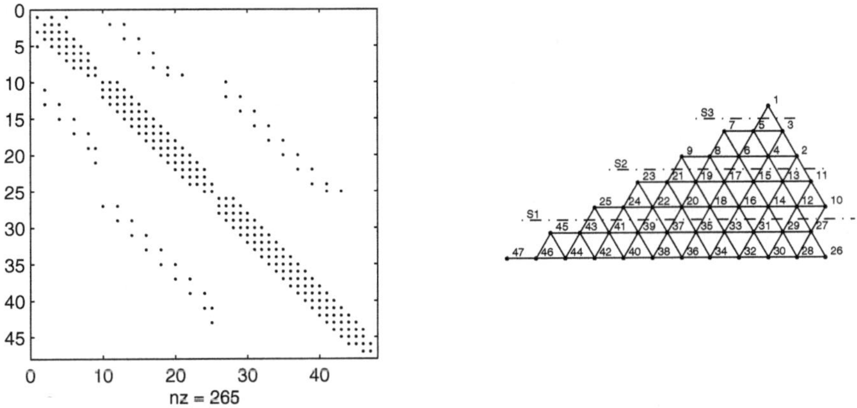


Figure 4. A pair separation: matrix structure and graph numbering.

The basic idea is the following: the graph of the matrix **A** is separated into pairs or triplets of the rows and the numbering within these pairs or triplets is defined so that the nonzero entries become as close as possible to the main diagonal. Examples of graph numbering using the pair, and the triplet separation with corresponding nonzero structure of the matrices are shown in Fig. 4 and Fig. 5, respectively.

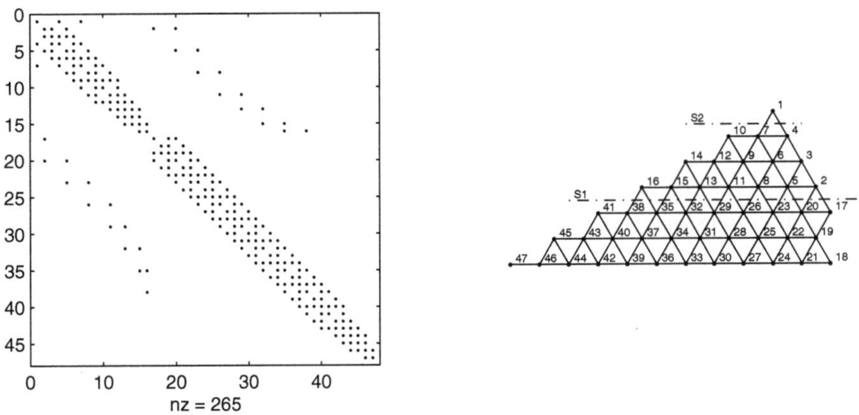


Figure 5. A triplet separation: matrix structure and graph numbering.

## Spectral properties of selected preconditioners

Let us consider a block tridiagonal positive definite M-matrix $\tilde{\mathbf{A}} \in \mathcal{E}^m$ ordered by means of the triplet separation technique, i.e., $\tilde{\mathbf{A}} = \mathbf{PAP}^{\mathrm{T}}$, where $\mathbf{P}$ is the corresponding permutation matrix. Let $\tilde{\mathbf{A}} = \mathbf{K} - \mathbf{R}$ be the splitting of the matrix $\tilde{\mathbf{A}}$, where $\mathbf{K} = \text{blockdiag}(\tilde{\mathbf{A}})$. According to Theorem 3.1, such a splitting is regular. The matrix $\mathbf{K}$ is symmetric and positive definite and $\lambda_{\min}(\mathbf{K}) > \lambda_{\min}(\tilde{\mathbf{A}})$. Let us set $p = \text{rank}(\mathbf{R})$. Then the matrix $\mathbf{K}^{-1}\tilde{\mathbf{A}}$ has unity as an eigenvalue of multiplicity $m - p$. Hence, the matrix $\mathbf{K}^{-1}\tilde{\mathbf{A}}$ has at most $p + 1$ different eigenvalues. Thus, using $\mathbf{K}^{-1}$ as a preconditioner, the preconditioned conjugate gradients will terminate at no more than $p+1$ steps, assuming no roundoff errors occur. The upper bound for the rank of the matrix $\mathbf{R}$ has been determined in Neumanová [9] as well as the bound for the extreme eigenvalue of the matrix $\mathbf{K}$.

Table 1. Extreme eigenvalues and condition numbers, 2-group test problem.

| $H/h$ | $m$ | 1st group | | | 2nd group | | |
|---|---|---|---|---|---|---|---|
| | | $\lambda_{\max}$ | $\lambda_{\min}$ | $\kappa$ | $\lambda_{\max}$ | $\lambda_{\min}$ | $\kappa$ |
| 2 | 233 | 8.038569 | 0.344414 | 23.339822 | 5.518094 | 0.402727 | 13.701823 |
| | | 1.523137 | 0.481229 | 3.165098 | 1.157471 | 0.865478 | 1.337378 |
| 3 | 502 | 7.536142 | 0.236699 | 31.838535 | 3.543457 | 0.230104 | 15.399375 |
| | | 1.698665 | 0.302132 | 5.622255 | 1.275276 | 0.753559 | 1.692337 |
| 4 | 875 | 7.362369 | 0.171035 | 43.045992 | 2.855009 | 0.165549 | 17.245660 |
| | | 1.810667 | 0.192257 | 9.417975 | 1.377968 | 0.655656 | 2.101663 |
| 5 | 1348 | 7.283288 | 0.125717 | 57.934079 | 2.535440 | 0.132503 | 19.134967 |
| | | 1.868197 | 0.133105 | 14.035528 | 1.466044 | 0.564624 | 2.596494 |
| 6 | 1923 | 7.240365 | 0.094686 | 76.467101 | 2.362193 | 0.112096 | 21.072996 |
| | | 1.903281 | 0.097151 | 19.590917 | 1.535615 | 0.482805 | 3.180608 |
| 7 | 2601 | 7.214685 | 0.073113 | 98.679147 | 2.257773 | 0.097832 | 23.078164 |
| | | 1.927666 | 0.072647 | 26.534693 | 1.596188 | 0.412062 | 3.873663 |
| 8 | 3381 | 7.198344 | 0.058167 | 123.753125 | 2.189996 | 0.086974 | 25.180015 |
| | | 1.954511 | 0.048785 | 40.063957 | 1.648276 | 0.352115 | 4.681076 |

Comparison of computed estimates of condition numbers and extreme eigenvalues of the original and preconditioned (via triplet separation technique) matrices are shown in Table 1.

## PCG algorithm

$x^0$ arbitrary, $r^0 = b - \mathbf{A}x^0$,
**Solve** $\mathbf{K}q^0 = r^0$ and set $p^0 = q^0$
*for* $k = 0, 1, \ldots, m-1$
    *if* $\|r^k\| > \text{EPSI}$
        $\alpha_k = \frac{(r^k, p^k)}{(\mathbf{A}p^k, p^k)}$
        $x^{k+1} = x^k + \alpha_k p^k$

$$
\begin{aligned}
&r^{k+1} = r^k - \alpha_k \mathbf{A} p^k && \text{(PCG)}\\
&\textbf{Solve} \quad \mathbf{K} q^{k+1} = r^{k+1}\\
&p^{k+1} = q^{k+1} - \tfrac{(q^{k+1}, \mathbf{A}p^k)}{(\mathbf{A}p^k, p^k)} p^k
\end{aligned}
$$

*else*

'END – required accuracy achieved.'

*end if*

*end for*

Within each step of the PCG algorithm we have to solve a system $\mathbf{K}q^k = r^k$. The matrix $\mathbf{K}$ is positive definite M-matrix, so, to determine $q^k$ Cholesky's factorization $\mathbf{K} = \mathbf{L}\mathbf{L}^{\mathrm{T}}$ has been used.

**Theorem 3.2** (Hackbusch [5]). *Let* $\mathbf{A},\mathbf{K} \in \mathcal{E}^m$ *be positive definite and let the matrix* $\mathbf{K}^{-1}\mathbf{A}$ *have its spectrum in the interval* $[\gamma, \Gamma]$, *where* $\Gamma = \lambda_{\max}(\mathbf{K}^{-1}\mathbf{A})$, $\gamma = \lambda_{\min}(\mathbf{K}^{-1}\mathbf{A})$, *and* $\kappa = \Gamma/\gamma$. *Then for the error* $e^i$ *of the iterates* $x^i$ *defined by the algorithm of preconditioned conjugate gradient method* (PCG) *the following error estimate holds:*

$$\|e^i\|_A \leq \frac{2(1-1/\kappa)^i}{(1+1/\kappa)^{2i} + (1-1/\kappa)^{2i}} \|e^0\|_A = \frac{2c^i}{1+c^{2i}} \|e^0\|_A, \tag{10}$$

$$\textit{where } c = \frac{\sqrt{\kappa}-1}{\sqrt{\kappa}+1} = \frac{\sqrt{\Gamma}-\sqrt{\gamma}}{\sqrt{\Gamma}+\sqrt{\gamma}}, \qquad \textit{and } \|e\|_A = \sqrt{(\mathbf{A}e, e)}.$$

## 4 Numerical results

The conjugate gradient method has been compared with the methods proposed in Section 3. A pair and a triplet separation technique and the corresponding PCG methods denoted by PCG3 and PCG4 were applied to a two group diffusion problem.

The results for a grid with the ratio $H/h = N = 3$ and $\mathrm{m} = 551$ in each group, and with prescribed accuracy $\varepsilon = 10^{-6}$ are shown in Table 2. $\lambda_{\min}$ and $\lambda_{\max}$ are the extreme eigenvalues of the original matrices (for CG method), or of the preconditioned matrices (for PCG3 and PCG4 method) in the first and the second energy group (the 1st and 2nd row of the table).

Table 2. A comparison of the methods.

| Method | Number of iterations | FLOPS | $\lambda_{\min}$ | $\lambda_{\max}$ |
|---|---|---|---|---|
| CG | 16 | 88270 | 1.2881<br>1.1470 | 41.2395<br>39.8241 |
| PCG3 | 9 | 67930 | 0.5489<br>0.3462 | 1.5698<br>1.5955 |
| PCG4 | 4 | 34100 | 0.5754<br>0.5761 | 1.4241<br>1.4230 |

A criticality problem of a nuclear reactor, modelled as a four group diffusion problem has been used for testing the PCG4 method. In Table 3 we explain some iteration characteristics in dependence on the number of equations in each energy group.

Table 3. A four group diffusion problem.

| $N$ | $m$ | $k_{\text{eff}}$ | INIT-first | INIT-last | NOP | OUTIT |
|---|---|---|---|---|---|---|
| 2 | 66 | 1.005485 | 7,6,5,6 | 4,4,3,3 | 1297 | 36 |
| 3 | 143 | 1.003135 | 9,7,7,8 | 5,4,4,4 | 2828 | 32 |
| 4 | 250 | 1.002087 | 12,9,8,10 | 6,5,4,5 | 5008 | 30 |

Table 4 shows a comparison of an error decrease of the standard CG and PCG4 method. The results there correspond to the the data for the 1st energy group of a four group diffusion problem, the number of equations is m = 975.

Table 4. An error decrease comparison.

| Number of iterations | 1 | 10 | 20 | 30 |
|---|---|---|---|---|
| CG | $1.459 \cdot 10^{+1}$ | $5.054 \cdot 10^{-1}$ | $4.659 \cdot 10^{-2}$ | $4.214 \cdot 10^{-3}$ |
| PCG4 | $1.385 \cdot 10^{+1}$ | $3.760 \cdot 10^{-2}$ | $1.376 \cdot 10^{-4}$ | $3.244 \cdot 10^{-7}$ |

**Acknowledgement.** This work has been partly supported by the Grant Agency of the Czech Republic, under grant No. 201/98/0528. The author wishes to thank Prof. Ivo Marek for his numerous inspiring ideas and comments given in connection with a numerical solution of the problem presented.

## References

[1] O. Axelsson, *Bounds of eigenvalues of preconditioned matrices*, SIAM J. Matrix Anal. Appl. **13** (1992), 847–862.

[2] J. Benda, E. Neumanová, *Some remarks on numerical solution of a nonlinear system*, in: Proc. of Conf. Numerical Modelling in Continuum Mechanics, Faculty of Mathematics and Physics, Charles University, Prague, 1995, pp. 10–17.

[3] G. Birkhoff, R. S. Varga, *Reactor criticality and nonnegative matrices*, SIAM J. Appl. Math. **6** (1958), 354–377.

[4] A. George, J. W.-H. Liu, *Computer Solution of Large Sparse Positive Definite Systems*, Prentice Hall, 1981.

[5] G. J. Habetler, M. Martino, *Existence theorems and spectral theory for multigroup diffusion model*, in: Proc. Symp. Appl. Math. 11, Nuclear Reactor Theory, Providence, 1961, pp. 127–139.

[6] R. Hackbusch, *Iterative Solution of Large Sparse Systems of Equations*, Springer-Verlag, 1994.

[7] L. A. Hageman, D. M. Young, *Applied Iterative Methods*, Academic Press, 1981.

[8] I. Marek, *Iterations of linear bounded operators in non self-adjoint eigenvalue problems and Kellog's iteration process*, Czechoslovak Math. J. **12** (1962), 536–554.

[9] I. Marek, *Matrices with positive entries and nuclear reactor criticality*, Pokroky Mat. Fyz. Astronom. **11** (1966), 121–134.

[10] E. Neumanová, *Numerical solution of ill-conditioned problems (in Czech)*, Ph.D. Thesis, Czech Technical University, Prague, 1999.

[11] R. S. Varga, *Matrix Iterative Analysis*, Prentice Hall, 1962.

# Index

Printing: Mercedes-Druck, Berlin
Binding: Stein + Lehmann, Berlin

# Scientific Computation

**A Computational Method in Plasma Physics**
F. Bauer, O. Betancourt, P. Garabechan

**Implementation of Finite Element Methods for Navier-Stokes Equations**
F. Thomasset

**Finite-Different Techniques for Vectorized Fluid Dynamics Calculations**
Edited by D. Book

**Unsteady Viscous Flows**
D. P. Telionis

**Computational Methods for Fluid Flow**
R. Peyret, T. D. Taylor

**Computational Methods in Bifurcation Theory and Dissipative Structures**
M. Kubicek, M. Marek

**Optimal Shape Design for Elliptic Systems**
O. Pironneau

**The Method of Differential Approximation**
Yu. I. Shokin

**Computational Galerkin Methods**
C. A. J. Fletcher

**Numerical Methods for Nonlinear Variational Problems**
R. Glowinski

**Numerical Methods in Fluid Dynamics**
Second Edition M. Holt

**Computer Studies of Phase Transitions and Critical Phenomena** O. G. Mouritsen

**Finite Element Methods in Linear Ideal Magnetohydrodynamics**
R. Gruber, J. Rappaz

**Numerical Simulation of Plasmas**
Y. N. Dnestrovskii, D. P. Kostomarov

**Computational Methods for Kinetic Models of Magnetically Confined Plasmas**
J. Killeen, G. D. Kerbel, M. C. McCoy, A. A. Mirin

**Spectral Methods in Fluid Dynamics**
Second Edition C. Canuto, M. Y. Hussaini, A. Quarteroni, T. A. Zang

**Computational Techniques for Fluid Dynamics 1** Fundamental and General Techniques Second Edition
C. A. J. Fletcher

**Computational Techniques for Fluid Dynamics 2** Specific Techniques for Different Flow Categories Second Edition
C. A. J. Fletcher

**Methods for the Localization of Singularities in Numerical Solutions of Gas Dynamics Problems**
E. V. Vorozhtsov, N. N. Yanenko

**Classical Orthogonal Polynomials of a Discrete Variable**
A. F. Nikiforov, S. K. Suslov, V. B. Uvarov

**Flux Coordinates and Magnetic Filed Structure: A Guide to a Fundamental Tool of Plasma Theory**
W. D. D'haeseleer, W. N. G. Hitchon, J. D. Callen, J. L. Shohet

**Monte Carlo Methods in Boundary Value Problems**
K. K. Sabelfeld

**The Least-Squares Finite Element Method**
Theory and Applications in Computational Fluid Dynamics and Electromagnetics
Bo-nan Jiang

**Computer Simulation of Dynamic Phenomena**
M. L. Wilkins

**Grid Generation Methods**
V. D. Liseikin

**Radiation in Enclosures**
A. Mbiock, R. Weber

**Large Eddy Simulation for Incompressible Flows** An Introduction Second Edition
P. Sagaut

**Higher-Order Numerical Methods for Transient Wave Equations**
G. C. Cohen

**Fundamentals of Computational Fluid Dynamics**
H. Lomax, T. H. Pulliam, D. W. Zingg

**The Hybrid Multiscale Simulation Technology** An Introduction with Application to Astrophysical and Laboratory Plasmas A. S. Lipatov

**Computational Aerodynamics and Fluid Dynamics** An Introduction J.-J. Chattot

# Scientific Computation

**Nonclassical Thermoelastic Problems in Nonlinear Dynamics of Shells** Applications of the Bubnov–Galerkin and Finite Difference Numerical Methods
J. Awrejcewicz, V. A. Krys'ko

**A Computational Differential Geometry Approach to Grid Generation** V. D. Liseikin

**Stochastic Numerics for Mathematical Physics** G. N. Milstein, M. V. Tretyakov

**Conjugate Gradient Algorithms and Finite Element Methods**
M. Křížek, P. Neittaanmäki, R. Glowinski, S. Korotov (Eds.)

MIX
Papier aus verantwortungsvollen Quellen
Paper from responsible sources
FSC® C105338

If you have any concerns about our products,
you can contact us on
**ProductSafety@springernature.com**

In case Publisher is established outside the EU,
the EU authorized representative is:
**Springer Nature Customer Service Center GmbH**
**Europaplatz 3, 69115 Heidelberg, Germany**

Printed by Libri Plureos GmbH
in Hamburg, Germany